Sustainable Development and Biodiversity

Volume 2

Series Editor
Kishan Gopal Ramawat
M.L. Sukhadia University Botany Department, Udaipur, Rajasthan, India

This book series provides complete, comprehensive and broad subject based reviews about existing biodiversity of different habitats and conservation strategies in the framework of different technologies, ecosystem diversity, and genetic diversity. The ways by which these resources are used with sustainable management and replenishment are also dealt with. The topics of interest include but are not restricted only to sustainable development of various ecosystems and conservation of hotspots, traditional methods and role of local people, threatened and endangered species, global climate change and effect on biodiversity, invasive species, impact of various activities on biodiversity, biodiversity conservation in sustaining livelihoods and reducing poverty, and technologies available and required. The books in this series will be useful to botanists, environmentalists, marine biologists, policy makers, conservationists, and NGOs working for environment protection.

More information about this series at http://www.springer.com/series/11920

Dilip Nandwani
Editor

Sustainable Horticultural Systems

Issues, Technology and Innovation

Springer

Editor
Dilip Nandwani
Agricultural and Environmental Sciences
Tennessee State University
Nashville
Tennessee
USA

ISSN 2352-474X ISSN 2352-4758 (electronic)
ISBN 978-3-319-06903-6 ISBN 978-3-319-06904-3 (eBook)
DOI 10.1007/978-3-319-06904-3
Springer Cham Heidelberg New York Dordrecht London

Library of Congress Control Number: 2014947972

Printed on acid-free paper

Springer is part of Springer Science+Business Media (www.springer.com)

To My Mother

Foreword

The *Sustainable Horticultural Systems: Issues, Technology and Innovation* book is a useful and diverse source of information on sustainable horticultural crop production on a global scale. Several of the authors are from a range of agricultural disciplines and from various countries, including developing countries from the tropics, which is a region highly impacted by climate change and natural disasters. Climate change poses many challenges to agricultural and horticultural production worldwide, such as the reductions in yields and quality of fruits and vegetables, increased irrigation water utilization, and increased incidences of weeds, insects, and plant diseases. Sustainable agriculture practices are resilient and becoming increasingly important due to pressing needs to protect the air, soils, and water; improve socio-economic conditions of farmers, farmworkers, and rural communities; and to provide healthy, safe, and nutritious horticultural products to a rapidly increasing world population. I am certain this book will contribute to the developing of more sustainable world horticultural production systems.

University of Georgia
Dept. of Horticulture
Sustainable/organic vegetable production

Juan Carlos Díaz-Pérez, PhD
Professor

Preface

Sustainable horticulture is gaining increasing attention in the field of agriculture as demand for the food production rises to the world community. Sustainable horticultural systems are based on ecological principles to farm, optimizes pest and disease management approaches through environmentally friendly and renewable strategies in production agriculture. It is a discipline that addresses current issues such as food security, water pollution, soil health, pest control, and biodiversity depletion. Novel, environmentally-friendly solutions are proposed based on integrated knowledge from sciences as diverse as agronomy, soil science, entomology, ecology, chemistry and food sciences. Sustainable horticulture interprets methods and processes in the farming system to the global level. For that, horticulturists use the system approach that involves studying components and interactions of a whole system to address scientific, economic and social issues. In that respect, sustainable horticulture is not a classical, narrow science. Instead of solving problems using the classical painkiller approach that treats only negative impacts, sustainable horticulture treats problem sources.

Because most actual society issues are now intertwined, global, and fast-developing, sustainable horticulture will bring solutions to build a safer world. This book gathers review articles that analyse current horticultural issues and knowledge and propose solutions. This book is the most up-to-date and comprehensive review of our knowledge on the use of innovative technologies and issues in sustainable horticultural systems with case studies from various regions of the world. It contains sixteen reviews written by leading international scientists from various countries. The reviews consider the production, management and issues in fruits and vegetable systems. The book has following sections:

- Section A: Sustainable Horticultural Systems
- Biodiversity in Sustainable Horticultural Systems
- Breeding and Improvement in Sustainable Horticultural Systems

The book is primarily designed for use by the undergraduates and post graduates studying horticulture, sustainable crop production, crop protection, agricultural sciences, plant pathology, and plant sciences. Horticulturists, vegetable specialists, plant and agricultural research scientists, crop protection, and in academia, will find much of great use in this book. Libraries in all universities and research establishments where agricultural and horticultural sciences are studied and taught should have multiple copies of this valuable book on their shelves. Editor wishes to thank all the contributors and staff of Springer for their cooperation in the completion of this book. Valuable suggestions and encouragement received from Prof. K.G. Ramawat is highly appreciated. Lastly, this journey would not have been possible without the support of my family, G. K. Nandwani (father), Varsha (wife), Gayatri (daughter) and Rahul (son).

Dr. Dilip Nandwani
March, 2014

Contents

Contributors

M.A. Adebayo Ladoke Akintola University of Technology, Ogbomoso, Nigeria

H. Adu-Dapaah CSIR-Crops Research Institute, Kumasi, Ghana

A. Agyeman CSIR-Crops Research Institute, Kumasi, Ghana

Raúl Alvarez-Mora Instituto Tecnológico Superior de Zacapoaxtla, Zacapoaxtla, Puebla, México

M.D. Asante CSIR-Crops Research Institute, Kumasi, Ghana

Tomas Ayala Silva United States Department of Agriculture, Agricultural Research Service, Subtropical Horticulture Research Station, National Germplasm Repository System, Miami, FL, USA

S Backiyarani Crop improvement, National Research Centre for Banana, (ICAR), Tiruchirapalli, TN, India

Saikat Kumar Basu Department of Biological Sciences, University of Lethbridge, Lethbridge, AB, Canada

D.C. Bhandari Division of Plant Improvement, Propagation and Pest Management, Central Arid Zone Research Institute, Jodhpur, Rajasthan, India

John Brown Agriculture and Life Science Division, College of Natural and Applied Sciences, University of Guam, Mangilao, GU, USA

William Cetzal-Ix Herbarium CICY, Centro de Investigación Científica de Yucatán, A. C. (CICY), Mérida, Yucatán, México

A Chandrasekar Crop improvement, National Research Centre for Banana, (ICAR), Tiruchirapalli, TN, India

Jaqueline Cosme-Pérez Instituto Tecnológico Superior de Zacapoaxtla, Zacapoaxtla, Puebla, México

Dana Jokela 145 Horticulture Hall, Department of Horticulture, Iowa State University, Ames, IA, USA

R. N. Kulloli Division of Integrated Land Use Management and Farming System, Central Arid Zone Research Institute, Jodhpur, Rajasthan, India

Ashwani Kumar Department of Botany, University of Rajasthan, Jaipur, India

Suresh Kumar Division of Integrated Land Use Management and Farming System, Central Arid Zone Research Institute, Jodhpur, Rajasthan, India

Noris Ledesma Fairchild Tropical Botanic Garden, Homestead, FL, USA

S. Lodha Division of Plant Improvement, Propagation and Pest Management, Central Arid Zone Research Institute, Jodhpur, Rajasthan, India

Rohanie Maharaj Biosciences, Agriculture and Food Technologies (BAFT), Eastern Caribbean Institute of Agriculture and Forestry (ECIAF) Campus, The University of Trinidad and Tobago, Centeno, Piarco, Trinidad and Tobago, W. I.

Mari Marutani Agriculture and Life Science Division, College of Natural and Applied Sciences, University of Guam, Mangilao, GU, USA

P.R. Meghwal Division of Integrated Land Use Management & Farming Systems, Central Arid Zone Research Institute, Jodhpur, Rajasthan, India

Michael Morgan Agricultural Experiment Station, University of the Virgin Islands, Kingshill, USA

Ajay Nair 145 Horticulture Hall, Department of Horticulture, Iowa State University, Ames, IA, USA

Dilip Nandwani Agricultural Experiment Station, University of the Virgin Islands, Kingshill, VI, USA

Velta Napoleon-Fanis Christiansted, VI, USA

Eliana Noguera-Savelli Francisco de Montejo, Mérida, Yucatán, México

M.K. Osei CSIR-Crops Research Institute, Kumasi, Ghana

Anjula Pandey Division of Plant Exploration and Germplasm Collection, National Bureau of Plant Genetic Resources, New Delhi, India

Gadi V. P. Reddy Western Triangle Agricultural Research Center, Montana State University, Conrad, MT, USA

M. S. Saraswathi Crop improvement, National Research Centre for Banana, (ICAR), Tiruchirapalli, TN, India

Dimple Singh-Ackbarali Biosciences, Agriculture and Food Technologies (BAFT), Eastern Caribbean Institute of Agriculture and Forestry (ECIAF) Campus, The University of Trinidad and Tobago, Centeno, Piarco, Trinidad and Tobago, W. I.

Stefan T. Jaronski USDA ARS Northern Plains Agricultural Research Laboratory, Sidney, MT, USA

Jennifer Tillman 145 Horticulture Hall, Department of Horticulture, Iowa State University, Ames, IA, USA

Janice Uchida Department of Plant and Environmental Protection Sciences, College of Tropical Agriculture and Human Resources, University of Hawaii at Manoa, Honolulu, HI, USA

S Uma Crop improvement, National Research Centre for Banana, (ICAR), Tiruchirapalli, TN, India

Koon-Hui Wang Department of Plant and Environmental Protection Sciences, College of Tropical Agriculture and Human Resources, University of Hawaii at Manoa, Honolulu, HI, USA

Shaohui Wu Western Triangle Ag Research Center, Montana State University, Conrad, MT, USA

Thomas W. Zimmerman Agricultural Experiment Station, University of the Virgin Islands, Kingshill, USA

About the Author

Dilip Nandwani (born 1965) did his M.Sc. (1987) and Ph.D. (1991) from the University of Jodhpur (Currently known as Jai Narayan Vyas University), Jodhour, India. He joined Tennessee State University as Associate Professor in 2014. Prior to joining TSU, he worked for the University of the Virgin Islands (2011–2013) as Research Associate Professor (Horticulture). He worked with Land Grant institutions in the American Pacific for a decade and served as Program Leader in Crop Production and Improvement. Dr. Nandwani has several years of research, extension and education experience in plant and horticultural sciences. He has published 103 articles (peer-reviewed, extension booklets, and conference proceedings) in internationally recognized journals. Dr. Nandwani is Certified Professional Horticulturist from the American Society for Horticultural Science, worked for the United Nations Environment Programme (UNEP-GEF) as Regional Advisor, International Agriculture Development, earned five awards, $ 2.2 M grants from the regional, national and international organizations, and presented research papers in over two-dozen countries.

Abbreviations

A	Acres
ACP	African, Caribbean and Pacific
AFLP	Amplified Fragment Length Polymorphism
AIMS	Africa, Indian Ocean, Mediterranean and South China Sea
AoA	Agreement on Agriculture
AOSIS	Alliance of Small Island States
BC	Before Christ
CA	California CAPs—cleaved amplified polymorphic sequence
CARICOM	Caribbean Community and Common Market
CASL	Community Adaptation and Sustainable Livelihoods
CAZRI	Central Arid Zone Research Institute
CDP	Committee for Development Policy
CET	Common External Tariffs
CFIA	Canadian Food Inspection Agency
CIEA	Centro De Investigacion y Estudios Avanzados
CIRAD	Centre for International Cooperation in Agronomic Research for Development
CRFM	Caribbean Regional Fisheries Mechanism
CRIDA	Central Research Institute for Dryland Agriculture
CSIDS	Caribbean Small Island Developing States
EU	European Union DNA—Deoxyribonucleic Acid
FAO	Food and Agricultural Organization
FAS	Foreign Agricultural Service
FEDA	Special Fund for Agricultural Development
FL	Florida
FOs	Farmers Organizations
FSMA	Food Safety and Modernisation Act
G	Guatemalan G × M—Guatemalan × MEXICAN
GATT	General Agreement on Tariffs and Trade
GDP	Gross National Product
GDEM	Global Digital Elevation Map
GIS	Geographic Information systems

GM	Genetically Modified
G × W	Guatemalan × West Indian
Ha	Hectare
IBA	Indole-3-Butyric Acid
ICAR	Indian Council of Agricultural Research
IFAD	International Fund for Agricultural Development
IPCC	Intergovernmental Panel on Climate Change
IPM	Integrated Pest Management
ISSR	Inter-Simple Sequence Repeat
LB	Pounds
M	Mexican
Mill	Miller
MDG	Millennium Development Goals
M × G	Mexican × Guatemalan
M × WI	Mexican × West Indian
MT	Metric Ton
MTT	Million Metric Tons
MY	Marketing Year
NASS	National Agricultural Statistics Service
NBPGR	National Bureau of Plant Genetic Resources
NCAP	National Centre for Agricultural Economics and Policy Research
NGB	National Gene Bank
NCGR	National Clonal Germplasm Repository
PFA	Pest Free Areas
PGR	Plant Genetic Resources
RAPD	Random Amplified Polymorphic DNA
RFLP	Restriction Fragment Length Polymorphism
SCAR	Sequence Characterized Amplified Regions
SDS-PAGE	Sodium Dodecyl Sulfate Polyacrylamide Gel Electrophoresis
SIDS	Small Island Developing States
SNP	Single-Nucleotide Polymorphism
SSR	Simple Sequence Repeat
UCR	University of California Riverside
UN	United Nations
UNDESA	United Nations Department of Economic and Social Affairs
UNEP	United Nations Environmental Program
USA	United States of America
USDA	United States Department of Agriculture
VGR	Vegetable Genetic Resources
WB	World Bank
WFP	World Food Program
WI	West Indian
WK	Weeks
WRCP	Wild Relatives of Crop Plants
WTO	World Trade Organization

Part I
Section A: Sustainable Horticultural Systems

Sustainable Vegetable Production: Caribbean Perspective

Velta Napoleon-Fanis and Dilip Nandwani

1 Value of Vegetables and Vegetable Production

Palada et al. (2006) defines a vegetable as a plant that is cultivated for an edible part such as the leaf, stem or root. Vegetables are very essential for the nutrition security of the people of the Caribbean and are absolutely important and vital for human health and not a luxury (USDA 1996). Vegetables are a rich source of many essential micronutrients, including vitamins C and K, foliate, thiamin, carotenes, several minerals, and dietary fiber. In fact, UN-SCN (2004) stated that vegetables are the most stable and sustainable sources of micronutrients. According to the overview statement of the Caribbean Agricultural Research and Development Institute (CARDI 2011), the Governments of the Caribbean islands have identified vegetables as part of the 'Regional Food Basket' and great efforts are being dedicated towards achieving 'food sovereignty'.

From 1987, CARDI in both Jamaica and Trinidad and Tobago have devoted a considerable amount of effort on research and development on leafy vegetables. Research included the Amaranth (*Amaranthus* sp), which is called 'callaloo' in Jamaica and cabbage and lettuce in Trinidad and Tobago (Chandler 1987). The CARDI factsheets for the region provides information of work that begun in the 1980's. The data provides information stating that the Trinidad and Tobago unit, together with CARDI St. Kitts and Nevis and St. Vincent and the Grenadines carried out studies on non-leafy vegetables like tomatoes, onions and sweet peppers; the Barbados unit worked on carrots and scallions (Hudson 1992); the St. Lucia unit studied chayotes also called 'cristophene' (Chase 1985); and CARDI Grenada researched eggplant (Buckmire 1980). Some of the work done have led to successful production of some vegetables. Take for example, in Jamaica, the Integrated Pest Management (IPM) in

V. Napoleon-Fanis (✉)
Sunny Isle, P.O. Box 6183, 00823 Christiansted, VI, USA
e-mail: veenapoleon@gmail.com

D. Nandwani
Agricultural Experiment Station, University of the Virgin Islands, 00851 Kingshill, VI, USA

D. Nandwani (ed.), *Sustainable Horticultural Systems,* Sustainable Development and Biodiversity 2, DOI 10.1007/978-3-319-06904-3_1

collaboration with its partners, has contributed significantly to the development of the export market for 'callaloo' and today, this vegetable enjoys preclearance status for shipments to the United States.

In order for there to be successful vegetable production, there is a need for radical changes in the Caribbean's vegetable production methods, due to the reliance on a large amount of chemical inputs to control pests and diseases, which are particularly aggressive in the region. This has been a major setback for sustainable vegetable production systems in the Caribbean (Chandler 1987).

According to Ali et al. (2002), vegetable production provides more jobs than other foods like cereal grains. These authors go on to say that vegetable production supports agribusiness and related service industries, thereby creating job opportunities. Vegetable production diversifies and generates farm income, usually at a higher rate than other agricultural crops (de Bon and Tran 2001). Two of the most important characteristics of vegetable production in the region are high labor intensity and high pesticide use, resulting in high chemical residues in the crops and the environment (Palada et al. 2006). The work done by CARDI currently addresses the growing demand for value added semi-processed vegetable products in some countries in the region.

According to a study by Food and Agriculture Organization (FAO), conducted by marketing consultant, Fitzroy James, in 2006, it was found that the total availability of fresh vegetable production in the Caribbean was 2.1 million metric t valued at US$ 1.6 million. This production represents the third largest sub-category accounting for 9% of total food crop production. The study also presented data which supports that the volume of vegetables produced in 2006 was estimated to be around 193,000 metric t. Although the Caribbean Islands produce mostly tropical vegetables, Fitzroy James found that the major crops were pumpkin (26%), tomatoes (19%), cabbages (18%), and cucumbers (13%), accounting for approximately 76% of the vegetables available for consumption.

The study by FAO. (2009) also determined that in terms of exports in 2006, the main vegetable products were mixed vegetables (71%), cabbages (8%) and peppers (7%) which were mainly produced and exported by Trinidad and Tobago and Guyana. Table 1 below shows a breakdown of production, import, export and availability of selected vegetable crops produced in the Caribbean (Table 1).

According to the FAO CARICOM Market Study Report (2009), the Caribbean region is effectively self-sufficient in pumpkins, squashes and gourds as can be seen by the production values stated in Table 1. These vegetables form a significant component of the vegetable sub sector and most if their production occurs in Jamaica followed by Guyana and Trinidad and Tobago (FAO CARICOM Report 2009). These crops are mostly produced by small and medium size growers in most of the region (Table 2).

In 2006, tomatoes were the second largest vegetable available in the region according to Table 1 above. Total production was estimated at around 35,000 metric t. Although the region is relatively self-sufficient in tomatoes, there is usually a seasonal shortfall in the rainy season from July to December. This shortfall is usually met by imports since it occurs when demand is at the highest, which is during

Table 1 CARICOM production and availability of fresh vegetable produce (2006)

Commodity	Production	Import	Export	Availability	Availability (%)
Pumpkins, squash and gourds	51,130	–	–	51,130	26
Tomato	35,755	1158	427	36,485	19
Cabbages and other brassicas	31,921	3122	538	34,504	18
Cucumbers and gherkins	25,823	63	191	25,695	13
Spinach	13,708	45	1	13,751	7
Lettuce and chicory	8491	–	–	8491	4
Eggplants (aubergines)	6462	34	293	6203	3
Okra	6095	–	–	6095	3
Cauliflower/broccoli	1925	3059	5	4979	3
Mixed vegetables	3658	3060	4858	1861	1
Lettuce	–	1713	–	1713	1
Maize, green	760	–	–	760	<1
Cabbage lettuce	–	708	1	708	<1
Mushroom	–	417	–	416	<1
Pepper (piper)	255	463	459	260	<1
Asparagus	–	99	6	93	<1
Brussels sprout	–	15	–	15	<1
Globe artichokes	–	8	24	–17	–
Total	*185,983*	*13,964*	*6804*	*193,143*	*100*

the tourist season for countries with well-developed tourist industries. Jamaica accounts for the bulk of production of cabbages and other brassicas and is the largest exporter of cabbage. Cucumbers and gherkins are produced throughout most of the region both by small farmers and under greenhouse conditions by medium scale producers in Jamaica and Trinidad and Tobago. They are produced both for the local domestic market and for the tourism sector and for pickle processing.

2 Research Studies Conducted on Vegetables and Vegetable Production

There have been numerous works by researchers across the Caribbean who seeks to find ways to increase vegetable production in the region. Researchers have given priority to developing pest management methods, by introducing companion plants in particular, substituting artificial fertilizer with locally available organic matter and also the breeding of resistant varieties of vegetables.

Table 2 Summarized findings of research studies on vegetable crop production from the Caribbean region (1964–2010)

Authors and year	Crop	Findings
Vlitos and Davies 1964	Tomato, Beet, Lettuce, Sweet Pepper, Carrot, Celery, Onion, Endive	The authors found that the vegetables grown in coconut fiber wastes were markedly superior to those grown in the soil and those grown in the fertilizer regime
Barrios 1965	Snap Bean	According to the author's findings, there is no relationship between magnesium deficiency, magnesium content and moisture in soil in snap bean varieties. The author believes that other factors may be involved
Irizarry et al. 1966	Tomato	The authors' findings indicate that mulching did not produce any significant yield, nor did it reduce soil fruit rot. However, it did suppress growth in 102 weed species
Charles 1966	Tomato	The author's research study showed that flower production, number of fruit harvested and fruit size indicated a considerable reduction in 'wet season' tomato yield. The author concluded that the unfavorable weather negatively influenced reproductive growth, flower setting and fruit development during the 'wet season'
Tai et al. 1968	Okra	The results of the study showed that the close spaced plots produced the highest yield, but harvest pods form them were difficult. According to the authors, loss due to difficulty of picking makes it more advisable to consider other spacing treatments
Kaan et al. 1969	Tomato	The authors obtained the varietal types of tomatoes that adapted to the West Indies environmental conditions, showed variability, resistance, and had superior fruit characteristics, precocity of floral induction and stylar length
Villanueva 1970	Sweet Pepper	Of the seven varieties tested by the author, the 'Cubanella' pepper variety was the highest yielding, followed by the 'Ruby Grant' and 'Florida King' varieties
de Jeffers 1970	Tomato, Sweet Pepper	The author found that there was increased production of tomatoes and sweet peppers. Onion was then no longer the main crop of specialization
Villanueva et al. 1970	Tomato	After 2 years of research on tomato varieties, the authors found that the outstanding varieties of tomatoes were 'Chico', 'Chico 111', 'Chico Grande' and 'Roma'

Table 2 (continued)

Authors and year	Crop	Findings
Kaan and Anais 1971	Cucumber	The authors could not recommend the best yielding selections because the varieties were either susceptible to powdery mildew or produced many distorted fruits. The best commercial cultivars were found to be 'F_1 Gemini', 'Saticoy' and 'Lehua'
Roach et al. 1971	Onion, Cabbage, Carrots, Lettuce, Snap Bean, Beet, Radish	The authors found the best cultivars from a selection of 50 or more varieties for each vegetable and selected these varieties for further trial
George 1972	Cucumber	The author's findings showed that there is a higher yield in cucumber production during the 'dry season' than during the 'wet season'
Forde 1972	Tomato	The author found the highest yields in the Leeward Islands to be during the 'dry-off season' period which is from the month of September to the month of December. High yields were also realized in St. Kitts during the 'wet-off season'
Smith 1973a	String Bean	Findings indicated that there is an increase in yield per plant if the crop is harvested frequently. According to the author, early removal of crop results in greater harvest by stimulating further flowering and fruit formation
Forde 1973	Sweet Pepper	According to the author's study, at low plant densities, the yield of fruit per plant was significantly higher. Satisfactory plant spacing was found to be 45.7×30.5 cm which gives an area of 710,736 plants per hectare
Smith 1973b	Tomato	The author found that in response to nitrogen applications, the 'Indian River' variety of tomato produced higher yields and had more vegetative growth than the 'Bounty' variety. In response to mulch cover, the Bounty' variety produced much larger and heavier fruit
Jackson and Sierra 1975	Tomato	The authors' findings indicate that at lower rates of metribuzin there was adequate control of weed species in tomatoes, especially in weed species like crabgrass, caltrop, purslane, jungle rice and pigweed

Table 2 (continued)

Authors and year	Crop	Findings
Mangual-Crespo et al. 1977	Onions	The authors found that there is a linear increase in yield with an increase in nitrogen (N) levels and that 267 kg of nitrogen per hectare was needed for normal growth and maturity in onions. Plants grown in less than 134 kg of N per hectare had deficiencies and did not mature properly
Jackson and Sierra 1979	Tomato	The results from the study indicated that 16 out of a 100 cultivars demonstrated no tolerance to herbicide applications. The authors found that one weed species, '*Echinochloa colonum*' demonstrated tolerance to all levels of application of Sencor (Metribuzin)
Tumuhairwe and Gumbs 1981	Cabbage	According to the authors, during the 'wet season', average head weights of cabbage harvested were highest on single and double-row ridges. This resulted in higher total yields per hectare in single and double-row ridges than in flatbed layouts. In the 'dry season', mulching along with irrigation improved both growth and yield
Gabriel et al. 1985	Tomato	The research indicated that the use of plastic cover over each plant increased vegetative growth but retarded fruiting and decreased yield. The authors found that the use of plastic mulch and plastic covered sheds both either equally increased yield or had no significant effect on yield
Sajjapongse and George 1985	Tomato	Open pollinated varieties and hybrids out-yielded the best variety 'Caraibe.' According to the authors, the yield produced from the 'Caraibe' variety was more than double and the fruit size and fruit numbers per plant were significantly larger
Martin 1985	Tomato	According to the author, segments with the characteristics of exceptionally long shelf life can be separated from pure lines. It was found however, that only a few of the genes that code for longer shelf life segregated during crossing over
Navarro and Newman 1987	Tomato	Research findings indicated that there was a significant increase in tomato yield for plants treated with higher irrigation rates; however, there was no significance in yield of the tomato planted in emitter placements

Table 2 (continued)

Authors and year	Crop	Findings
Collingwood et al. 1989	Cucumber	The authors found that mulched plots produced higher yields and used less water than the plots without the mulch
Wessel-Beaver et al. 1990	Tomato	According to the study, heat tolerant varieties 'Heatwave' and 'Captain' maintained excellent yields in the months of April and May, and good yields in June. The authors found that the use of heat tolerant varieties is essential for successful summer tomato production
Crossman and Collingwood 1991	Thyme	The researchers found that fresh and dry matter yields of thyme produced from fertilizer treatment levels of 112 and 169 kg if nitrogen per hectare were superior to thyme yields of fertilizer treatments of 0 kg of nitrogen per hectare
Cooper and Gordon 1992	Hot Pepper	Pepper seed extraction using trisodium phosphate treatment showed satisfactory germination for up to 6 months following treatment, according to the authors' findings
Collingwood et al. 1992	Tomato	The authors found that the cultivars 'Celebrity Floridae' and 'UH-N69' displayed more vigorous growth than the other cultivars. They had larger fruits and were most suitable for the Virgin Islands fresh market
Chinnery et al. 1993	Cabbage	Findings indicate that pests inhibited the production of cabbage in Barbados and there was limited resistance to the insecticides tested (DDT, DIPEL, Jupiter, IGR). The authors found that there is need for a pool of efficacious insecticides that can be rotated in order to have suitable increase in cabbage production
Palada et al. 1994	Thyme	The authors' results showed that urea and cow manure were found to be the best sources of nitrogen to increase yields in thyme production
Crossman et al. 1995	Onion	Of the 16 short-day cultivars studied, the authors found that the mean yield was significantly higher for the 'Texas Grano 1015'. This cultivar also produced the largest bulbs
Palada et al. 1996	Cucumber	Authors found that the use of black mulch for the thyme variety 'Calypso' showed higher yield than when using straw mulch in 1995. There was no significant difference with the use of mulch with the variety 'Dasher IT' in 1996; however, yield was much higher in 'Dasher IT' than in 'Calypso' yield

Table 2 (continued)

Authors and year	Crop	Findings
Dupuy et al. 1996	Tomato	The authors found that correct planting time, use of resistant and early varieties, application of legal measures combined with production of tomato plantlets under mosquito net tunnels reduced the early infection of tomato crops with gemini-virus (TYLCV), and lead to an increase in yield and greater outcomes to tomato farmers
Palada et al. 1997	Chive, Parsley	The results of the study indicated that green manure crops had a significant effect on some yield parameters of both crops. Authors also found that some culinary herbs can benefit when grown with green manure crops
Rhoden et al. 1997	Hot Pepper	The findings showed that hot pepper 'Scotch Bonnet' grown in the shade had increased vegetable growth but decreased fruit production and biomass accumulation
Biney 1998	Onion	Both planting methods, transplanting and direct seeding had similar yields, however, direct seeding recorded the least double bulbs and higher plant survival. Authors found that direct seeding was more advantageous and cost effective then transplanting
Palada and Crossman 1998a	Okra	The authors' findings indicated the planting density of 98,522 plants per hectare would be optimum for maximum yield in okra
Palada and Crossman 1998b	Parsley, Chive	The authors established that the use of locally available grass as mulch has potential for increasing production of parsley and chives in the U.S. Virgin Islands
Palada and Crossman 1999	Chive, Basil, Sage, Parsley, Cilantro	Findings indicate that without chemical fertilizers, legume green manure crops, particularly 'sun-hemp' and 'hyacinth bean' increase yields in culinary herbs in a crop rotation system
Crossman et al. 1999a	Hot Pepper	According to the results, 'Scotch Bonnet' peppers are the most adapted and promising cultivar in the Virgin Islands, according to the results. According to the authors, the 'West Indian Red' also has good potential in the Virgin Islands

Table 2 (continued)

Authors and year	Crop	Findings
Crossman et al. 1999b	Spinach	The results from findings showed that the optimum cow manure application rate for leaf yield is 10–20 treatment per hectare. The authors observed significant and quadratic responses in the number of stems harvested, length of the longest stem, fresh plant weight, leaf and stem as well as dry weight of plant, stem and leaf
Palada et al. 2000a	Eggplant	The authors found that the cultivars suitable for the sustainable production of eggplant using an organic management system in the Virgin Islands were found to be: 'Black Nite', 'Black Beauty', 'Black Bell', 'Italian Pink Bicolor' and 'Hybrid Beauty'
Palada et al. 2000b	Chive	According to the authors' findings, organic mulch such as shredded paper and grass straw improve yield and income in chives production in the Virgin Islands
Adams et al. 2001	Hot Pepper	The results of the study showed that higher plant population densities produced the highest yields. The authors found that there was no effect on size, shape and weight due to higher plant population densities
Gardner and Queely 2001	Hot Pepper	The researchers established that plants grown on organic mulches performed significantly better and had higher yields than those grown on synthetic mulches
Palada and O'Keefe 2001	Hot Pepper	The research findings indicate that the cultivars, 'Habanero' and 'West Indies Red' produced more fruit and larger fruit than the cultivars 'Red' and 'Yellow Scotch Bonnet.' The authors found that irrigation rate has no effect on hot pepper yield, and that hot peppers can be produced with minimum irrigation
Palada et al. 2001	Tomato	The results of the study showed that the cultivars suitable for production using an organic management system are: 'Bonita', 'Empire', 'Keepsake', 'Merced', 'Mountain Pride' and 'Sunmaster'
O'Keefe and Palada 2002	Hot Pepper	The authors established that spacing influenced yield based on cultivar. The best yields were attained by the 'Jamaica Scotch Bonnet' and the 'Yellow Scotch Bonnet'

Table 2 (continued)

Authors and year	Crop	Findings
Palada et al. 2003	Sweet Pepper	According to the authors, an irrigation regime does influence the yield of 'Puerto Rico Sweet Pepper' in terms of the number of fruits and marketable yield. Increase in the moisture in the soil increased marketable yield
Palada and Mitchell 2004	Cucumber	The authors found that although the new cultivars 'Cobra' and 'Olympian F_1' produced yields that were not significantly different to the yields of common cultivars 'Calypso' and 'Dasher 2,' cultivars 'Cobra' and 'Olympian F_1' are suitable for farmers who wish to replace old and common cultivars
Palada et al. 2006	Hot Pepper	The authors' findings indicate that in the early stages of hedgerow establishment, tree cropping competition does not cause a reduction in growth and yield of hot pepper
Allen 2004	Broccoli, Cabbage, Lettuce, Onions, Sweet Pepper, Collard Green, Snap Bean, Tomato	According to the author, vegetable crop varieties grown in soil amended with composted yard waste and cow manure produced yields that were equal to and sometimes exceeded that which were produced by plots amended with commercial fertilizer
Santos et al. 2007	Tomato	Results of the study suggested that fertilization with sulfur had a significant effect on the yield of tomato
Gardner et al. 2008	Cucumber	The authors found that there are no advantages in using trellises over using conventional system where vines are allowed to run freely on the ground
Chapman 2009	Lettuce	According to the author, well decomposed poultry and sheep manure could be used as an alternative source of organic matter to replace filter press mud in lettuce production
Weekes 2009		The author's findings showed that the yields obtained from planting on plastic mulch were increased by 50 % which led to a reduction in total cost of labor for weeding and irrigation
Etienne 2011	Tomato	According to the author, the preliminary results indicate that plants grown in the organic compost had more vigorous growth, flowered earlier, had better fruit set and produced heavier fruits

Table 2 (continued)

Authors and year	Crop	Findings
Nandwani 2012	Sweet Pepper	According to the author, the sweet pepper cultivar that matured the fasted and produced the highest marketable yield was the 'White King.' Findings also indicated that the largest fruits were produced by the 'Aristotle' cultivar
Napoleon-Fanis and Nandwani 2014	Watermelon	According to the findings, bensulide and halosulfuron used only as pre-emergence herbicides do not provide the much needed control of grasses, sedges and broadleaf weeds in transplanted watermelon production. The authors suggest that combinations of both pre and post emergence herbicides should be used in transplanted watermelon
Nandwani and Bailey 2013	Cucumber	The authors studied ten varieties of cucumber (*Cucumis sativus*) evaluated in the aquaponic system in the US Virgin Islands. Cucumber var. "Speedway", "Fanfare" and "Sweeter Yet" yielded 24 fruits per m^2 per planting. The study showed that "Palace King" produced the greatest mass (6.2 kg/m^2/crop). Varietal differences influence the mass produced for each crop

Conclusion

The role of CARDI, the work done by governments of the Caribbean islands, CARICOM, and also the numerous research studies that have been conducted over the past 50 years in order to promote and enhance vegetable production in the Caribbean region are very important. These efforts and studies have improved vegetable varieties and ameliorated sustainable vegetable production practices that are currently adopted and practiced by producers and farmers of the region. In addition, these labors have impacted vegetable production to a great extent in that there is now increased productivity and many of the islands can boast the export of vegetables to other Caribbean islands as well as the United States.

Quality is an important aspect of the vegetable industry, whether the vegetable crop is sold locally, regionally or internationally. Quality and productivity are intertwined and obtaining the most excellent quality of vegetables is the ultimate goal for vegetable producers of the Caribbean region. Studies on sustainable vegetable production continue across the region today, which in future seek to benefit the Caribbean region through the expansion of better vegetable crop production, and through improved technologies that will augment the global export of quality and more nutritious vegetables.

Acknowledgments The senior author would like to thank University of the Virgin Islands Agriculture Experiment Station to conduct research with the Horticulture Program and program staff Vanessa Forbes, Shamali Dennery, Paulino Perez, Victor Almodovar and Stafford M. A. Crossman.

References

Adams HV, Laukner FB, Sisnett DO (2001) Effect of high plant population densities on yield, plant and fruit characteristics of the hot pepper cultivar 'West Indies red'. Proc Caribb Food Crops Soc 37:197–201

Ali M, Farooq U, Shih YY (2002) Vegetable research and development in the ASEAN region: a guideline for setting priorities. In: Kou CG (ed) Perspective of ASEAN cooperation in vegetable e-research and development. AVRDC, Shanshua, pp 20–64

Allen JR (2004) Differences in vegetable crop yield response to composted waste as the soil amendment in urban gardening. Proc Caribb Food Crops Soc 40:32–40

Bailey D, Nandwani D (2013) Production of cucumber (*Cucumis sativus*) in the UVI aquaponic system. Proc Caribb Food Crops Soc. Trinidad p. (in press)

Barrios L (1965) The influence of magnesium on leaf composition, yield and quality of snap beans (*Phaseolus vulgaris* L.). Proc Caribb Food Crops Soc 3:106–107

Biney EA (1998) Comparison of transplanting and direct seeding technology for four onion cultivars. Proc Caribb Food Crops Soc 33:88–95

Buckmire KU (1980) Eggplant production for Grenada. CARDI Factsheet PT.1/80. St. Michael, Barbados. Caribb Agri Res Dev Inst

CARDI (2011) Caribbean Agricultural Research and Development Institute. http://www.cardi.org/commodities-themes/fruits-vegetables/. Accessed 26 Sept 2013

CARICOM (2006) Fresh and processed produce trade tariffs and statistics 2001–2006. Caribbean Regional Negotiating Machinery

Chandler F (1987) Efficient crop management for quality vegetable production. CARDI Factsheet CP-F/ 9.87. St. Augustine, Trinidad

Chapman I (2009) Evaluation of productivity of loose-leaf lettuce (*Lactuca sativa*) using different medium materials. Proc Caribb Food Crops Soc 45:161

Charles WB (1966) Evaluation of tomato varieties in a breeding program. Proc Caribb Food Crops Soc 4:2–42

Chase V (1985) Eat St. Lucian produce: Christophene. Castries, St. Lucia, Carib Agri Res and Dev Inst

Chinnery LE, Gibbs IH, Chandler LR (1993) Decline in cabbage production in Barbados: cause and solution. Proc Caribb Food Crops Soc 29:67–80

Collingwood C, Crossman CS, Navarro A (1989) Effect of black plastic mulch on cucumber yields, water use and economic returns. Proc Caribb Food Crops Soc 25:211–216

Collingwood C, Crossman CS, Palada MC (1992) Tomato germplasm evaluation for growth and productivity in the U.S. Virgin Islands. Proc Caribb Food Crops Soc 28:232–238

Cooper B, Gordon M (1992) Production of West Indian hot pepper seed. Proc Caribb Food Crops Soc 28:22–231

Crossman CS, Collinwood C (1991) Effect of varying levels of applied nitrogen on thyme production. Proc Caribb Food Crops Soc 27:165–170

Crossman SMA, Palada MC, Kowalski JA (1995) Evaluation of onion germplasm for yield characteristics in the U.S. Virgin Islands. Proc Caribb Food Crops Soc 31:111–116

Crossman CS, Palada MC, Davis AM (1999a) Performance of West Indian hot pepper cultivars in the Virgin Islands. Proc Caribb Food Crops Soc 35:169–177

Crossman CS, Palada MC, Davis AM (1999b) Growth and yield response of malabar spinach to levels of dehydrated cow manure application. Proc Caribb Food Crops Soc 35:178–182 de Bon H, Tran KT (2001) The response of peril-urban agriculture to urbanization in Vietnam. FAO Paper EC600E

Dupuy JW, Sencion M, Febello J, Ferreiras V (1996) Alternative methods for producing industrial tomato plants by small farmers. Proc Caribb Food Crops Soc 32:104–105

Etienne D (2011) Yield performance of tomato (*Lycopersicon esculentum*) grown in various media under greenhouse in Dominica. Proc Caribb Food Crops Soc 47:50

Forde CM (1972) Factors affecting tomato production in the Leeward Islands: time planting. Proc Caribb Food Crops Soc 10:45–51

Forde CM (1973) The influence of plant density on sweet pepper (*Capsium annum*) yields in St. Kitts. Proc Caribb Food Crops Soc 11:456–459

Gabriel L, Bridgelal P, Bart A, Wilson H, Ramdeensingh S (1985) Effect of plastic cover on tomato growth in the rainy season. Proc Caribb Food Crops Soc 21:22–24

Gardner CS, Queeley GL (2001) Effect of organic and synthetic mulch (reflective and non-reflective) on yield of Scotch Bonnet hot pepper. Proc Caribb Food Crops Soc 37:175–177

Gardner CS, Queeley GL, Grant KT, Brown BG, Hylton T (2008) Production of table cucumber (*Cucumis savita*) on two trellis systems in northern Florida. Proc Caribb Food Crops Soc 44:393–397

George G (1972) Evaluation of slicing cucumber varieties at Union agricultural station St. Lucia. Proc Caribb Food Crops Soc 10:52–55

Hudson C (1992) Tyre gardens. CARDI factsheet CP-F/21.92. St. Michael, Barbados: Carib Agri Res Dev Inst

Irizarry H, Azzam H, Woodbury R (1966) Evaluation of black polyethylene mulch paper for tomato production in Puerto Rico. Proc Caribb Food Crops Soc 4:75–79

Jackson GC, Sierra C (1975) The tolerance of tomato cultivars to post-emergent application of lexone (metribuzin). Proc Caribb Food Crops Soc 13:183–189

Jackson GC, Sierra C (1979) The tolerance of tomato cultivars to post-emergent application of sencor (metribuzin). Proc Caribb Food Crops Soc 16:113–122

James F (2009) CARICOM/CARIFORUM fresh and processed produce market-opportunity study. A report prepared for the Food and Agriculture Organization (FAO) of the United Nations under the project-promoting CARICOM/CARIFORUM food security. Market opportunity study report

Jeffers W de C (1970) Tomato and pepper production in Barbados. Proc Caribb Food Crops Soc 8:38–41

Kaan F, Anais G (1971) A comparative trial of cucumber varieties and hybrids in Gaudeloupe. Proc Caribb Food Crops Soc 9:64–70

Kaan F, Beramis M, Messeaen CM (1969) Research of tomato varieties of the Antilles. Proc Caribb Food Crops Soc 7:173–181

Mangual-Crespo G, Orengo E, Cruz-Perez L (1977) Effect of plant spacing and fertilizer levels on yield and dry weight of onions (*Allium cepa* L.) 'Texas grano 520'. Proc Caribb Food Crops Soc 14:553–557

Martin F (1985) Genetic strategies for breeding long-lasting tomatoes. Proc Caribb Food Crops Soc 21:5–9

Nandwani D (2012) Growth and yield response of sweet pepper cultivars in the U.S. Virgin Islands. Proc Caribb Food Crops Soc 48:231–237

Napoleon-Fanis V, Nandwani D (2014) The effect of pre-emergence herbicides on the quality and yield of transplanted watermelon (*Citrullus lanatus*). Proc Carib Food Crops Soc (in press)

Navarro A, Newman J (1987) The effect of drip irrigation rates and two emitter placements on tomato production. Proc Caribb Food Crops Soc 23:101–106

O'Keefe DA, Palada MC (2002) In-row plant spacing effects on growth and yield on four hot pepper cultivars. Proc Caribb Food Crops Soc 38:162–168

Palada MC, Crossman SMA (1998a) Planting density affect growth and yield of bush okra. Proc Caribb Food Crops Soc 34:52–57

Palada MC, Crossman SMA (1998b) The influence of mulch type on yield of parsley and chives in the U.S. Virgin Islands. Proc Caribb Food Crops Soc 34:40–45

Palada MC, Crossman SMA (1999) Production of culinary herbs in rotation with green manures in the Virgin Islands. Proc Caribb Food Crops Soc 35:155–162

Palada MC, Mitchell JM (2004) On-farm evaluation cultivars for summer production in the U.S. Virgin Islands. Proc Caribb Food Crops Soc 40:16–21

Palada MC, O'Keefe DA (2001) Response of hot pepper cultivars to levels of drip irrigation. Proc Caribb Food Crops Soc 37:190–196

Palada MC, Crossman SMA, Kowalski JA (1994) Growth and yield response of thyme (*Thymus vulgaris* L.) to sources of nitrogen fertilizer. Proc Caribb Food Crops Soc 30:58–64

Palada MC, Crossman SMA, Kowalski JA (1995) Evaluation of onion germplasm for yield characteristics in the U.S. Virgin Islands. Proc Caribb Food Crops Soc 31:111–116

Palada MC, Crossman SMA, Kowalski JA (1996) Effects of black plastic mulch and straw mulch on yield and water use efficiency of cucumber. Proc Caribb Food Crops Soc 32:83–93

Palada MC, Crossman SMA, Kowalski JA (1997) Growth and yield of chive and parsley grown in rotation with tropical green manure crops in the Virgin Islands. Proc Caribb Food Crops Soc 33:201–207

Palada MC, Crossman SMA, Davis AM (2000a) Yield performance of eggplant cultivars grown under organic management system. Proc Caribb Food Crops Soc 36:87–92

Palada MC, Crossman SMA, Davis AM, Chichester E (2000b) Evaluation of organic mulches for chive production in the Virgin Islands. Proc Caribb Food Crops Soc 36:68–72

Palada MC, Davis AM, Crossman SMA (2001) Yield performance of tomato cultivars grown under organic management system. Proc Caribb Food Crops Soc 37:154–160

Palada MC, Mitchell JM, O'Keefe DA (2003) Growth and yield response of Puerto Rico sweet pepper to levels of drip irrigation in the Virgin Islands. Proc Caribb Food Crops Soc 39:272–277

Palada MC, Mitchell JM, Becker BN (2004) Growth and yield of hot pepper in hedgerow intercropping with *Morinda citrifolia* L. during early establishment. Proc Caribb Food Crops Soc 40:32–40

Palada MC, Thomas JK, Thomas AL (2006) The world vegetable center in enhancing and promoting vegetable production in the tropics. HortScience 4(3):556–560

Rhoden EG, Gueye M, Ekuka G, Basquiat P, Khan KA, Aglan H (1997) Use of shade in vegetable production: hot pepper. Proc Caribb Food Crops Soc 33:88–95

Roach FL, Hoad RM, Headley W, Eavis MT (1971) Vegetable cultivar screening trials: a progress report. Proc Caribb Food Crops Soc 9:42–51

Sajjapongse A, George C (1985) Evaluation of tomato varieties in St. Lucia. Proc Caribb Food Crops Soc 21:17–21

Santos BM, Esmel CE, Moratinos H, Rechcigl JE (2007) Using sulfur fertilizer sources for tomato production. Proc Caribb Food Crops Soc 43:119

Singh RH, Rankine LB, Seepersaud G (2007) Competitiveness study, policy proposals: enhancing international competiveness, the CARICOM regional transformation program for Agriculture, CARICOM

Smith L (1973a) Investigation on the effect of different frequencies of harvest on the yield of string beans. Proc Caribb Food Crops Soc 11:167–173

Smith L (1973b) An investigation on yield response of two tomato varieties on nitrogen and potash fertilizer and to a grass cover. Proc Caribb Food Crops Soc 11:153–160

Tai EA, Kanhai S, Gardner-Brown T (1968) Spacing and fertilizer trial with okra. Proc Caribb Food Crops Soc 6:62–66

Tumuhairwe J, Gumbs FA (1981) Studies on soil management for dry and wet season production of cabbage (*Brassica deracea* cv. Capitata) in Trinidad. Proc Caribb Food Crops Soc 17:1–10

United Nations System Standing Committee on Nutrition (UN-SCN) (2004) Fifth report on the world nutrition situation. Nutrition for improved development outcomes. World Health Organization, UN-SCN, Geneva

U.S. Department of Agriculture (1996) The food guide pyramid. Control nutrition policy promoting. Washington, DC. Home Gdn. Bul. 252

Villanueva G (1970) Ensayo comparative de variedades en el cultivo del pimento. Proc Caribb Food Crops Soc

Villanueva G, Montas Vizcaino F, Saito T (1970) Influencia de epoca de siembra en la produccion de variedades de tomate. Proc Caribb Food Crops Soc 8:46–49

Vlitos AJ, Davies WNL (1964) Coconut fiber waste as a basic medium for the production of vegetable crops. Proc Caribb Food Crops Soc 2:95–122

Weekes T (2009) Analysis of cost of production of two major vegetable crops under various mulching regimes in St. Kitts. Proc Caribb Food Crops Soc 47:50

Wessel-Beaver L, Fornaris G, Armstrong A, Caraballo E, Velasquez L (1990) Extending the tomato season in Puerto Rico: problems of summer production. Proc Caribb Food Crops Soc 26:568–573

Horticulture Based Production Systems in Indian Arid Regions

D. C. Bhandari, P. R. Meghwal and S. Lodha

1 Introduction

The Indian arid zone covers around 12% of country's geographical area occupying 31.8 million ha of land (Fig. 1). It covers parts of Andhra Pradesh, Gujarat, Haryana, Karnataka, Maharashtra, Punjab and Rajasthan states of India (Fig. 2). These areas experience an annual rainfall between 100 and 500 mm with a coefficient of variation varying from 40 to 70%. The region is characterized by low and erratic rainfall with extremes of temperature (1–48 °C), high wind velocity and sandy soils. Vegetation constitutes primary source of life support where animal husbandry being major vocation of people that depends entirely on natural vegetation, besides being of direct economic relevance, provides stability to wind prone sandy friable surface covering nearly two-third of the region. Further, inhospitable climate, too deep or too shallow soils with low moisture and poor fertility, deep underground water, which is often brackish or saline, coupled with intense biotic pressure permits specialized plants, which are well adapted to these climatic, edaphic and biotic adversities and fluctuations. With the increasing biotic pressure, most of the arid and semi-arid regions are confronted with the challenges of producing more per unit land with uncertain and dwindling supplies of water.

To sustain livelihood, desert dwellers have tested many crops for the last several centuries. During this exercise, they have identified crops, which can be cultivated under the harsh climate of the region with scarce water through rainfall or in the form of deep underground water. This traditional knowledge of hardy crops, cultivation practices and land races was inherited generation after generation to

D. C. Bhandari (✉) · S. Lodha
Division of Plant Improvement, Propagation and Pest Management,
Central Arid Zone Research Institute, Jodhpur 342003, Rajasthan, India
e-mail: bhandaridc@yahoo.com

P. R. Meghwal
Division of Integrated Land Use Management & Farming Systems,
Central Arid Zone Research Institute, Jodhpur 342003, Rajasthan, India

D. Nandwani (ed.), *Sustainable Horticultural Systems*, Sustainable Development and Biodiversity 2, DOI 10.1007/978-3-319-06904-3_2

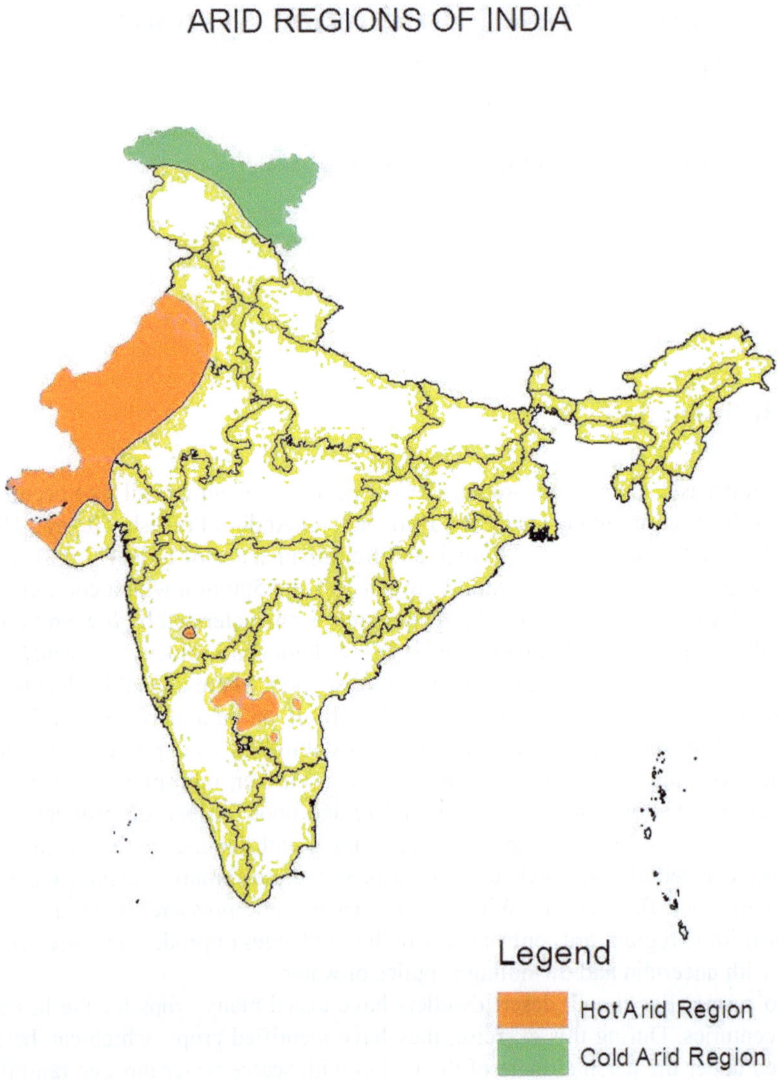

Fig. 1 Hot and cold arid regions in India

the extent that 'Thar desert' is considered as most thickly populated desert of the world because of time tested water harvesting structures providing livelihood security. The principle rainfed crops include pearl millet (*Pennisetum glaucum* (L.) R. Br.), guar or cluster bean (*Cyamopsis tetragonoloba* (L.) Taub.), moth bean (*Vigna aconitifolia* Jacq. Marechal), green gram (*V. radiata* (L.) Wilczek), cowpea (*V. unguiculata* (L.) Walp.), sesame (*Sesamum indicum* L.), sorghum (*Sorghum vulgare* Pers.), etc. These crops are cultivated during rainy season (July–October) at the

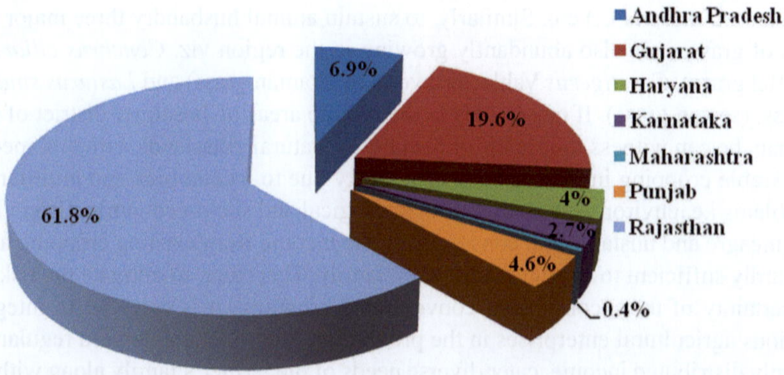

Fig. 2 Area under hot arid region in different states of India

onset of the monsoon. Depending on distribution of rainfall, fluctuations in temperature and other climatic features and selection of varieties, these crops mature in a period of 60–100 days. Produce of these crops are the main source of food for the people, which they consume for a year, keep as a security for next year (may be a drought year) and remainder is sold in the market. Certain time tested land races were evolved, which can produce seed as well as fodder for the livestock maintained by the farmers.

During winter season, wherever irrigation facilities are available, farmers grow wheat (*Triticum aestivum* L.), mustard (*Brassica juncea* (L.) Czern & Coss.), chickpea (*Cicer arietinum* L.) cumin (*Cuminum cyminum* L.), blond psyllium or isabgol (*Plantago ovata* Forsk.), onion (*Allium cepa* L.), garlic (*A. sativum* L.), etc. Among these, cumin and isabgol are also monopoly crops of this region in India owing to favorable soil and climatic conditions. These are considered as 'cash crops' because even under limited irrigation farmers can earn substantial amount of income from one hectare of land. In spite of being risk oriented, farmers are traditionally cultivating these crops in their fields by adopting a sound crop rotation.

In addition to above annual crops, some important tree species are also an integral part of arid agriculture, which provide alternative means of security to the farmers. One of the most important trees is the Indian mesquite locally called as khejri (*Prosopis cineraria* (L.) Druce), also known as 'Kalpavriksh' of the Indian desert. This tree is the most important component of the agro-forest in agricultural fields. Farmers are maintaining 30–60 trees of khejri in a hectare as it improves the growth of companion annual crops. Being a legume, it also fixes nitrogen in the soil. All parts of tree are used in one or other way as fodder or vegetable. Other trees of importance are marwar teak (*Tecomella undulata* D. Don), gum Arabic or kumat (*Senegalia senegal* Britton), Israeli babool (*Acacia tortilis* Hayne), etc. The region is also bestowed with drought hardy horticultural and medicinal plants. Some of these are Indian jujube (*Ziziphus mauritiana* Lamk.), kair (*Capparis decidua* (Forsk.) Edgew), Bengal quince or bael (*Aegle marmelos* (L.) Corr. Serr.), henna

(*Lawsonia inermis* L.) etc. Similarly, to sustain animal husbandry three major species of grasses are also abundantly growing in the region viz. *Cenchrus ciliaris* L. (buffel grass), *C. setigerus* Vahl. (birdwood or dhaman grass) and *Lasiurus sindicus* Boiss. (sewan grass). If one travels in the remote areas of Jaisalmer district of Rajasthan, he can witness thousands of hectares of natural grasslands with this species.

Arable cropping in the region is often risky due to its complex and multifarious problems i.e. environmental, biotic, technological and socio-economic. Crop yields are meagre and unstable and consequently the income from existing cropping alone is hardly sufficient to sustain the farmers' family. Therefore, to mitigate the risk and uncertainty of the income from conventional cropping, it is essential to integrate various agricultural enterprises in the production programme that yield regular and evenly distributed income, cater diverse needs of the farmer's family along with imparting sustainability through conservation and improvement of natural resources in fragile arid ecosystem. These regions are endowed with appreciable agro-ecological diversity and hence various components viz. crop, animal, tree, grass, fruit tree can be integrated in production system for livelihood security. Moreover, simple and high input packages designed for mono crop systems do not fit well either with the complexity and diversity of resource poor farming community or with their poor access to agricultural and risk prone environment.

2 Importance of Horticultural Crops

Over the past 40 years, the world's agricultural systems have been changing in response to population pressures (Waterlow et al. 1988). Population growth and local economics are driving both the intensification of agriculture and its extensification in to the marginal lands, where risks of crop failure and environmental degradation are high. As Lal (1987) points out, 'subsistence farmers, who face famine, would consider a successful technology to be one that produces some yield in the worst year rather than one that produces high yields in the best'. Horticulture based production systems are now considered to be the most ideal strategy to provide food, nutrition and income security to the people (Chundawat 1993; Chadha 2002). Integration of annual crops with fruit trees yields multiple outputs that ensure production and income generation (Osman 2003).

The importance of horticulture in improving the productivity of the land, generating employment, improving economic conditions of the farmers and entrepreneurs, enhancing exports and above all, providing nutritional security to the desert dwellers, can hardly be overemphasized. Horticulture has assumed significant importance in the crop diversification in recent years, which has become essential to arrest serious land degradation and enhancing the farm income. In fact, the horticulture has also gained commercial importance with a very significant share in the economy of the region. Diversification of agriculture from traditional land use with predominantly cereal/legume-based cropping systems to more productive and remunerative one has become a milestone to be achieved. Horticulture provides one

Table 1 Improved cultivars, propagation method, plant geometry and yield potential of different fruit crops

Fruit crops	Improved cultivars	Propagation method	Yield (kg plant^{-1})
6×6 m spacing			
Bengal quince	Dhara Road, Faizabadi local, NB5, NB9, Pant urvashi, Pant aparna, Pant shivani and Pant sujata	Patch budding (May–July)	30–60
Indian cherry	–	Seeds and budding	40–150
Indian gooseberry	Chakiya, Kanchan, NA7, Krishna, Anand 2	Patch budding (July–August)	40–150
Indian jujube	Gola, Mundia (early) Seb, Banarasi, Kaithli, Goma keerti (medium), Umran, Illaichi, Tikdi (late)	I-budding (July–August)	40–100
Orange	Kinnow	Budding (March–April)	30–50
Sweet orange	Mosambi and blood red	Budding (March–April)	30–50
5×5 m spacing			
Pomegranate	Jalore seedless, Ganesh, G137, G131, P26, P 23, Mridula, Araktha, Bhagwa, etc.	Hard wood cutting and air layering (July–August)	15–25
Sour lime	Kagzi lime, Vikram, Pramalini, etc.	Air layering and budding	20–30
4×4 m spacing			
Karonda	Pant manohar, Pant suvarna, Pant sudarshan	Seeds and cutting (August)	10–20

of the few viable and most attractive alternative land use system. Apart from their contribution to the total agricultural production, their potential for providing much higher income to the farmers has been another major factor for favoring these crops in this campaign. Awasthi and Pareek (2008) have reviewed the horticulture based cropping system for arid region. The improved cultivars of prevalent fruit crops with their propagation method, spacing and yield potential was worked out during past four decades (Table 1).

2.1 Horticultural Crops

2.1.1 Indian Jujube (*Ziziphus mauritiana* Lamk.)

The Indian jujube (ber) of family Rhamnaceae is one of the most ancient cultivated fruit trees in north Indian plains. It grows even on marginal lands or inferior soils where most other fruit trees either fail to grow or give very poor performance. It is regarded as the king of arid zone fruits and also as poor man's apple. There are three main species found in the country. The *Z. mauritiana* is the main species of commercial importance with its several varieties. *Z. nummularia* is prized for its leaves (rich in protein) which provide fodder (Pala) for livestock. The third one, *Z. rotundifolia* also bears edible fruits but of smaller size. It is used as rootstock for commercial Indian jujube. The seeds contain saponins, jujubogenin (Kawai et al. 1974) and obelin lactone. Jujube fruits contain fairly high amount of vitamin C, besides vitamin A, B, protein, calcium and phosphorus (Jawanda and Bal 1978). It is a perennial hardy fruit tree which gives income from multiple products such as fruits, fodder and fuel wood even in severe drought conditions to the resource deficient farmers. It is the only fruit crop which can give good returns even under rainfed conditions and can be grown in a variety of soils and climatic conditions ranging from sub-tropical to tropical.

2.1.2 Indian Gooseberry (*Emblica officinalis* Gaertn.)

The Indian gooseberry (aonla) of family Euphorbiaceae is being cultivated in India since Vedic Era. As a result of intensive research and development, it has attained commercial status and also proved to be potential fruit crop for arid ecosystem. It is hardy, prolific bearer and highly remunerative even without much care and can be grown in variable agro-climatic and soil conditions. The fruits are recognized for their nutritive, medicinal and therapeutical values and are rich source of vitamin C (4–9 mg g^{-1}), pectin, iron, calcium and phosphorus. The fruit is the main ingredient in Chayvanprash and triphala used in Ayurvedic medicine.

2.1.3 Indian Cherry (*Cordia myxa* L.)

Indian cherry of family Boraginaceae, locally known as lasoda is another important fruit plant suitable for arid and semi-arid regions of India. Its fruits and other parts have multiple uses in human health, nutrition and other uses. Green unripe fruits are important as fresh vegetable and pickles during April–May when availability of conventional vegetables is scarce. The species is also important ecologically in providing vegetative cover as tree component of arid farming system, preventing soil erosion and promoting biodiversity. The advantage with this species for agro forestry system is that it offers least competition with rainy season crops since its fruiting season is during summer season when main crops are already harvested. This plant also offers scope in using harvested rain water for fruit production since it requires irrigation only for 2–3 months period during summer season (April–June).

2.1.4 Pomegranate (*Punica granatum* L.)

Pomegranate (anar) of family Lythraceae is an economically important commercial fruit crop of arid and semi-arid regions. Commercial plantations of pomegranate exist in Maharashtra, Gujarat, Rajasthan, Andhra Pradesh and Karnataka owing to its preference for arid climate. Its xerophytic characteristics and hardy nature makes it suitable crop for dry, rainfed, pasture and undulating land, where other fruit crops cannot grow successfully. Besides, being a favorite table fruit it is also used for preparation of juice and squash. Dried seeds give an important condiment coined as anardana. It also has medicinal value and rind is being used for dyeing cloths.

2.1.5 Indian Mesquite (*Prosopis cineraria* (L.) Druce)

Indian mesquite (khejri) of family Fabaceae is an important component of farming system and plays significant role in the economy of Indian desert. It is found growing in the arid and semi-arid parts of Rajasthan, Gujarat, Haryana, Punjab, Delhi and some parts of southern India. This tree grows well in all sorts of climatic constraints which is evidenced by the fact that new foliar growth, flowering and fruiting occur during extreme dry months (March–June) when most other trees of the desert remain leafless or dormant. Because of its multiple economic value and suitability in agro forestry systems, it is conserved in arable land where its population is regulated by the farmer (Saxena 1977). All arid land forms except hills and saline depression receiving an average rainfall of 150–500 mm are having good density of the tree. The immature pods are rich in crude protein, carbohydrates and minerals. Duhan et al. (1992) recorded 18 % crude protein, 56 % carbohydrates, 0.4 % each of phosphorus and calcium and 0.02 % iron in immature pods, which are used as vegetable both fresh as well as after dehydration, while ripe dried pods having 9–14 % crude protein and 6–16 % sugar (Arya et al. 1991) can be powdered and used in the preparation of bakery items such as biscuits and cookies. The variability in pod quality traits is most important from the horticultural quality point of view. Diversity was observed in pod characteristic such as taste, tenderness, fiber content, color, length, thickness, seed number, seed size, protein content and mineral constituents (Dwivedi et al. 1997; Pareek 2002).

2.1.6 Kair (*Capparis decidua* (Forsk.) Edgew)

Kair is a multipurpose, perennial, woody shrub or small tree of family Capparaceae which grows widely without much care in the Thar Desert of western Rajasthan. It is much branched, leafless bushy and thrives well in the most adverse climatic conditions and in the soils of poor fertility. It is highly suitable for stabilizing sand dunes and controlling soil erosion by wind and water. Due to its xerophytic adaptive nature, the plant grows successfully under harsh climatic conditions. Its berry-shaped unripe fruits are rich in carbohydrates, proteins and minerals used as fresh vegetables and in the preparation of pickles. Dehydrated fruits are used in the off season as vegetable either alone or in combination with other dried vegetables. In general, it is highly valued by inhabitants of hot arid areas. Natural propagation occurs through seeds and root suckers, though vegetative propagation through hard wood cuttings has been tried (Meghwal and Vashishtha 1998).

2.1.7 Karonda (*Carissa carandas* L.)

Karonda is an evergreen spiny shrub or a small tree up to 3 m height and suitable for arid tropics and sub-tropics. It grows successfully on marginal and wastelands. The plant is also useful for making attractive thorny dense hedge around any fruit orchard. It yields a heavy crop of attractive berry like fruits which are edible and rich in vitamin C and minerals especially iron, calcium, magnesium and phosphorus. Mature fruit contains high amount of pectin and, therefore, besides being suitable for making pickle, it can be exploited for making jelly, jam, squash, syrup and chutney, which are of great demand in the international market. Its main flowering season is March–April with fruits maturing during August–September which enables the plants to make best use of monsoon rain. However, some varieties/plant types also flower during October–November.

2.1.8 Bengal Quince (*Aegle marmelos* (Linn.) Correa)

Bengal quince (bael) of family Rutaceae is an indigenous hardy fruit crop and can be grown successfully in dry areas. It is well known for its nutritional and therapeutic properties. The ripe fruits are laxative and unripe ones are prescribed for diarrhea and dysentery and are in great demand for native system of medicine such as Ayurvedic. Various chemical constituents, viz. alkaloids, coumarins and steroids have been isolated and identified from different parts of bael tree such as leaves, wood, root and bark by various workers. The marmelosin content of fruit is known as the panacea of the stomach ailments.

2.1.9 Kinnow (*Citrus reticulata* Blanco)

Kinnow mandarin is a hybrid cultivar of citrus developed by crossing King (*Citrus nobilis*) with Willow leaf (*Citrus deliciosa*) and is extensively grown in Punjab and Rajasthan states of India. This easy peel citrus developed has assumed special economic importance and export demand due to its high juice content, special flavor, and as a rich source of vitamin C. Its beautiful golden orange color, abundant juice, excellent aroma and taste have contributed greatly to the success of this fruit.

2.1.10 Kachri (*Cucumus callosus* (Rottl.) Cogn)

Kachri, a drought hardy cucurbit grows naturally and is also cultivated with rainfed crops. It is a short duration crop; flowering starts just after 30–35 days of sowing and produces 2–8 kg per vine small sized edible fruits of high nutritive value with little care. The fruits are rich in minerals specially calcium and used fresh in garnishing of vegetables. Most desert inhabitants store fruits round the year as dried slices, whole dried or in powder form for use. This is generally grown in all farming systems in combination with other rainy season crops without any competition for available soil moisture.

3 Traditional Production Systems

Arable cropping enterprise in dry lands of arid and semiarid regions of India is often un-remunerative due to aberrations of monsoon like late onset, prolonged dry spell, early withdrawal and unequal distribution of rainfall (Gill et al. 1998). Hence an integrated approach of land management is essential for efficient utilization of natural resources to meet the requirements of farmers without deteriorating the land productivity and also to stabilize the income (Gill and Deb Roy 1998). Even the traditional cropping system has multi species character example pearl millet, green gram, guar, moth bean, sesame between Indian mesquite and between fruit trees like Indian cherry, jhar ber, kair, pilu (*Salvadora oleoides*). Leaves of these fruit trees are lopped and fed to livestock during the lean period. The prevalence of this system reveals its potential advantage in production, stability, resiliencies and ecological stability. Farmers are often seen mixing the seeds of above arable crops altogether for sowing in their fields as mixed cropping with an optimism to harvest at least some produce even when the conditions are not congenial for one or another crop. Under such a situation, horticultural based production systems are considered an ideal strategy for the overall development and wellbeing of its farming community of the region.

3.1 Agri-horti System

Intercropping of agricultural crops with woody species is an age old practice in traditional farming systems in our regions. Growing of arable crops in association with horticultural plants is one of the viable approaches to sustain the agricultural production and stabilize the rural economy in the region. Suitability of intercrops depends primarily on soil and climatic conditions, however, compatibility aspects deserve prime consideration. Crops of competitive nature are largely not preferred and such crops are usually not grown as intercrop. Exhaustive crops like maize, wheat, sugarcane, cotton, etc. are not worth cultivation with horticultural plants (Krishnamurthy 1959; Naik 1963). By contrast, crops with companion and synergistic attributes are considered compatible to accentuate early income, optimize land use efficiency, facilitate better harness of solar energy, reduce the soil erosion, increase biological efficiency both in time and space dimensions in horticulture/fruit based production systems (Table 2).

Jujube plants after planting cover the entire inter-row spaces in a period of about 5 years under rainfed conditions. During this period, considerable loss in soil health particularly soil erosion occurs from these barren interspaces. By inclusion of suitable intercrops during rainy season, these losses can be minimized and additional income may be generated. Inter cropping in newly developed jujube orchard had no adverse effects on plant growth up to 5 years. However, in subsequent years, tall growing crops like pearl millet and guar were adversely affected and the effect was perceptible up to 2.5 m distance. Studies revealed that the yields of both jujube and

Table 2 Recommend agri-horti crops component for Indian arid regions

Growing conditions	Horticultural component			Crop component
	High storey	Medium storey	Ground storey	
Rainfed (150–300 mm)	Bordi and Indian mesquite	Jhar ber	Cucurbits and guar	Guar, moth bean, pearl millet and sesame
Rainfed (300–500 mm)	Indian cherry, Indian jujube and Indian mesquite	Jhar ber	Cowpea, cucurbits, guar and Indian bean	Cowpea, guar, green gram, moth bean, pearl millet and sesame
Irrigated	Bengal quince, Indian gooseberry Indian jujube and Indian mesquite	Guava, kinnow, karonda, lime, pomegranate and sweet orange	Brinjal, chilli, cole crops, cucurbits, garlic, okra, onion, peas, root/leafy vegetables and tomato	Chickpea, green gram, groundnut mustard and seed spices

the intercrops were higher than in monoculture (Singh 1997). Intercropping of guar, green gram and sesame with jujube (cv. *Seb*) increased the fruit yield to 14.8 from 5.2 kg tree^{-1}. Cultivation of guar in this system gave additional advantage of 782 kg seed yield, which was higher than that obtained from green gram and cowpea in drought years. However, Gupta et al. (2000) reported that 3-year old plantation of jujube at a density of 400 plant ha^{-1}in association with green gram performed well with seasonal rainfall of 210 mm. Intercropped green gram yielded only 160 kg ha^{-1} as against 620 kg ha^{-1} from pure crop, but the fruit yield from intercropped system increased the net profit to INR 2886. The economic analysis of cost and returns of jujube and Indian goose berry based agri-horti systems revealed that highest net profit of INR 13,487 from one hectare was realized from jujube with guar followed by 11,700/- from jujube with green gram (Meghwal and Henry 2006). Jujube, being a multipurpose species provided better income and interspaces were best utilized for growing of leguminous crops. Least gross return and negative profits were obtained from Indian gooseberry with guar, since fruit plants were just 3 years old without bearing of fruits; nevertheless an additional income of INR 3720 from a hectare was obtained just by growing guar, which met out 79 % of the expenditure on maintaining this system. The highest B:C ratio (2.15) was achieved in jujube and guar followed by sole plantation of jujube (2.10) and jujube with green gram (2.02), which was higher than the B:C ratio obtained by either of sole crops. This shows that agri-horti system minimizes the risk and helps in imparting economic stability. Cover cropping with lobia (*Dolichos biflorus*) was found to increase water holding capacity of light soils as a result of increased organic carbon content in these regions (Pareek and Vishal Nath 1996). The legume, Caribbean stylo (*Stylosanthes hamata*) is a good cover crop in the semi-arid regions (Raturi and Chadha 1993). Studies on standardization of jujube based production system revealed that 5 × 5 m spacing (400 plants ha^{-1}) for cultivation of jujube (cv. *Gola*) was ideal for legume intercropping. To develop viable agri-horti model for the Kachchh region of Gujarat state, a study was conducted with Indian gooseberry, jujube and pomegranate as fruit crops and cowpea (cv. *GC3*), guar (cv. *Maruguar*) and moth bean (cv. *CZM2*) as annual crops (Dayal et al. 2009). The results revealed that intercropping gave higher yield of annual crops and within fruit crops yield promotion was higher with jujube as compared to Indian gooseberry and pomegranate.

Under irrigated conditions, horse gram (*Macrotyloma uniflorum*), cumin, chilli (*Capsicum annum* L) and other vegetables are more profitable as intercrops with jujube (Pareek 1983). In the semi-arid regions, vegetable guar gave the highest net returns per hectare followed by cowpea, okra and brinjal as intercrops with jujube (Raturi and Chadha 1993). Similarly, intercropping of pea in jujube orchard did not affect the growth and fruit yield and was found more economic and suitable to increase per unit crop yield (Singh and Vishwanath 1995). However, strong competition has been observed between jujube and crops like wheat and chickpea (Khan 1993). He suggested that the cost of maintaining jujube trees was more than the value forgone by growing these crops. Therefore, such crops should not be grown as intercrop in jujube orchards.

Using sprinklers, studies were conducted in a jujube based cropping system having rainy and winter season annual and perennial crops on a typical sand dune of undulating topography (Saroj et al. 2003). Jujube (cv. *Gola*) was planted in three spacing (6×6, 8×8 and 16×4 m) as over storey component while groundnut and vegetable guar as rainy and wheat and mustard as winter season crops and Indian aloe as perennial under storey components. Studies revealed that during initial stage of jujube establishment, there was no competition between both the components for resources. Instead the inputs applied for the ground storey components enhanced the vegetative vigor of jujube plants than its sole plantation. It appears that management practices played an important role than land use system for better establishment and growth of jujube. The growth of jujube with Indian aloe was more vigorous due to higher level of inputs, particularly the frequency of irrigation (9). In the third year of plantation, jujube plants came in bearing and yielded 1440, 813 and 823 kg fruits ha^{-1} under 6×6, 8×8 and 16×4 m spacing, respectively. Among under storey components, highest yield of jujube fruits was recorded with Indian aloe (1355 kg ha^{-1}), while with groundnut-wheat and guar-mustard, the yield level of jujube fruits was almost equal (962 and 980 kg ha^{-1}, respectively) and minimum was recorded under sole plantation (809 kg ha^{-1}). The spacing of jujube had marginal effect on yield of ground storey components during initial stage of establishment. The average yield of ground storey crops over 2 years was recorded as 348, 5097, 1005, 1149 and 208 kg ha^{-1} in groundnut, guar, wheat, mustard and Indian aloe, respectively. These studies demonstrated that guar-mustard and Indian aloe can be integrated as a compatible ground storey component with jujube as compared to groundnut-wheat, which is a prevalent rotation under irrigated conditions of the region. Incidentally, guar-mustard rotation required only four additional irrigations compared to 12 in other rotation. By adopting such a system, even marginal lands can also be brought under cultivation as evident by the fact that yield levels during second year of winter season cultivation were better owing to stabilization of field.

Introduction of canal irrigation in certain arid districts has resulted in crop diversification particularly cultivation of cash crops like groundnut. However, the untimely availability of canal water along with reduced quantity has compelled for inclusion of low water requiring, perennial crops and fruit trees in the cropping system for its sustainability under adverse conditions. Intercropping of annual crops with the perennial trees may provide the extra income to the growers when the fruit trees are in their juvenile phase along with assured production from the system in canal command area. Therefore, an experiment was conducted under extreme arid situation (<250 mm rainfall) to find out the productivity of different intercrops with fruit trees. Three crops viz. moth bean, guar and groundnut were intercropped in interspace of 3-year old jujube, bael and kinnow orchards. Crop diversification in these cropping systems revealed that maximum plant height of moth bean and groundnut was recorded under the intercropping with bael, whereas in jujube plantation maximum plant height was recorded in guar (Yadava et al. 2006). Moth bean inter cropping with bael produced 14.6 and 27.2 % higher total dry matter and grain yield, respectively than its intercropping with jujube. Among the entire intercrops, highest mean root dry matter addition to soil after crop harvest was observed in groundnut, which was significantly higher over the intercropping of moth bean and guar. However, the mean highest system productivity (871 kg ha^{-1}) was observed under the intercropping of groundnut with the fruit trees, which was 15.1 and 9.8 % higher over intercropping of moth bean and guar, respectively. Highest total income and net profit was realized with bael with groundnut followed by jujube with groundnut and kinnow with groundnut intercropping. The highest B:C ratio was recorded with bael and guar followed by bael and moth bean (Fig. 3). Among the fruit trees, bael appeared most promising with highest B:C ratio irrespective of any annuals. The highest B:C ratio though was obtained with groundnut (2.56) but almost equal performance of guar (2.48) in association with any fruit trees provides scope of this industrially important seed legume more than groundnut in extreme arid conditions as it is less water requiring, labor intensive, less exhaustive and is having low requirement of plant protection measures.

A study on kinnow based system under irrigated conditions of north-western arid part was conducted on pre-bearing (3–5 years) and bearing (8–10 years) orchards for two consecutive years (Bhatnagar et al. 2007). During rainy season, cotton and green gram, while in winter barley and chickpea and their different combinations were grown in both the types of orchard. Their effects on tree growth parameters, leaf nutrient status, number of fruits, fruit yield and physico-chemical parameters were studied. Significant effect of intercrop on the growth of trees in pre-bearing orchard was observed only during first year. Available nitrogen content in pre-bearing and bearing kinnow leaves was lower in the intercrop than the normal leaves and it was also lower in pre-bearing trees. This may be due to low requirements and consequently less absorption of nitrogen during pre-bearing stage of the orchard, which was also true for P and K contents. Effect of intercrops did not show any significant effect on total soluble solid contents, acidity and ascorbic acid content of fruits. Number of fruits per plant varied from 219 (cotton) to 299 (without intercrop) in different combinations. The less number of fruits/plant under cotton intercropping might be due to its exhaustive nature resulting in nutrient competition with the main crop. The fruit per plant under intercrops was maximum (293) in sole cropping of

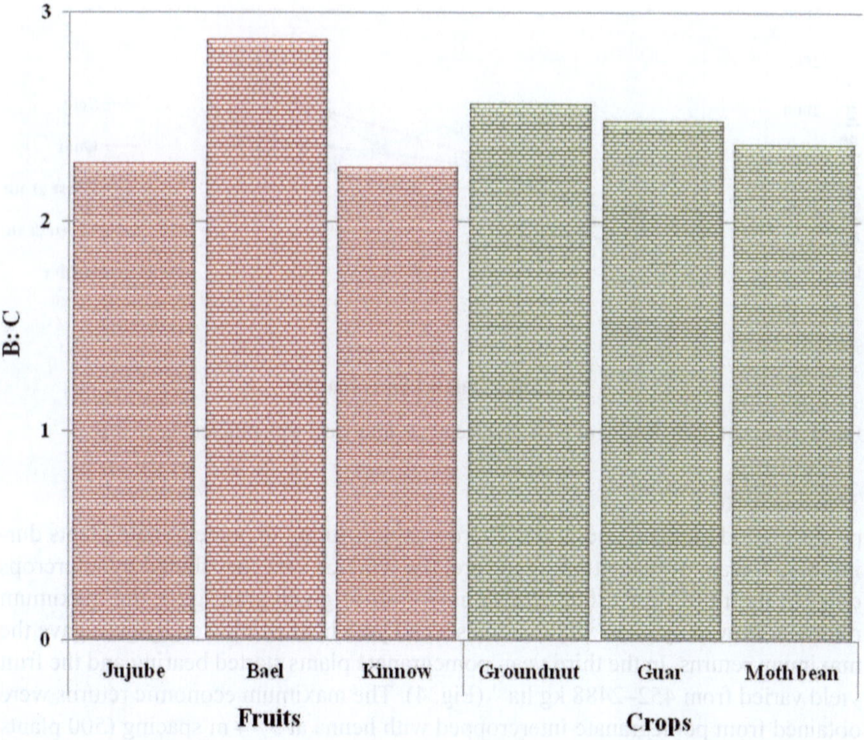

Fig. 3 Economic evaluation (B: C) of intercrops with jujube, bael and kinnow orchards

green gram followed by cotton-chickpea (280) and cotton-barley (277). The plant yield varied from 43.8 kg in cotton to 57.5 kg under cotton-chickpea intercrop. However, the plant yield was maximum (58.8 kg) in sole cropping of green gram and minimum under cotton (43.8 kg) indicating the beneficial association of legume with tree component. Singh et al. (1999) also reported the positive effect of legumes in mandarin orchards. By inclusion of cotton in production system, fewer yields of fruit was recorded, which can be attributed to competitive interaction of cotton (Reddy and Reddy 2005).

In pomegranate based cropping system, inclusion of guar, horse gram, green gram and henna improved the profitability over sole pomegranate cultivation on medium soils (Lal 2008). In 2003–2004, the highest biomass and grain yield among pulses intercropped between rows of pomegranate plants was in green gram followed by guar and horse gram. The gross returns were highest in green gram followed by horse gram, guar and henna, and lowest in sole crop. In second year, there was a terminal drought and the yields from intercrops were very low except in henna and guar. In the third year, there was an intermittent drought leading to low yields of intercrops except in henna and horse gram. The performance of horse gram was found to be better as the crop received rainfall in the later stage of growth, therefore, could

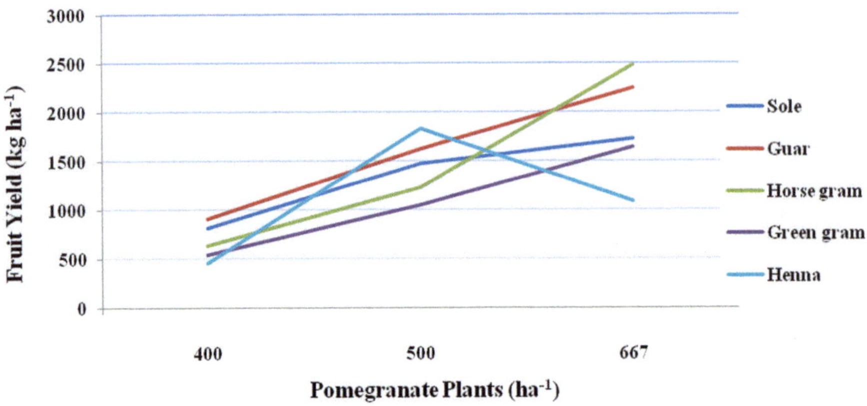

Fig. 4 Effect of plant density of pomegranate on its fruit yield with different intercrops

produce good biomass and grain. There was no fruiting in pomegranate plants during first 2 years of plantation; therefore, the produce was contributed by intercrops only. In the third year (2003–2004), intercrop of green gram gave the maximum returns followed by guar, while in the second year henna followed by guar gave the maximum returns. In the third year, pomegranate plants started bearing and the fruit yield varied from 452–2488 kg ha^{-1} (Fig. 4). The maximum economic returns were obtained from pomegranate intercropped with henna at 5×4 m spacing (500 plants ha^{-1}). However, in other crops the maximum returns were obtained in plots where these were intercropped with pomegranate at 5×3 m spacing (667 plants ha^{-1}). Further, pomegranate was found compatible with pearl millet, green gram, isabgol, sorghum and cumin in traditional pomegranate producing areas (Gupta 2000).

Indian gooseberry fruit based multi-storey production system comprising Indian gooseberry-jujube-brinjal-moth bean-fenugreek and Indian gooseberry-bael-karonda-moth bean-chickpea showed that ground storey crops did not affect growth of over storey crops and vice versa and these systems have been found promising under arid conditions (Awasthi et al. 2005, 2008). The net return obtained from above cropping through ground storey crops during first year was to the tune of INR 23,614 and 25,662 per hectare, respectively which increased up to 20 and 15%, respectively, in the second and third years. This indicates that during juvenile phase of fruit tree, there are ample scopes for raising annual, biennial and perennial crops, which can meet diversified needs of farmers.

In continuation of this study, in newly established Indian gooseberry orchard, experiment was conducted on performance of rainy and winter season intercrops (Awasthi et al. 2009). During rainy season, moth bean was a common rainfed intercrops, while in winter season fenugreek, chickpea, mustard and cumin were grown. Crop residues obtained after the harvest of moth bean, fenugreek, chickpea, mustard and cumin were recycled in same plots in all the treatment combinations. Studies revealed that intercrops grown in association with Indian gooseberry did not offer any competition on its annual growth and development due to different rooting behavior of companion crops. Significant increase in plant height and other growth parameters of fruit crop was recorded over period in all the intercrop combinations, however, it was more pronounced (> 100 %) when grown in association with moth bean-chickpea and moth bean-fenugreek crop rotations as compared to its sole plantation. This was attributed to inputs given to intercrops and their leguminous nature compared to sole fruit crop. Kumar and Pandey (2004) also reported better growth of Indian gooseberry when intercrop with leguminous crops. Saroj et al. (2003) reported the positive influence of intercrops on growth and vigor of jujube in the similar climatic conditions. The total biomass of both rainy and winter season crops increased over the years and was significantly higher when grown as intercrop with fruit than their sole cropping, which was ascribed to relatively less competition among companion crops due to deep rooting behavior of fruit crop. Incorporation of crop residues in the soil improved nutrient status of N (0.9–4.5 %), P (12.5–27.5 %) and K (2.45–14.7 %). Decomposition of residues and improved organic matter was attributed for better nutrient status (Kumar et al. 2008). Economic analysis showed that the intercrops gave higher returns compared to sole cropping, maximum in the sequence involving moth bean and cumin followed by moth bean-chickpea and moth bean-mustard. Higher returns due to cultivation of cumin were owing to its high market price. Improved economic returns were also reported in jujube and Indian gooseberry (Dhandar et al. 2004) and in Indian mesquite (Kaushik and Kumar 2003) based cropping systems.

These studies conclusively demonstrated that in arid regions, agri-horti system under rainfed and irrigated conditions provide assured sustainability due to additional gains from under storey crops, better utilization of natural resources like water and soil, recycling of on-farm wastes, improving soil fertility, preventing soil erosion and land degradation. The system was found more remunerative particularly during juvenile stages of the horticultural crops. However, for long term sustainability, choice of crops for intercropping is more important, which are photo insensitive, less exhaustive and economical. In general, legumes were found more compatible with horticultural plants under rainfed conditions, while under irrigated conditions cumin and mustard were considered more remunerative. Samadia et al. (2004) proposed suitable horticultural crops for agri-horti production systems for rainfed and irrigated areas of arid region.

3.2 Agri-horti-silvi System

Introduction of canal irrigation and development of tube wells has facilitated double cropping in many pockets in the region but needs judicious use of water. However, the region is prone to wind erosion often making agri-horti system also risky. To avoid the soil erosion and evaporative water losses through speedy winds from field, it has been scientifically advised to adopt the planting of perennials as shelter belts in or around the field. Such systems were termed as agri-horti-silvi system to provide more security, a symbiotic affinity which leads to the higher production of fruit, fodder, fuel and timber wood per unit area. Drought hardy fruit crops can survive and provide income to the farmers even under severe drought. Silvicultural plantations would check the drift of sand, provide forage, fuel and timber wood and would help in creating favorable micro-climatic conditions. To manage the arid lands in judicious way, a combination of horticultural crops, silvicultural plantations and fodder crops has been suggested.

To get maximum biomass and production per unit area from agri-horti-silvi production system, a study was conducted under extreme arid condition (Yadava et al. 2013). The green gram and guar were grown as intercrops under this system with 9 years old plantation of citrus (*Citrus aurantifolia*), mopane (*Colophospermum mopane*) and shisham (*Dalbergia sissoo*). Plant height of citrus and mopane was maximum with cluster bean, whereas shisham attained higher height with green gram. Highest stem girth and canopy of citrus and mopane was observed with green gram, which was 45.24 and 18.18 % (stem girth) and 39.48 and 19.54 % (canopy) higher over sole plantation. The effect of both the intercrops on growth of trees was almost equal. The mopane and shisham adversely affected the total chlorophyll content of intercrops. The highest chlorophyll content in guar (2.44 mg g^{-1}) and green gram (2.62 mg g^{-1}) and chlorophyll a:b ratio was recorded highest in intercropping with citrus and lowest with mopane. Among trees, the total chlorophyll content was highest in mopane (2.69 mg g^{-1}) followed by shisham (2.31 mg g^{-1}) and in citrus (1.58 mg g^{-1}), whereas chlorophyll a:b ratio was found to be non-significant. Highest biological and seed yield of green gram (1412.5 and 471.3 kg ha^{-1}) and guar (1352 and 419.8 kg ha^{-1}) was recorded with citrus over intercropping with shisham and mopane, respectively. Greater leaf area per plant and chlorophyll content in guar (2.44 mg g^{-1}) and green gram (2.62 mg g^{-1}FW) were attributed for more seed yield. Highest leaf water potential in guar (−3.18 MPa) and green gram (−3.28 MPa) was recorded in intercropping with mopane, whereas among trees it was highest in mopane (−6.99 MPa) followed by citrus (−5.51 MPa) and shisham leaves (−3.58 MPa). The study revealed that with use of minimum irrigation through sprinkler system, intercropping under horti-silvi system can be successful with established citrus plantations without any reduction in productivity. In agri-silvi system, mopane proved to be most exhaustive for moisture having more negative leaf water potential than shisham. Therefore, some cultural intervention (manuring, mulching or tillage operations) will be required for getting better yields from intercrops.

In an another study having wheat as winter season intercrop, irrigation and fertilizer for wheat promoted the growth of horti and silvi species compared to sole plantation (Singh et al. 2012). *C. mopane* with *C. myxa* combination was found more competitive because of vigorous growth of these species by utilizing above and below ground resources extending competition with the wheat reducing yield by 25 %. However, a combination of *P. cineraria* and *Z. mauritiana* appeared to be the best where yield reduction in wheat was bridged down to 6 %. This combination needs validation and promotion for greater benefits in terms of fodder, fruit and food production in the arid environment. Many crops such as vegetables, pulses, and fodder crops have been found suitable for intercropping in citrus orchards (Bajwa and Jawanda 1954). Further, other tree species and suitable arable crops should also be explored for better sustainability.

3.3 Horti-pastoral System

Growing grasses for in the fruit orchard for rearing cattle is pronounced as horti-pastoral system, an old and traditional farming system mainly adopted in dry areas. This system is also adopted as an alternative land use system in northern regions of India. Farmers in these areas grow fruit trees like guava, mango, citrus, pomegranate, jujube, Indian gooseberry, etc. and rear small cattle that graze on pastures. In modern times, hybrid varieties of fruit trees are being grown and people inhabiting these areas earn money from selling fruits, milk and other different products. In horti-pastoral system, the inter spaces between fruit trees are utilized for the cultivation of grasses and grass legume mixtures. In this system, fruit trees usually form the first tier whereas grasses are grown as ground storey crop. Only during dormant season of the fruit tree, the livestock are allowed to graze on the available pasture for a period of 3–4 months in a year depending on stocking rate. This system is more suitable for rainfed conditions where animal husbandry is also a source of livelihood. In practice, desert inhabitants having land and dependent solely on rains also maintain small ruminants like goat and sheep. In the integrated farming system model with a land holding of 7–8 ha, 10–15 % of land is allocated for this system.

Z. nummularia (jhar ber) is an important top feed species commonly found in farmers' fields. During drought years when crops fail, this plant comes to the rescue of the farmers and provide them fodder, fuel wood and fruits. Sharma and Vashishtha (1985) evaluated the performance of two species of *Ziziphus* viz. *Z. nummularia* and *Z. rotundifolia* in an established buffel grass pasture. During two years of establishment, 80 and 60 % mortality was recorded in *Z. nummularia* and *Z. rotundifolia*, respectively due to poor and slow root development in the transplanted seedlings, which could not compete with vigorous root system of grass component. Different populations (70–280 plants ha^{-1}) of both the species did not affect the pasture production once established. The study concluded that jujube species can be successfully grown in buffel grass pasture provided pasture is established after two years of the establishment of the jujube plants. The plant to plant and row to row spacing should be kept at 6 × 6 m in order to get 280 plants ha^{-1} of jujube. Economics of the system over a period of 9 years was worked out by method of discounting (Sharma and Diwakar 1989).The benefits accrued in the first year were less than the cost of production due to gestation period involving various operations like grubbing, ploughing, sowing, harvesting, etc. and the material costs like grass seeds, seedlings of *Ziziphus* spp., manure, etc. However, second year onwards, the benefits accrued were substantially higher than the production costs like gap filling, interculture, harvesting, etc. and other inputs (seeds, seedlings, water). The cost was positively related with the yield of forage and its seed production throughout the life of the system except the year of establishment. However, the benefits accrued varied with the rainfall pattern of each year (Fig. 5). The economic parameters delineated clearly the economic soundness of the system even on the higher costs (25 %). Therefore, this system may be adopted on large scale not only on common property resources (CPRs) but also at farmers' fields particularly on marginal lands under the similar agro-climatic conditions. This will generate additional income as well as help farmer in rearing small ruminants.

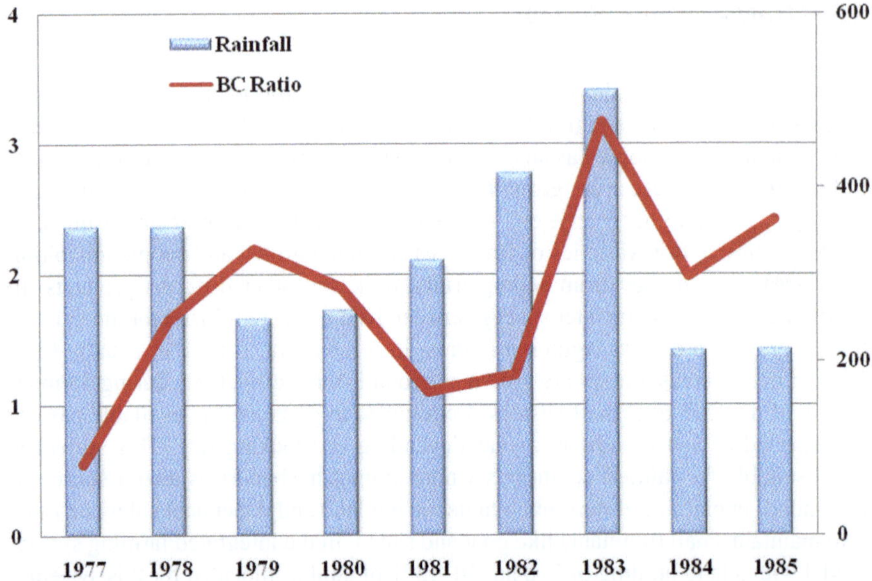

Fig. 5 Relationship between amount of rainfall and economic returns (B: C) from horti-pastoral system

In an experiment, where *C. ciliaris* was introduced in an 18 years old jujube orchard no adverse effects were observed on jujube yield but additional 2.5 t ha^{-1} dry forage and 25–30 kg grass seed yield was achieved, besides average jujube fruit yield of 7.5 t ha^{-1} (Vashishtha and Prasad 1997). In semi-arid conditions also, *C. ciliaris* and *Stylosanthus hamata* in various combinations were grown in the interspace of jujube (cv. *Gola*). Introduction of pasture was reported to have no adverse effect on the growth parameters of jujube during the establishment phase of first five years (Kumar et al. 2002). Forage production during the initial three years ranged between 3–7 t dry matter $ha^{-1}$and thereafter, a decline in the production was recorded.

An effort was made to compare economic viability of horti-pastoral system with arable farming (Gajja et al.1999). In horti-pastoral system, *Z. mauritiana* (cv. *Gola*) was taken as a major component with a population of 250 plants ha^{-1}. The inter row space between jujube was used for planting pasture grass, *C. ciliaris*. In this system, jujube fruits and grass were sold in the market, since no grazing was allowed during the study period. The cost of the system was more during the year of establishment and in the second year. However, from third year onwards returns were higher than the cost and system became static from sixth year onwards except in tenth year when the pasture was re-sown. The return from this system was obtained from sale of jujube fruits, grass fodder, pasture seed, leaf fodder of jujube and twigs (pruned wood) for fuel. At the terminal stage, returns were also available from wood of jujube plants. However, 90 % return each year was exclusively obtained from the sale of jujube fruits.

The total area sown with arable crops (pearl millet, guar, green gram, cowpea, castor and moth bean singly or in combination) varied from 0.4 to 2.65 ha for a period of 7 years. The variable cost, gross income and profit data indicated that highest variable cost was recorded in green gram followed by guar and lowest in castor. The gross income as well as profit over variable cost was higher in guar followed by cowpea. Therefore, in the changing economic scenario, guar should be cultivated on a large scale.

The discounted measures of both systems fail to take into consideration adequately the costs and benefits in different timings. However, by discounting technique, future cash flows can be reduced to the present worth estimation of benefits and costs to arrive at net-present-worth (NPW), benefit-cost ratio (B-CR), and annuity value (AV). Sensitivity analysis showed that how the value of criteria changes with the change in value of any variables in discounted cash flow. In these systems, AV and B-CR were higher in horti-pasture system compared to only arable farming. Comparing the AV of individual crop with alternate land use system, it was observed that horti-pasture was more beneficial than arable farming. In case of 25 % less rainfall than normal (330 mm) during cropping season, 30% less profit would be obtained, while returns from horti-pasture system will still remain unaffected. The sensitivity analysis revealed that horti-pasture system at 20 % discount rate was viable.

In Indian gooseberry based system, marvel grass (*Dicanthium annulatum*) as under storey forage species with different doses of nitrogen was assessed for ten consecutive years (Kumar et al. 2009). The pasture production was marginally higher with association of trees as compared to pure pasture. Further, high dose of nitrogen (60 kg ha^{-1}) gave significantly higher dry pasture as well as fruit yield compared to sole pasture. The mean B:C ratio over ten years was 1.85 for pure pasture and 3.7 with this system. The employment generation was 2.07 and 4.74 man days per month in pure pasture and in horti-pasture system, respectively. These studies demonstrated that adoption of horti-pastoral system in rainfed agriculture is economically viable, ecologically feasible and socially accepted to minimize the migration from rural areas for the hunt of jobs.

3.4 *Horti-silvi-pastoral System*

The concept of horti-silvi-pastoral production system is based on both biological and socio-economic considerations. The former includes all the advantages of fruit and trees on the soil and environment, such as close and efficient nutrient cycle, maintenance of organic matter, prevention of run off and soil erosion, regulation of microclimate, and above all adaptability of trees to soils that are incapable of sustaining annual crops. The socio-economic factors that substantiate the potential value of this production system are the burgeoning human and animal population pressure and lack of resources due to which poor marginal farmers are compelled to cultivate unproductive soils not suitable for agriculture and practice land management systems that may have disastrous consequences. Livestock population of the

Indian arid region having a man: animal ratio ranging from 1:1.3 to 1:5 depending upon rainfall and other conditions against national average of 1:0.5 ratio. Total human population in the region is around 50 million (Census 2011). Food and forage are the basic requirements for the health of desert dwellers and their animals. The consumption of fruits in the daily diet of rural mass is almost negligible as against the recommended requirement of 100 g/day (NIN 2010). Animal husbandry being major source of income, the availability of fodder in adequate quantity is all the more important. Therefore, this system has great potential to provide a sustainable land use system in arid regions.

The selection of fruit trees would depend upon the agro-climatic conditions and availability of irrigation facilities. In the region, there are certain pockets where irrigation facilities are available either due to underground resources, time tested water harvesting structures or canal irrigation. Recognizing the importance of trees as top feed during lean periods, it is useful to know the potential top feed species along with their palatability and nutritive value. As far as possible, multipurpose species would be an ideal choice under the given conditions. Ganguli et al. (1964) identified 16 top feed species based on palatability ratings for leaves in the order of preference as *A. tortilis*, *P. cineraria*, *S. oleoides*, *Z. nummularia*, *Acacia senegal*, *Albizia lebbeck*, *Anogeissus rotundifolia*, *A. pendula*, *Calligonum polygonoides*, *Azadirachta indica*, *Grewia tenax*, *Gymnosporia spinosa*, *P. juliflora* and *T. undulata*. These top feed species occur in distinct habitats and remain green even during hot summer months. To provide balanced nutritive fodder to the animals, both grasses and legumes in the pasture are advocated. The selection of the species would, however, depend upon the soil and climatic conditions. Based on the rainfall and edaphic factors suitable grasses are classified, which include *L. sindicus*, *Panicum antidotale*, *Saccharum bengalensis*, *C. ciliaris* for sand dunes and sandy plains; *D. annulatum*, *Heteropogon contortus* for sandy loam and sandy clay loam soils; *Sehima nervosum*, *D. annulatum* for sand stones and rocky sites; *C. ciliaris*, *C. setigerus* for well drained sandy alluvial soils; and *Sporobolus marginatus*, *Chloris virgata* for low lying saline soils. This classification is based on the edaphic factors (Bhati 1984), but the selection of grass species will also depend on the rainfall.

The management of trees, fruit crops and pasture grasses will involve planting all the three components in such a way so as to avoid competition for light, nutrition and water. The tree species should be planted on the periphery of the orchard in south-west direction, which would act as shelterbelt and wind break. While planting trees/shrubs for shelterbelt, selection of species should be on the basis of height. The tall species should be planted on the boundary, the middle row should consist of shrub, having low height and again the third row should be of tall species. These should also be staggered rather than in a straight row in order to minimize wind velocity and the damage caused by hot and cold winds prevalent in the region. The choice of the tree/shrub species will be governed by the rainfall conditions. In rainfall zone of 300–400 mm, *A. tortilis*, *P. cineraria*, *T. undulata* can be selected as tall species whereas *C. mopane*, *Z. nummularia*, *C. polygonoides* etc. can form less high bush component. The distance of planting of tree/shrub should be in such a way so that it forms an effective shelterbelt. The tree should invariably be planted at a distance of 4–5 m whereas shrubs at 2–3 m. Fruit plants require more care, therefore, selected cultivars of fruit plants should be planted at recommended distance, which would enable the pasture to grow in between rows of the fruit trees.

In horti-silvi-pasture system, silvicultural plantation (trees/bushes) should be planted as shelterbelt and fruit trees can be planted simultaneously. The sowing/planting of grass should be taken up 1 or 2 years after planting the trees/shrubs and fruit plants. The planting of fruit trees should be done during rainy season. In case of budded plants, in situ budding should be invariably followed. The root stock should be raised and can be budded after 1 year. The plants raised through cutting, layering etc. should be planted during rainy season and irrigated for establishment during first year. The fruit plants should be well trained during initial three years so as to provide them a definite shape suiting to the farming system. In case where the animals like sheep and goat are allowed for grazing in the pasture, the budding should be done at the height of 1 m and plants should be trained so that the animals do not harm the fruit trees. In case of Indian jujube, the leaves can be fed to the animals after the harvest of fruits. The pomegranate plants should be trained on single stem and branching should be allowed at 1 m height. The fruit plants should be provided required nutrition. The requirement of N, P and K will depend upon the species and age of the plant. The doses of N, P and K are constant after 6 year of age of the plants. The doses of nitrogen should invariably be given in three splits and that of phosphorus and potash should be given once a year. While planting/sowing grass or pasture legume, leaving 50 cm space from the basin of the fruit plant on each side will reduce the competition for nutrients and water in both the components. Grass extends its runners every year which may reach the basin of fruit plant therefore the suggested distance from basin to be maintained every year. In an ongoing experiment on horti-pastoral system at CAZRI, Jodhpur, *C. ciliaris* was introduced in an established orchard of Indian jujube in 1993, in between rows of 6 m distance. During third year, 3.29 t h^{-1}grass was harvested with 25 kg h^{-1} grass seed in addition to 29 kg $tree^{-1}$ jujube fruits. There was no adverse effect on jujube yield with the introduction of grass. Korwar et al. (1988) conducted work on horti-pastoral system at Hyderabad (Andhra Pradesh) taking guava and custard apple as fruit crop and *S. hamata* and *C. ciliaris* as grasses. Plant height of guava was more (1.78 m) with *C. ciliaris* whereas in custard apple it was more with control. The yield of *C. ciliaris* in first year was 5–9.6 t ha^{-1}.

The horti-silvi-pastoral system can produce more biomass per unit area, which also meets the basic requirement of fruit, fuel and fodder for the rural masses of the region. Not much work has been done on this multidimensional aspect and hence there is a need to conduct intensive research on multi-location to generate information for this composite system. This system will not only provide the judicious use of available land resources but will also generate additional employment opportunities for rural people.

4 Support System for Sustainable and Integrated Production Systems

The majority of the population in arid and semiarid areas depends on agriculture and pastoralism for subsistence. These activities face many constraints due to predominance of erratic rainfall patterns, torrential rainfall majority of which is lost due to runoff, high rate of evapotranspiration further reducing yields, weeds growing more vigorously than cultivated crops and competing for scarce reserves of moisture, low organic matter levels and high variables responses to fertilizers (CASL 2006). Hence, there is a need of a more efficient capture and use of the scarce water resources to recycle for its utilization. An optimization of the rainfall management, through water harvesting in sustainable and integrated production systems can contribute for improving the small-scale farmers' livelihood by upgrading the rainfed agriculture production. A method of water collection that has, historically, been "applied in arid and semiarid regions where rainfall is either not sufficient to sustain a good crop and pasture growth or where, due to the erratic nature of precipitation, the risk of crop failure is very high," (Prinz 1996) water harvesting is now being employed all over the world. Water harvesting has been defined as the collection of runoff for its productive use (Critchley and Siegert 1991), yet this definition is too general and can be more accurately defined as the process of collecting and concentrating water from runoff into a run-on area where the collected water is either directly applied to the cropping area and stored in the soil profile for immediate use by the crop (Prinz and Singh 2000). In Indian arid regions, the gap between water supply and demand necessitates harnessing of available water with efficient in situ and ex situ water conservation and proper recycling at critical stages of crop growth. The traditional water harvesting structures are of immense use; however, blending it with modern technology can make it more effective. Reducing water demand by changing water delivery scheme from supply to demand driven basis, improving the efficiency of water use through advanced irrigation technologies, improved soil and water management practices can play a great role in alleviating the adverse effects of water scarcity. The construction of farm ponds and tanks with appropriate catchment area helps in runoff collection during periods of high intensity rainfall. The water collected is used for drinking purposes and supplemental irrigation during periods of drought and for the establishment of the nursery plants and trees. The traditional methods of water harvesting still prevalent in the region are tanka (sump wells), nadis (village pond), and khadin system (runoff farming). In some of the arid districts of Rajasthan state about 42 % of the population is dependent upon nadis, 35 % on tanks, 15 % on open wells and tube wells whereas other sources account only for 7–8 % of the total requirement (Singh and Awasthi 2006). Studies conducted at CAZRI and elsewhere show that application of water at critical stages of tree/crop growth increases the yield substantially. Harvested rainwater can be used to provide supplemental/lifesaving irrigation, particularly to trees and crops during long dry spells within a season. Khan (1995) observed that with supplemental irrigation (60 l irrigation^{-1} plant^{-1}) the fruit yield of jujube and pomegranate increased substantially. In comparison to no irrigation, increase in fruit yield with 2, 4 and 6 irrigations for jujube was 46.4, 80.3 and 124 %, respectively, whereas in pomegranate it was 69.8, 112.5 and 191.7 %, respectively.

5 A Success Story

A progressive farmer having 4.5 ha of land was adopted by the Central Arid Zone Research Institute to demonstrate the jujube based horti-agri-silvi-pastoral production system. The problem of insufficient, uneven and erratic rains leading to short and long durations of moisture stress was overcome by adopting rain water harvesting system enabling the successful establishment and growth of jujube. The production system had 800 plants of jujube planted in 6 m row-width, native grass, interwoven with 414 multi-purpose tree species (MPTS) like *P. cineraria*, *T. undulata*, *E. officinalis*, *A. lebbeck* and *A. tortilis*, which contributed significantly to the improvement of environment and soil quality and reduced soil erosion and high wind velocity. Jujube is prone to wild animals, therefore a trench of 1.5 m depth and 2 m width was dug out around the periphery of the whole farm. Outside this trench, fast growing thorny species like *A. tortilis* and *P. juliflora* were planted in close spacing. A thermometer driven alarm bell was installed to give signal when temperature falls below 1 °C to apply measures to check frost injury. On an average fruit production per tree was 20 kg, which on hectare basis came to 5.36 t. These were graded into two categories. Large sized fruits were sold through fruit shops and departmental stores, while small sized were sold in fruit market for juice preparation. A wide publicity through press and media was given about this production system. Various farm operations (pruning, manuring, weeding, bird scaring, sprays, watering, harvesting and packing) were done by the trained villagers in neighborhood giving employment to the tune of 750 man days in a year. At the time of pruning, 1.8 t dry fuel and 940 kg ha^{-1} dry leaves were also obtained, which fetched INR 3800. Buds were also provided to various nursery growers resulting in rapid spread of improved jujube in arid and semi-arid regions. The waste land around the farm was utilized to grow indigenous grass *C. biflorus* and *C. ciliaris*. On an average, 600 kg of grass was harvested every year fetching INR 1000. MPTS provided fuel wood and pods from *P. juliflora* and *A. tortilis* used as excellent fodder and pods of *P. cineraria* as vegetable for human consumption. Integrating silviculture system in jujube orchard played significant role in changing the microclimate, soil conservation, reducing wind velocity and enhancing farm income and providing employment to the local people in lean period. In this diversified production system, six goats and two cows were reared and fodder was made available from farm. On an average 1.5 and 3 l milk was obtained from goats and cow, respectively. Waste recycling was done by making compost from fallen leaves, damaged jujube and rotten fruits, droplets of goat, cow dung, pods of leguminous trees and fruits of neem in order to get compost having disease suppressive characteristics besides improving soil fertility (Lodha et al. 2002; Bareja et al. 2013). A trench of 40 m length and 3 m width provided around 25 t of compost in 120 days. Amending soil with this compost made this farm almost completely organic.

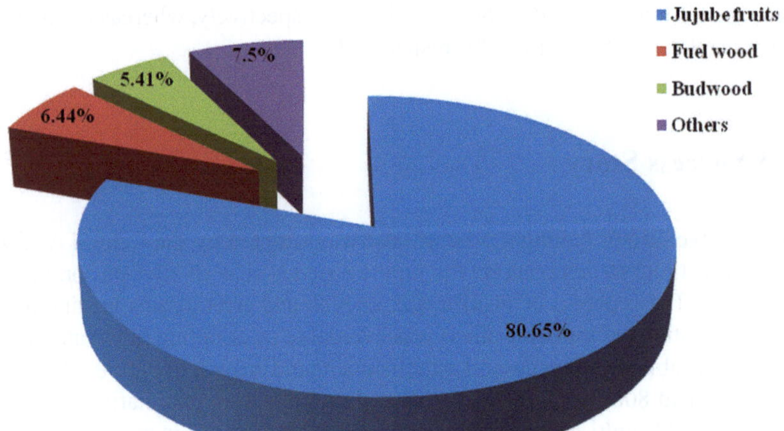

Fig. 6 Economic returns from horticulture based integrated production system at grower's orchard

A nursery was established for raising grafted jujube, which was the first private nursery in the region. Seeds of wild jujube were sown in last week of March for root stock that provided a chance of re-sowing in April in case of any mortality. During 1979–1982, budded plants of jujube (26,292) were sold to various agencies. As an impact of this nursery, many other grafted jujube nurseries came in to existence in next one decade and around 13 million saplings were sold for plantation in many states of India. The economic evaluation revealed that the maximum return was from the sale of jujube fruits; however the contribution of other components over period was also visible (Fig. 6). Thus, this jujube based diversified production system was considered as a role model for reproducing in other parts of arid region resulting in large acreage put under this system benefitting the local population. Moreover, the risk of failure of arable crops was averted and a sound system was developed. This progressive farmer received a national award (N. G. Ranga Farmer Award) from the ICAR, New Delhi in 2004 for diversified agriculture.

Conclusion

The adoption of above production systems has resulted in congenial microclimate reducing the aridity and crop losses and increased the productivity of the farm due to:

6.1 Efficient Land Use

The horticulture based production systems can be practiced both in rainfed and irrigated conditions for enhanced productivity, profitability and sustainability. These systems are also useful for those marginal and degraded lands where arable farming is either not possible or uneconomical. Adopting horti-pastoral or horti-silvi-pastoral system can ameliorate these lands in a long term due to soil conservation and improvement in fertility levels. Windbreak effect due to live fences and tree plantation leading to better soil moisture retention and reduced transpiration will be an additional advantage.

6.2 Socio-Economic Issues

Adoption of horticulture based production systems can improve the socio-economic conditions of resource deficient farming community. These will provide an assured source of livelihood as perennial component will generate farm produce even during low rainfall or drought conditions. Many of these underutilized fruit species play an important role in the social economy and livelihoods of tribal, small, marginal and landless farmers. Produce of trees provides additional income to these farmers and substantial livelihood support in addition to the nutritional security to the children and women. Organized production and processing for value addition of products would further enhance income of small and marginal farmers and also help in on-farm conservation of valuable germplasm (Malik et al. 2013). Increased biomass availability will help in integration of livestock in to the farming system. Most of the perennial components also provide fodder for huge animal population under adverse conditions. Nearly every part of *Ziziphus* plants can be utilized. Due to the high dry weight protein content, leaves are an important source of protein for animals (Arndt 2001; Dalziel 1937; Dastur 1952; Ngwa et al. 2000). Leaves of *Ziziphus* spp. are readily eaten by camels, sheep, goats and cattle (Tewatia and Khirwar 2002). Adoption of these systems in irrigated pockets will improve social conditions of the desert inhabitants owing to additional returns from per unit land.

6.3 Employment Generation

These production systems will certainly open an opportunity for additional employment for rural youth during most of the months compared to arable farming, which is exclusively dependent on rains and confined to a limited period. These systems involve collection of fodder from grass and top feed tree component, raising nurseries, grafting of fruit trees, training, application of agronomical and plant protection measures, irrigation, grading, marketing, pruning, and managing farm produce for compost preparation or recycling in soil.

6.4 Nutrition Component

The rural masses in arid regions do not have adequate fruits and vegetables in their diet. With large scale plantation of horticultural crops in their farm under integrated system, the inhabitants of region will certainly get nutrition in the form of vegetables and fruits. In Zimbabwe, the dry jujube powder is used in baking and to prepare jam (Maposa and Chisuro 1998) and a traditional loaf (Kadzere 1998) and kachaso, a crude spirit (Arndt 2001).

6.5 Entrepreneurship

Adoption of these systems can also help in establishing village based cottage industries for making jam, murabba, pickles, jelly, juice, squash, etc. This will help in generating additional employment to school drop outs and women folk in addition to a regular source of income. Such agro-based industries have already come in to existence in the region for the last one decade. In the lean period, same industries can prepare pickles from chilli and sauce from tomato cultivated as horticultural crops in the region under irrigated conditions. Cakes are made out of dried and fermented pulp from jujube in western Sudan (Dalziel 1937) and in Zambia (Kalikiti 1998). Such initiatives with newer products for developing entrepreneurship and to generate additional employment can also be started in this region. Jujube trees are considered amongst the best for rearing lac insects (Hussain and Khan 1962). A lac yield of 1.5 kg $tree^{-1}$ $year^{-1}$ was obtained by collection during October–November in India. It is a chief host plant of *Kerria lacca* and *K. sindica* (Li and Hu 1994). By using 6–8, 2–3 m long shoots of 2–3 cm thickness on a stump for inoculation by lac insects, a yield of 3–6 kg raw lac can be obtained in 3 years. When jujube trees are used for rearing lac insects, then their use for fruits is not viable. Jaiswal et al. (2002) collected socio-economic data on lac production from Jharkhand, Orissa and West Bengal. In the Islamabad area of Pakistan, *Z. jujuba* flowers attract honeybees, which can contribute to conservation and the economic stability of the people in the local area (Chemas and Gray 1991; Fatima and Ramanujam 1989; Muzaffar 1998). In India and Queensland, the flowers of *Z. mauritiana and Z. jujuba* are rated as a minor source of nectar for honeybees (Dash et al. 1992). The honey is light and of fair flavour (Morton 1987). In certain parts of India, *Ziziphus* leaves have been used as a food for silkworms. *Ziziphus* spp. have been recorded as a secondary food plant for rearing Indian tasar silkworm, *Antheraea mylitta* Drury, larvae and proved to be better than the primary food plant *Shorea robusta* (Dash et al. 1992).

6.6 Environment Issues

Improvement in soil health due to better fertility and moisture holding capacity, soil conservation, aeration and porosity, check in land degradation and soil erosion are the major advantages in horticultural based production systems. In general, there is a drop in ambient temperature during summer months and frost protection to annual crops in winters. Shade effects also lead to drop in soil temperature and improvement in soil moisture for better growth of intercrops. One of the prominent fruit crop, *Z. mauritiana* plays an important role in the conservation of soil because of its abundant and vigorous root systems even in Sahelian climate (Depommier 1988; Arndt 2001). Soil erosion in desert areas is largely due to the removal of structure less topsoil by wind and rain. This can largely be checked by planting wind breaks, creating shelterbelts and stabilizing sandy tracts and dunes with adapted grasses and shrubs like *Ziziphus* (Khoshoo and Subrahmanyam 1985). *Z. nummularia* shrubs have been shown to effectively check wind erosion, help in deposition of soil, and bring about a change in the microhabitat, causing favorable conditions for the appearance of successional species such as perennial grasses. Several species of *Ziziphus* can endure extreme stress caused by drought, salinity, and in some cases waterlogging. This makes the cultivated jujube ideal for planting on marginal or degraded lands provided the right genotypes are selected for alkali-sodic soils (Dagar et al. 2001; Hebbara et al. 2002) thereby controlling desertification.

6.7 Policy Related Issues

Arid regions share more than 10 % of the geographical area of the country, but harbor about 4 % human population only owing to many inherent reasons on one hand but carries a vast animal population on the other, due to animal husbandry based livelihood. Efforts to enhance arid zone development are seriously hampered by poverty, which goes hand in hand with household food insecurity and environmental degradation. Eradicating poverty, attaining food security and at the same time conserving the environment are seen as contradictory objectives to achieve. This is because the majority of the rural poor destroy their environment in order to earn a living. State and central government policy planners in the country are concerned with a situation when during complete and partial drought conditions, region needs support from neighboring states for food, fodder and other life supporting materials. For instance, in the 2 drought years (1987 and 2002), even water was supplied by public transport means to many severely affected districts, besides food and fodder. Large scale mortality was witnessed in animal population due to starvation, though the region is having large self-sown natural pastures. Introduction of canal irrigation in recent years has no doubt changed the agricultural scenario in the region but crop diversification and excess use of irrigation has also brought newer environmental problems including waterlogging. Frequent irrigation in sandy soils of arid regions always results in buildup of salinity thereby making the land unproductive. Therefore, the situation warrants the judicious use of irrigation and adoption of integrated farming system with component of perennial species (fruits, trees and grasses) to provide long term solution and sustainability in the region. Climate change is yet another issue for environmentalists and policy planners in the recent time, which is bound to adversely affect the productivity of region. Therefore, improvement in vegetation cover is also a matter of serious concern for all those who are concerned for arid zone development. Expansion of perennial component particularly those plants which can also provide livelihood security and nutrition to desert dwellers is required. In view of this need, National Horticulture Mission has given special emphasis for the promotion of horticultural crops, pressurized system of irrigation and other soil moisture conservation practices so that sizeable area can be brought under horticulture based production systems for judicious use of irrigation water in the region. At present 83 % water is being used for irrigation, out of which at least 25 % is going waste due to percolation and other losses. The aim of policy planners is to save at least 20 % of this water for its use in other productive purposes. Migration in the hunt of food and jobs is also a regular phenomenon, which has received due attention of policy planners and generating employment through village based cottage industries has been given greater emphasis in recent years. Number of self-employment schemes involving agro-based industries has been floated by state and central governments for rural youth in order to check the migration. These will keep farmers occupied with land based activities for extended periods, thus, staying back on their land almost 12 months in a year and hence taking better care of land.

References

Arndt SK (2001) http://chemsrv0.pph.univie.ac.at/ska/ziplant.htm

Arya S, Toky OP, Bisht RP, Tomar R (1991) *Prosopis cineraria*-promising multipurpose tree for arid lands. Agroforest Today 3(4):13–14

Awasthi OP, Pareek OP (2008) Horticulture based cropping system for arid region. Range Manage Agroforest 29(2):67–74

Awasthi OP, Singh IS, Dhandar DG (2005) Aonla based multitier cropping system for sustainable production in western Rajasthan. In: Proceedings of national seminar on globalization of aonla organized by Aonla Grower Association of India, Madurai, India, 7–9 Oct 2005

Awasthi OP, Saroj PL, Singh IS, More TA (2008) Fruit based cropping system for arid regions. Central Institute for Arid Horticulture, Bikaner. Technical Publication No. 25, p 18

Awasthi OP, Singh IS, More TA (2009) Performance of intercrops during establishment phase of aonla (*Emblica officinalis*) orchard. Indian J Agric Sci 79(8):587–591

Bajwa BS, Jawanda JS (1954) Intercropping suitability in citrus orchards. Indian J Hortic 11:13–15

Bareja M, Mawar R, Mathur M, Lodha S (2013) On-farm waste-based composts suppressive to *Macrophomina phaseolina* inciting dry root rot of guar in arid environment. Aust J Plant Pathol 42:9–16

Bhati TK (1984) Grass component in silvipastoral system with special reference to Indian arid zone. In: Shankarnarayan KA (ed) Agro-forestry in arid and semi-arid zone. CAZRI, Jodhpur, pp 150–155

Bhatnagar P, Kaul MK, Singh J (2007) Effect of intercropping in kinnow based production system. Indian J Arid Hortic 2(1):15–17

CASL (2006) Arid and semi-arid lands: characteristics and importance. Community adaptation and sustainable livelihoods programme. International Institute for Sustainable Development, Canada

Census (2011) Government of India, New Delhi

Chadha KL (2002) Diversification to horticulture for food, nutrition and economic security. Indian J Hortc 59:209–229

Chemas A, Gray VR (1991) Apiculture and management of associated vegetation of the Maya of Taxicacaltuyub, Yucatan, Mexico. In: Iqbal I, Hamayun M (eds) Studies on the traditional uses of plants of Malam Jabba Valley, District Swat, Pakistan. Agroforest Syst 13(1):13–26

Chundawat BS (1993) Intercropping in orchards. Adv Hortic 2:763–775

Critchley W, Siegert K (1991) A manual for the design and construction of water harvesting schemes for plant production. 17/1991, FAO Corporate Document Repository. http://www.fao.org/docrep/U3160E/u3160e00.htm#Contents

Dagar JC, Singh G, Singh NT (2001) Evaluating forest and fruit trees for rehabilitation of semiarid alkali-sodic soils in India. Arid Land Res Manage 15:115–133

Dalziel JM (1937) The useful plants of west tropical Africa. Crown Agents. 2nd Reprint, London 1955

Dash AK, Nayak BK, Dash MC (1992) The effect of different food plants on cocoon crop performance in the Indian tasar silkworm, *Autheraea mylitta* Drury (Lepidoptera: Saturniidae). J Res Lepidoptera 31(1–2):127–131

Dastur JF (1952) Useful plants of India and Pakistan. A popular handbook of trees and plants of industrial, economic and commercial utility. D. B. Tarporewala Sons & Co, Bombay, p 252

Dayal D, Shamsudheen M, Ram B (2009) Alternate farming systems suitable for Kachchh region of Gujarat. CAZRI, Regional Research Station, Kukma p 22

Depommier D (1988, Publ 1990) *Ziziphus mauritiana*: cultivation and use in Kapsiki country, northern Cameroon. Bois et Forêts des Tropiques No. 218, 57–62, CTFT-CIRAD, ICRAF, Nairobi, Kenya

Dhandar DG, Saroj PL, Awasthi OP, Sharma BD (2004) Crop diversification for sustainable production in irrigated hot arid ecosystem of Rajasthan. J Arid Land Stud 148:37–40

Duhan A, Chauhan BM, Punia D (1992) Nutritional value of some non-conventional plant foods of India. Plant Foods Hum Nutr 42:193–200

Dwivedi NK, Bhandari DC, Verma SK (1997) Collecting *Prosopis cineraria* (L.) Mac bride germplasm in western India. IPGRI Newsletter for Asia. The Pacific and Oceania No 22, p 19

Fatima K, Ramanujam CGK (1989) Pollen analysis of two multifloral honeys from Hyderabad, India. J Phytol Res 2(2):167–172

Gajja BL, Bhati TK, Harsh LN, Khan MS (1999) Comparative economics of silvopasture, hortipasture and annual crops on marginal agricultural lands of arid zone of Rajasthan. Ann Arid Zone 38(2):173–180

Ganguli BN, Kaul RN, Nambiar TN (1964) Preliminary studies on a few top feed species. Ann Arid Zone 3(2):33–37

Gill AS, Deb Roy R (1998) Tree/crop interface competition studies under semiarid conditions-relative yield of wheat. Range Manage Agroforest 19(2):154–164

Gill AS, Deb Roy R, Chaudhary BL (1998) Effect of multipurpose tree species on pulse production under semi-arid conditions. Indian J Agron 43(3):650–565

Gupta JP (2000) Technology approach for greening degraded arid lands. CAZRI, Jodhpur, p 51

Gupta JP, Joshi DC, Singh GB (2000) Management of arid agro-ecosystem. In: Yadav JSP, Singh GB (eds) Natural resource management for agricultural production in India. Indian Society of Soil Science, New Delhi, pp 553–668

Hebbara M, Manjunatha MV, Patil SG, Patil DR (2002) Performance of fruit species in saline-waterlogged soils. Karnataka J Agric Sci 15:94–98

Hussain AKM, Khan TA (1962) Lac cultivation in east Pakistan. Pak J Forest 12(3):220–222

Jaiswal AK, Sharma KK, Bhattacharya A (2002) Household survey for assessing utilization of conventional lac host trees for lac cultivation. New Agric 13:13–17

Jawanda JS, Bal JS (1978) The ber-highly paying and rich in food value. Indian Hortic 23(3):19–21

Kadzere I (1998) Role of *Ziziphus mauritiana* in the livelihood of some communities in Zimbabwe. In: International Workshop on *Ziziphus mauritiana*, Harare, Zimbabwe, 13–16 July 1998

Kalikiti F (1998) *Ziziphus mauritiana* in Siavonga district, Zambia. In: International Workshop on *Ziziphus mauritiana*, Harare, Zimbabwe, 13–16 July 1998

Kaushik N, Kumar V (2003) Khejri (*Prosopis cineraria*) based agroforestry system for arid Haryana. J Arid Environ 55(2):218–221

Kawai K, Toshiyuki A, Yukio O, Shibata S (1974) A new sapogenin in the saponins of *Zizyphus jujuba*, *Hovenia dulcis* and *Bacopa onniera*. Phytochemistry 13(12):2829–2832

Khan FS (1993) Bioeconomic interaction of single scattered trees on farmlands. Pak J Agroforest 43(3):173–180

Khan MA (1995) Importance of water resources management in arid Rajasthan. J Nat Herit Inst 4:53–59

Khoshoo TN, Subrahmanyam GV (1985) Ecodevelopment of arid lands in India with non-agricultural economic plants—a holistic approach. In: Wickens GE, Goodin JR, Field DV (eds) Plants for arid lands. Unwin Hyman, London

Korwar GR, Osman M, Tomar DS, Singh RP (1988) Dryland horticulture (Extension Bulletin No. 4). CRIDA, Hyderabad

Krishnamurthy S (1959) Effect of intercrops on the citrus health. Indian J Hortic 16(4):221–227

Kumar D, Pandey V (2004) Vegetative growth of aonla as influenced by intercrops under rainfed conditions of Agra. Orissa J Hortic 32:109–111

Kumar S, Kumar S, Baig MJ, Choubey BK, Sharma RK (2002) Effect of nitrogen on the productivity of aonla based hortipastoral system. Indian J Agroforest 4:94–97

Kumar B, Gupta RK, Bhandari AL (2008) Soil fertility changes after long term application of organic manures and crop residues under rice–wheat system. J Indian Soc Soil Sci 56(1):80–85

Kumar S, Kumar S, Choubey BK (2009) Aonla-based hortipastoral system for soil nutrient build-up and profitability. Ann Arid Zone 48(2):153–157

Lal R (1987) Managing the soils in sub-Saharan Africa. Science 236:1069–1076

Lal G (2008) Pomegranate-based cropping system for sustainable production under arid ecosystem. In: Narain P, Singh MP, Kar A, Kathju S, Kumar P (eds) Diversification of arid farming systems. Arid Zone Research Association of India and Scientific Publishers,, Jodhpur, pp 56–60

Li J, Hu X (1994) Rejuvenation and utilization of wild *Ziziphus mauritiana*. Forest Res 7(2):224–226

Lodha S, Sharma SK, Aggarwal RK (2002) Inactivation of *Macrophomina phaseolina* propagules during composting and efficacy of composts on dry root rot and seed yield of cluster bean. Eur J Plant Pathol 108(3):253–261

Malik SK, Bhandari DC, Kumar S, Dhariwal OP (2013) Conservation of multipurpose tree species to ensure ecosystem sustainability and farmers livelihood in Indian arid zone. In: Nautiyal S, Rao KS, Harald K, Raju KV, Ruediger S (eds) Knowledge systems of societies for adaptation and mitigation of impacts of climate change, Environment Science and Engineering. doi:10.1007/978-3-642-36143-2_16. Springer, Berlin

Maposa M, Chisuro D (1998) Importance of *Ziziphus mauritiana* (masau) in the Mukumbura area of Zimbabwe-from a farmer's and extensionist's point of view. In: International Workshop on *Ziziphus mauritiana*, Harare, Zimbabwe, 13–16 July

Meghwal PR, Henry A (2006) Economic evaluation of different agro-horticulture systems under rainfed condition of arid zone. In: Kumar D, Henry A, Vittal KPR (eds) Legumes in dry areas. Indian Arid Legume Society and Scientific Publishers, Jodhpur, pp 299–303

Meghwal PR, Vashishtha BB (1998) Effect of time of planting and auxin on rooting of cuttings of ker (*Capparis decidua* L. Forsk). Ann Arid Zone 37(4):401–404

Morton J (1987) Indian Jujube. 272–275. http://www.hort.purdue.edu/newcrop/morton/indian_jujube.html

Muzaffar N (1998) Beekeeping in Islamabad area. Islamabad Horticultural Society 16th Annual Spring flowers, fruit and vegetable show. In: Iqbal L, Hamayun M (eds) Studies on the traditional uses of plants of Malam Jabba valley. district Swat, Pakistan p 44

Naik KC (1963) South Indian fruits and their culture. P Verdachari & Co, Madras

Ngwa AT, Pane DK, Mafeni JM (2000) Feed selection and dietary preferences of forage by small ruminants grazing natural pastures in the Sahelian zone of Cameroon. Anim Feed Sci Technol 88:253–266

NIN (2010) Dietary guidelines for Indians—a manual. National Institute of Nutrition, Hyderabad, p 115

Osman M (2003) Alternate land use systems for sustainable production in rainfed areas. In: Pathak PS, Newaj R (eds) Agroforestry-potential and opportunities. Agrobios, Jodhpur, pp 177–181

Pareek OP (1983) The Ber. Indian Council of Agricultural Research, New Delhi

Pareek OP (2002) Top-working wild khejri trees. Indian Hortic 4(1):11–12

Pareek OP, Nath V (1996) Coordinated fruit research in Indian arid zone-a two decade profile. NRC for Arid Horticulture, Bikaner

Prinz D (1996) Water harvesting: past and future. In: Pereira LS (ed) Sustainability of irrigated agriculture. Proceedings, NATO Advanced Research Workshop, Vimeiro, 21–26 March, Balkema, Rotterdam, pp 135–144

Prinz D, Singh AK (2000) Technological potential for improvements of water harvesting study for the world commission on dams, Cape Town, South Africa. Report: dams and development.

Raturi GB, Chadha KL (1993) Horti-silvi-pastoral system. In: Chadha KL, Pareek OP (eds) Advances in horticulture. vol 2, no 2. Malhotra Publishing House, New Delhi, pp 805–811

Reddy TY, Reddy GHS (2005) Cropping system. Principles of agronomy. Kalyani Publishers, Ludhiana, pp 468–489

Samadia DK, Vashishtha BB, Raturi GB (2004) Prospects of vegetables production in the arid zone of India. In: Kumar S, Joshi RK, Pal S (eds) Impact of vegetable research in India. NCAP, ICAR, New Delhi, pp 101–114

Saroj PL, Dhandar DG, Sharma BD, Bhargava R, Purohit CK (2003) Ber (*Ziziphus mauritiana* Lamk.) based agri-horti system: a sustainable land use for arid ecosystem. Indian J Agroforest 5:30–35

Saxena SK (1977) Vegetation resources in Indian arid zone. In: Proceedings of summer institute on resource inventory and land use planning. CAZRI, Jodhpur, pp 277–284

Sharma SK, Diwakar GD (1989) Economic evaluation of horti-pastoral system in arid region of western Rajasthan. Indian J Range Manage 10(2):119–122

Sharma SK, Vashishtha BB (1985) Evaluation of jujube-buffel grass hortipastoral system under arid environment. Ann Arid Zone 24(3):303–309

Singh RS (1997) Note on the effect of intercropping on growth and yield of ber (*Z. mauritiana*) in semi-arid region. Curr Agric 21(1–2):117–118

Singh IS, Awasthi OP (2006) Managing scarce water resources in hot arid regions: a review. Indian J Arid Hortic 1(2):32–40

Singh RS, Vishwanath S (1995) Performance of pea varieties as intercrop in ber orchard in semi-arid region. Indian J Hortic 52(2):137–140

Singh S, Srivastava AK, Shivankar VJ, Singh S (1999) Citrus based cropping system-a review. Agric Rev 20(2):123–128

Singh B, Bishnoi M, Baloch MR (2012) Tree growth and wheat yield in agri-horti-silvi system in the arid region of Rajasthan. Indian Forester 138:726–732

Tewatia BS, Khirwar SS (2002) Utilization of ber (*Ziziphus mauritiana* Lamk.) leaves, hay and silage in goats. Indian J Anim Nutr 19:329–333

Vashishtha BB, Prasad RN (1997) Horti-pastoral systems for arid zone. In: Yadav MS, Singh M, Sharma SK, Tewari JC, Burman U (eds) Silvipastoral systems in arid and semi-arid ecosystems. CAZRI, Jodhpur, pp 277–288

Waterlow JC, Armstrong DG, Fowden L, Riley R (eds) (1988) Feeding a world population of more than eight million people: a challenge to science. Oxford University and Rank Prize Funds, Oxford

Yadava ND, Beniwal RK, Soni ML (2006) Crop diversification under fruit based cropping system in arid zone of western Rajasthan. Indian J Arid Hortic 1(1):20–22

Yadava ND, Soni ML, Nathawat NS, Birbal (2013) Productivity and growth indices of different intercrops in agri-horti-silvi system in arid western Rajasthan. Ann Arid Zone 52 (in press)

Principles and Practices of Sustainable Vegetable Production Systems

Ajay Nair, Dana Jokela and Jennifer Tillman

1 Introduction

Much of our modern agricultural production systems have lost the balance needed for long-term sustainability (Kimbrell 2002). With excessive dependence on fossil fuels and external inputs, most production systems are overusing and degrading the soil, water, and other natural resources upon which agriculture has always relied. Intensive agriculture has increased crop yields, but posed severe environmental problems (Pimentel et al. 1995). Agriculture is at a pivotal stage in terms of not only meeting the growing demand for food, but also improving sustainability of current production systems, under rising production costs, increasingly scarce natural resources, and changing climate. For nearly a decade there has been increased awareness of food quality, safety, health standards and global environmental issues. Never before have we observed such intense scrutiny of our production systems with respect to how they affect the environment and our natural resources. Agricultural production practices are being modified to minimize any deleterious effects to the environment and society. Sustainable agriculture has since emerged as a powerful tool in providing adequate quantity and quality of food to the growing world population with minimum footprint on the environment. Sustainability in agricultural production systems could be defined as practices, techniques, and approaches that contribute to food security and safety, sustained economic viability, enhanced environmental quality, and higher quality of life for farmers, farm workers, and society as a whole (National Research Council 2010). Environmental quality, natural resource conservation, profitability, preservation of farming communities, productivity, and human health are all interrelated aspects of sustainable agriculture (Sitarz 1998).

A. Nair (✉) · D. Jokela · J. Tillman
145 Horticulture Hall, Department of Horticulture,
Iowa State University, Ames, IA 50011, USA
e-mail: nairajay@iastate.edu

D. Nandwani (ed.), *Sustainable Horticultural Systems,* Sustainable Development and Biodiversity 2, DOI 10.1007/978-3-319-06904-3_3

Vegetable production systems are embracing principles and practices of sustainability at a rapid pace. In recent years, there has been a growing consumer demand and enthusiasm for locally produced foods which has spurred growth in local food systems. Synonymous to local food is the term 'sustainability'. The local food movement reflects an increasing interest by consumers in supporting local farmers, and in better understanding the origin of their food (Ilbery and Maye 2005). Consumers who are willing to pay higher prices for locally produced foods place importance on product quality, nutritional value, methods of crop production and the effects of production techniques on the environment (Martinez et al. 2010). Most of the fruit and vegetables sold through community supported agriculture programs, farm stands, and farmer's market are sustainably grown with minimal impact on the environment. Growers are interested in developing an integrated whole system approach by adopting production techniques that enhance soil health, crop productivity, and improve long-term farm profitability.

Vegetable production systems demand high labor inputs, include complex production techniques, advanced cultural practices, and crop rotations, however, sustainability can be integrated into many facets of vegetable production, starting from planting to harvest. This chapter will discuss and highlight production techniques and practices that are being utilized by vegetable growers in an attempt to move their production systems along a trajectory towards environmental, economic, and social sustainability.

2 Integration of Cover Crops in Vegetable Production Systems

Sustainable vegetable production places great emphasis on farming practices that safeguard soil on a long-term basis. One of the major tenets of sustainable production is to build soils that provide essential nutrients for crop growth, support a diverse and active biotic community, and exhibit a good soil structure. There is a growing interest among growers to utilize production techniques that reduce soil erosion, minimize nutrient leaching, suppress weed emergence, and build soil quality and organic matter. Cover crops are now being widely used by conventional, sustainable, and organic vegetable growers, to accomplish these tasks and also to maintain high soil fertility (Russo and Webber III 2007; Snapp et al. 2005; Ngouajio et al. 2003). A cover crop is a crop planted either before during, or after the vegetable crop to effectively manage soil quality and fertility, water quality, weeds, pests, and diseases. Cover crops are often terminated before the planting of the vegetable crop but there are production systems that use cover crops as living mulch between rows of crops. Cover crops have profound impact on soil quality as they add to the soil organic matter pool, enhance soil structure and fertility, improve soil water holding capacity, reduce the loss of nutrients and sediments in surface run-off, and suppress weed populations (Carrera et al. 2007; Clark 2007;

Snapp et al. 2005). Additional benefits from cover crops, when incorporated or used as mulch include reduction of insect and disease spread (Andow 1991), enhanced microbiological attack of soil pathogens (Griffin et al. 2009; Thurston 1992), and increased soil microbial biomass and biological activity (Buyer et al. 2010; Nair and Ngouajio 2012). Cover crops help support diverse and active soil biotic communities that serve as a foundation for agricultural sustainability. Biological activity in soil is the driving force behind decomposition processes that break down complex organic compounds and convert them to forms that are taken up by plants (Friedel et al. 2001).

2.1 *Types of Cover Crops*

There are many cover crop species that can be utilized in vegetable cropping systems. Selection of a cover crop depends on when it can be planted and the goal for its use. Non-legume cover crop species recycle existing soil nitrogen and other nutrients and can reduce leaching losses whereas legume cover crops are primarily used to fix atmospheric nitrogen (Sainju et al. 1998). Cover crops are divided into three broad categories based on growth habits:

Perennial Perennials persist for several years without replanting but can also be grown as a cover crop for one season. If they are not terminated/killed before incorporating into the soil, they can potentially regrow and become weeds in the Midwestern states of United States.

Summer Annual Summer annuals are seeded in spring or summer and grown until the end of the season. Although a few annuals will survive frosts, most will winterkill.

Winter Annual Winter annuals are normally seeded in the fall, 40–45 days before a killing frost. Most winter annuals are tough and will survive the winter. Depending on the time of seeding, some growth will occur in the fall after which the plant goes dormant for the winter. Growth resumes early in the spring.

Selecting the best cover crop for any vegetable production system will largely depend upon climatic conditions, planting window, end use of the cover crop, and resources available to establish and terminate the cover crop. Some cover crops that are frequently used in vegetable production systems include:

Cereal rye (*Secale cereale*)—Cereal rye is a versatile winter annual cover crop adaptable to most regions in United States. It can be easily established after fall vegetable harvest and provides ample ground cover to suppress weeds and reduce soil erosion over the winter. The plant has a deep and extensive root system that enables it to scavenge nutrients from the soil profile.

Annual ryegrass (*Lolium multiflorum*)—Similar to cereal rye, annual ryegrass can also be seeded in the fall. The crop produces sufficient top growth and provides good ground cover in the winter, although, it may winter-kill in USDA Plant Hardiness Zone 4 and below.

Oat (*Avena sativa*)—Oat is an upright, annual grass, that thrive under cool, moist conditions on well-drained soil. The crop has a fibrous root system, takes up excessive soil nutrients, provides quick weed suppression, and decent biomass that could be tilled-in to increase soil organic matter. This cover crop winterkills in USDA Plant Hardiness Zone 6 and below.

Sorghum-sudangrass (*Sorghum bicolor x S. bicolor*)—This is a summer annual cover crop that produces significant amount of biomass (4000–5000 lb. dry matter/acre). Due to its high biomass production capability, this cover crop is largely recommended to improve worn-out soils. Besides smothering weeds and adding organic matter to the soil, the extensive root system of the plant can penetrate compacted subsoil.

Buckwheat (*Fagopyrum esculentum*)—It is a summer or cool season cover crop which suppresses weeds and attracts beneficial insects and pollinators with its abundant blossoms. This cover crop has a quick turnaround as it takes only 45–50 days from seeding to termination. Buckwheat's strong weed suppressing ability makes it ideal for smothering warm season annual weeds.

Red clover (*Trifolium pratense*)—It is a short-lived perennial that is winter hardy in much of the U.S. (Hardiness Zone 4 and warmer). Red clover provides many benefits such as nitrogen fixation, weed suppression, soil protection, and can also be utilized for forage purposes. Among all clovers, red clover is the most adapted to many soil types and climatic niches.

Crimson clover (*Trifolium incarnatum*)—This cover crop is grown mostly as a winter- or summer-annual depending upon prevalent climatic conditions. In hardiness Zone 7 and above, this crop will overwinter. Crimson clover is gaining popularity as a winter-killed annual in USDA Hardiness Zones 5 and colder. Crimson clover provides good groundcover and weed control and can contribute 70–150 lb/A of nitrogen.

White clover (*Trifolium repens*)—It is a perennial cover crop and a preferred choice for "living mulch" systems planted between rows of vegetable crops, fruit bushes or trees. Once established, it provides long-term cover, weed suppression, and soil health improvement while attracting pollinators. This cover crop is tolerant to wet conditions and is an excellent choice for seeding in areas where there is intensive field traffic.

Hairy vetch (*Vicia villosa*)—Hairy vetch is widely used by vegetable growers as a winter hardy cover crop as it is widely adapted through USDA Hardiness Zone 4. Hairy vetch is often planted as a companion crop with cereal rye. It fixes atmospheric nitrogen, helps improve soil tilth, and provides weed control during its vigorous growth in the spring. It can also be grazed or harvested as forage. Nitrogen credit from this cover crop can range from 70–150 lb/A. It has the disadvantage of producing a significant percentage of hard seed that do not germinate the first year, but will often germinate later. This issue, however, can be addressed by killing the plant before seed maturation.

Mustard (*Brassica* spp.)—There are number of crops in the Brassicaceae family that share the name Mustard. Few examples include white or yellow mustard

Fig. 1 Yellow mustard (**a**) and buckwheat (**b**) cover crop in June 2013

(*Sinapis alba)*, brown mustard (*Brassica juncea*), and black mustard (*B. nigra)*. These crops are known for their rapid fall growth, great biomass production and nutrient scavenging ability. These cover crops have been shown to produce compounds called glucosinolates that are biotoxic and have broad activity against bacteria, fungi, insects, nematodes, and weeds when the crop is mowed and incorporated into the soil (Sarwar et al. 1998).

Oilseed radish (*Raphanus sativus*)—This cool season cover crop is in the Brassicaceae family and has the ability to produce large amounts of biomass, recycle soil nutrients, suppress weeds and pathogens, break-up compaction, and reduce soil erosion. It is becoming a popular cover crop in vegetable production systems due to its ability to absorb excess nutrients left at the end of the fall season, thus preventing leaching or runoff of nutrients into water systems (Fig. 1). This cover crop will winter kill in regions which receive freezing temperatures of 20 °F and below. Oilseed radish also produces glucosinolates and is widely acclaimed for its biofumigation property.

2.2 Soil Erosion and Water Quality

The primary benefit of cover crops is reduction of soil erosion by wind and water runoff (Langdale et al. 1991). Soil is susceptible to erosion when there is no vegetative ground cover or plant residue on the surface. A cover crop provides vegetative cover during periods when a crop is not present and reduces the impact of falling raindrops, which otherwise would detach soil particles and increase erosion (Dabney et al. 2001). It also slows the rate of runoff, thus improving moisture infiltration into the soil. No-tillage and other conservation tillage practices combined with cover crops have shown to significantly reduce runoff and soil erosion losses. Cover crops have also shown to improve water quality by suppressing nitrate leaching. Wyland et al. (1996) found that nitrate leaching was reduced by 65–70 % in a broccoli system with a cover crop compared to a fallow rotation.

Table 1 Nitrogen fixation by leguminous cover crops

Cover crop	Nitrogen contributed lb/A
Crimson clover	70–150
Hairy vetch	100–120
Red clover	75–130
White clover	80–130

2.3 Soil Fertility

One of the approaches widely practiced by sustainable vegetable growers is the use of cover crops to improve soil fertility. The use of leguminous cover crops in vegetable production systems represents an import of N into the system via fixation of atmospheric N. Cover crops such as red clover, hairy vetch, white clover, and crimson clover have been shown to fix significant amount of atmospheric nitrogen (Snapp and Fortuna 2003). Table 1 shows the amount of nitrogen that is fixed by the legume cover crops (Clark 2007).

When planted after a leguminous cover crop, increases in yields in tomato (*Lycopersicon esculentum*), lettuce (*Lactuca sativa*), and eggplant (*Solanum melongena*) has been reported (Abdul-Baki and Teasdale 1993; Shennan 1992; Sainju et al. 2002; Stirzaker and White 2005). Increase in yields have been contributed to the increased N supply by the legume (Abdul-Baki et al. 1996). Because of the low C/N ratio, leguminous cover crops rapidly decompose in the soil after termination and release N which is available for uptake by the succeeding crop within 2–4 weeks (Kuo et al. 1997; Wilson and Hargrove 1986). Legume cover crops have also shown to enhance root growth of vegetable crops. Sainju et al. (2001) reported enhanced root growth in tomato due to increased N supply by hairy vetch and crimson clover. The increase in root growth due to hairy vetch and crimson clover was similar to that produced by N fertilization rates of 90 and 180 kg ha^{-1}.

2.4 Nutrient Scavenging

Besides adding N, as in the case of legumes, some cover crops scavenge residual N, conserve it, and make it available to successive crops. After the fall crop harvest, significant amounts of N fertilizer applied to the summer crop is left as residual N in the soil (Hallberg 1989). This N has the potential to leach or denitrify. In order to minimize N loss due to leaching or denitrification, vegetable growers often seed winter cover crops that scavenge residual N and moisture after crop harvest in the fall. Examples of few cover crops that are seeded in fall to scavenge N include cereal rye, annual ryegrass, oats, yellow mustard, and oilseed radish. A Study conducted by McCracken et al. (1994) reported 94 % reduction in nitrate-N leaching by using cereal rye as a cover crop as compared to a no-cover crop control. Although rye is the most widely grown winter cover crop for its N scavenging capacity, brassica cover crops can capture equal or sometimes even higher amount of soil N (Dean and Weil 2009; Kristensen and Thorup-Kristensen 2007). Oilseed radish also is an excellent scavenger of nitrate from deeper soil layers after harvest of the cash crop. The oilseed radish cultivar Renova, for example, was shown to recycle more than 140 pounds of nitrogen per acre in a growing season (Ngouajio and Mutch 2004).

2.5 Weed Management

Weed management is achieved in conventionally grown crops mainly through chemical practices. In sustainable production systems, more emphasis is placed on cultural, mechanical, and biological control methods to manage weeds. With proper choice and management of cover crops and plant residues, it may be possible to reduce herbicides use in vegetable production systems. Cover crops have been shown to effectively suppress weeds in vegetable cropping systems (Teasdale 1996). Researchers have effectively used cover crops of wheat (*Triticum aestivum* L.), barley (*Hordeum vulgare* L.), oats (*Avena sativa* L.), rye (*Secale cereale* L.), and sudangrass (*Sorghum sudanese* L.) to suppress weeds, primarily annual broadleaf weeds (Putnam et al. 1983). Fall planted mixtures of cover crops that included rye, crimson clover, barley, and hairy vetch, suppressed weeds in a processing tomato production system (Creamer et al. 1996). Similar results were reported from a muskmelon study that used rye and hairy vetch mixture in the fall (Nair et al. 2012). In Iowa, fall planted cereal rye that was rolled and crimped in the spring before planting of summer squash, effectively reduced weed emergence in a summer squash production system (Tillman et al. 2013). Summer cover crops such as sorghum sudangrass, buckwheat, and cowpea significantly reduced populations of both broadleaved and grass weeds in a lettuce study (Kruse and Nair 2013). Legume cover crops such as crimson clover effectively suppressed common lambsquarter emergence and growth in a sweet corn study without affecting sweet corn yield (Dyck and Liebman 1994), however, there are studies that have shown negative interaction of cover crops on crop yield. Burgos and Talbert (1996) tested several cover crop mixtures in a sweet corn study and found wheat, rye, and rye + hairy vetch to effectively suppress weeds, but also reduce sweet corn emergence, height, and yield. Though, cover crops can certainly serve as a sustainable weed management tool, they will often be a component of a weed management system, not a stand-alone weed management strategy. Growers should integrate other mechanical and cultural techniques, along with cover crops, to effectively manage weeds.

2.6 Pest and Disease Management Using Cover Crops

Non-chemical approaches to manage pest and diseases include techniques such as fertility management, choice of irrigation methods, crop rotation, and organic matter management. Cover crops are also turning out to be an effective tool in managing pest and diseases. Cover crops have been shown to provide habitats for beneficial insects and pests and suppress diseases (Costello and Dane 1998; Griffin et al. 2009). Brassica crops, such as oilseed radish (*Raphanus sativus* L.), yellow mustard (*Sinapis alba* L.) and rape (*Brassica napus* L.), are increasingly being used as cover crops in temperate regions of North America for biofumigation. Biofumigation refers to the incorporation of plant residues that contain biocidal compounds, some of which are toxic to plant pathogens, and others that are allelopathic to weeds (Snapp et al. 2005). All members of the Brassicaceae family examined to date contain glucosinolates-sulfur-containing compounds that are not toxic but are enzymatically hydrolyzed to yield a variety of biologically active products, including isothiocyanates, ionic thiocyanates, nitriles, oxazolidinethiones, organic cyanates and epithionitriles (Mithen 2001). These compounds, especially isothiocyanates, have the ability to suppress seed germination and seedling growth of many weed species, in the laboratory, greenhouse, and field studies (Krishnan et al. 1998). Besides weeds, brassica cover crops also suppress phytophagous insects, nematodes, and other harmful pests (Haramoto and Gallandt 2004). Brassica species are widely used as a disease suppressing cover crops and have shown promise in controlling plant parasitic nematodes (Zasada and Ferris 2004). The approach of using cover crops to suppress diseases is sustainable; however, the efficacy of control varies based on cover growing conditions, biomass, varietal differences in isothiocyanate levels, sensitivity of the target pathogen, and ability of soil microorganisms to rapidly breakdown the active biocidal compounds (National Research Council 2010). For example, certain cultivars of oilseed radish such as Adagio and Ultimo are reported to give better nematode suppression (especially cyst nematodes) than other oilseed radish cultivars (Ngouajio and Mutch 2004). There is more research needed to document the efficacy and reliability of using brassica cover crops as an effective pest management tool for its widespread adoption.

3 Crop Rotation

Crop rotation is an underpinning for any sustainable production system. It is a planned system of growing different crops in recurrent succession on the same land. Benefits of crop rotation include weed, pest, and disease management, increased soil moisture and nutrients, and higher yields (National Research Council 2010). Growers practice crop rotation to maintain soil productivity, improve soil health, and reduce pests and diseases. Most vegetable farms grow many different crops and crop families, however, not all crops are equally profitable. The challenge of a good crop rotation system is to grow the type and quantity of crops needed to ensure the farm's profitability while continually building soil quality for long-term productivity (Mohler and Johnson 2009). Choice of crops or the sequence of crops within a crop rotation depends upon a number of factors, including botanical family, market

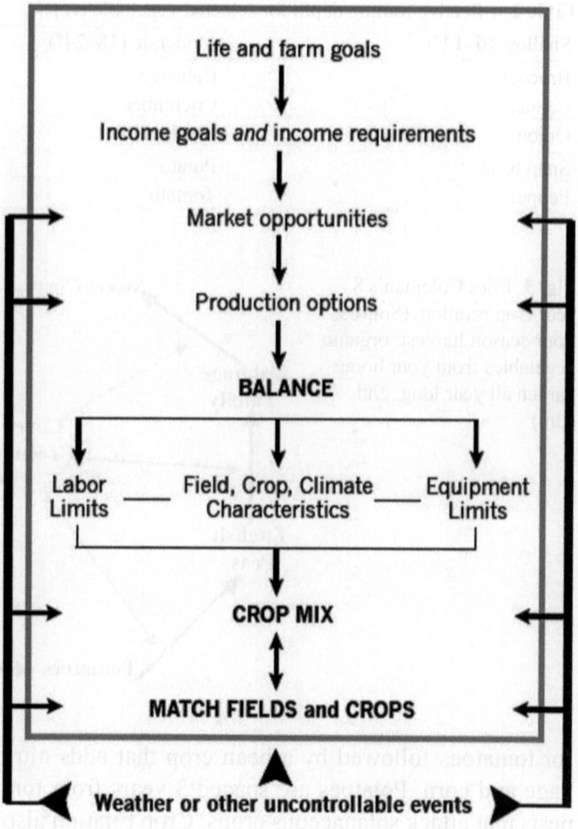

Fig. 2 Factors affecting crop rotation planning. Information obtained from an expert panel of 12 organic farmers from Northeast U.S. (Source: Mohler and Johnson 2009.)

demand, season of planting, labor availability, ability of the crop to compete with weeds, and nutrient requirement of the crop (Fig. 2).

Crop rotation has a positive effect on crop yield and performance. The reason for increased yields is not due to a single cause, but multiple factors that are affected by crop rotation. For example, the yield increase could be due to changes in soil fertility, reduction in pest populations, or improvement of soil health. Walters and Eckenrode (1996) found significantly reduced onion maggot populations in rotated compared to continuous onion production. One of the crop rotation fundamental vegetable growers implement is the alternate planting of deep-rooted crops and shallow rooted/fibrous-rooted crops (Table 2). Deep-rooted species benefit the shallow rooted species by bringing up nutrients from deep in the soil. Eliot Coleman, author of the "Four-Season Harvest", developed an eight crop rotation that is popular among vegetable growers (Fig. 3). In this rotation, Coleman reasons that both potatoes and squash are good "cleaning" crops, meaning that they cover the soil and suppress weed seed germination and growth. These plants are then followed by root crops. The pea crop in the rotation adds nitrogen to the soil

Table 2 Effective rooting depth for selected vegetable crops

Shallow (6–12″)	Moderate (18–24″)	Deep (> 36″)
Broccoli	Cabbage	Asparagus
Lettuce	Cucumber	Lima bean
Onion	Muskmelon	Parsnips
Snap bean	Potato	Sweet potato
Pepper	Tomato	Watermelon

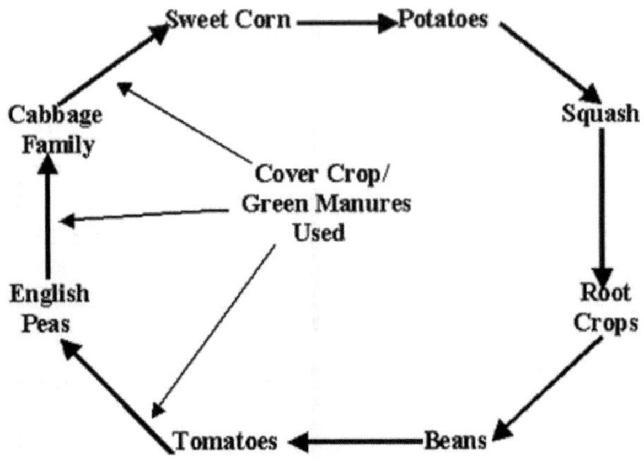

Fig. 3 Eliot Coleman's 8 year crop rotation. (Source: Four-season harvest: organic vegetables from your home garden all year long, 2nd edn.)

for tomatoes followed by a bean crop that adds nitrogen for heavy feeders cabbage and corn. Potatoes are spaced 3 years from tomatoes to reduce build-up of pests that attack solanaceous crops. Crop rotation also helps to manage soil-borne plant pathogens. By planting non-related crops in the same area in a rotation, the pathogen life cycle can be disrupted. If a particular pathogen attacks many host species, rotation may not be as successful, whereas if a pathogen attacks one or a few species rotation can be used successfully (Zadoks and Schein 1979). Some pathogens may last longer in soil than others thus requiring longer crop rotations. Table 3 lists major vegetable crops and their pathogens, and the time required to rotate those crops to eliminate those pathogens.

On vegetable farms, practicing crop rotation can sometimes be challenging due to diverse crops, planting schedule, weather, or market demands. In addition, crop rotation requires increased management skills because of the complexity in finding the right set of crops to improve yield, while also potentially reducing input expenses. Crop rotation is the foundation of a sustainable production system and is an irreplaceable component of the sustainability matrix. Rotating crops increases soil biodiversity and nutrient cycling capacity, reduces buildup and carryover of soil-borne diseases, breaks disease and pest cycles, and helps create favorable growing conditions for healthy, well-developed crop root systems (University of Minnesota Extension 2005).

Table 3 Major vegetable diseases and rotation time required to reduce pathogen populations to non-harmful levels

Vegetable	Disease	Rotation years required (years)
Asparagus	Fusarium rot	8
Cabbage	Clubroot	7
Cabbage	Black rot	2–3
Eggplant	Phomopsis rot	3
Muskmelon	Fusarium wilt	5
Potato	Blackleg	>3
Tomato	Bacterial speck	>3

4 Fertility Management

Sustainable vegetable crop production mandates comprehensive plans for managing nutrients. The most commonly used fertility inputs in vegetable production are chemical fertilizers. Sustainable vegetable production involves judicious use of chemical inputs to minimize any deleterious effects of excessive fertilization. High rates of nitrogen fertilization can reduce mycorrhizal activity and impair the ability of nitrogen-fixing bacteria to colonize roots of legumes (Amarasiri 1990). Timing and placement are critical when using synthetic fertilizers that are readily available to plants. Vegetable growers often apply nitrogen and potassium fertilizers in split applications. This significantly improves nutrient use efficiency and reduces loss due to runoff and leaching. Additionally, placing the fertilizer just to the side and below the crop row, is a far more efficient use of fertilizer as compared to broadcasting. Slow release formulations of fertilizers offer another alternative that reduce the number of applications and time spent on fertilizer applications. Sustainable vegetable production emphasizes a balanced use of synthetic fertilizers with other sources of nutrient inputs such as legumes, animal manure, and compost.

Legumes, as mentioned in the section pertaining to cover crops, form a symbiotic relationship with Rhizobium, root nodule bacteria that fix atmospheric nitrogen for the plants to use. Crop rotations that include legumes can reduce synthetic fertilizer inputs as the fixed nitrogen is returned to the soil with incorporation of the legume crop residue. Legumes such as sunn hemp (*Crotalaria juncea* L.) and cowpea (*Vigna unguiculata* L.) have been shown to increase tomato marketable yields. Campiglia et al. (2009) reported similar yields for potatoes that were fertilized with synthetic fertilizer and potatoes that followed a subclover or hairy vetch cover crop.

Sustainable production systems rely heavily on organic inputs such as composts and manures. Composts have favorable effects on crop productivity. Delate et al. (2008) showed that growth and yields of peppers grown using compost-based organic fertilizer can surpass those of conventionally grown peppers. Organic fertility amendments like compost and manures have been shown to improve the physical, chemical and biological properties of soil and produce yields equivalent to conventional cropping systems (Bulluck et al. 2002; Drinkwater et al. 1998). Application of compost enriches soil food web (bacteria, fungi, protozoan and nematode density) and also positively affect a number of soil characteristics, including soil organic matter, and soil respiration (Carrera et al. 2007; Lundquist et al. 1999; Treonis et al. 2010). A drawback of using compost is the unpredictability of the timing of nutrient release from the compost. Nutrient mineralization depends on quality of compost, soil moisture, temperature, and microbial activity in the soil. A Study conducted by Evanylo et al. (2008) reported that supplemental fertilizer was needed to attain high yields in sweet corn despite using high rates of compost. This was attributed to compost mineralization occurring after the period of peak crop demand.

5 Weed Management

Weeds can be detrimental to crop yields due to competition for water, light, and nutrients as well as their tendency to promote insect and disease issues and interfere with harvest (Finney and Creamer 2008). There are two main goals for controlling weeds. The first goal is to limit weed pressure on crops during the critical period in which weeds exert most of their influence on the crop. The critical period depends on the type of crop, time of planting, weed competition, and growing degree days (Garvey et al. 2013; Swanton et al. 2010; Adkins et al. 2010). Because of this, a higher level of weeds will not always mean a lower yield (Sanchez et al. 2008). Table 4 provides critical weed-free periods for some of the most commonly grown vegetable crops. The second goal in weed management is to reduce the weed seed bank in order to exert long-term weed control. This involves controlling weeds even if they are not in crop fields to limit all weed seed propagation (Grubinger 1999).

There are many strategies to manage weeds in vegetable production, many of which can be used together for maximum weed control. There are two general categories: methods that prevent weeds from emerging (pre-emergence control) and methods that kill weeds once they have emerged (post-emergence control).

5.1 Pre-emergence

5.1.1 Cover Crops

In some weed/crop interactions, cover crops have been shown to suppress weeds without affecting crop growth. This is attributed to allelopathic effects of certain cover crops. Crimson clover residue can suppress lambsquarter emergence without negatively affecting sweet corn growth (Dyck and Liebman 1994). Residue from sunn hemp can reduce the germination of pigweed, but extracts made from sunn hemp may not significantly affect the germination rates of sweet corn or cucumber (Skinner et al. 2012). These effects are highly dependent on the crop, however, as sunn hemp was shown to drastically reduce the germination of lettuce, onion, bell

Table 4 Critical weed-free periods for selected horticultural crops (Finney and Creamer 2008).

Crop	Critical weed-free period after seeding or planting	Weed species
Bean, snap	Unifoliate stage to flowering	Common cocklebur
Carrot	3–5 wk	Purple nutsedge
Cucumber	2–5 wk	Mixture of common lambsquarters, common ragweed and longspine sandbur
Muskmelon	4–6 wk 0–3 wk	Mixture of two pigweed species Smooth pigweed
Okra	37 wk	Purple nutsedge
Squash	4–6 wk	Mixture of quackgrass, horsenettle, common lambsquarters, and common ragweed
Sweet potato	2–6 wk	Mixture of sicklepod, redroot pigweed and yellow nutsedge
	0–4 wk 1–8 wk	Mixture of purple nutsedge, yellow cleome, large crabgrass, three-lobe morningglory, spreading dayflower, itchgrass, goosegrass, bermudagrass, and sour paspalum Mixture of green kyllinga, wild poinsettia, common purslane, garden spurge, cogongrass, arrowleaf sida, giant sensitive plant, purple nutsedge and goosegrass
Tomato	4–5 wk 3–5 wk	Mixture of common lambsquarters, common ragweed and longspine sandbur Purple nutsedge
Watermelon	0–6 wk	Large crabgrass

pepper, and turnip (Skinner et al. 2012). Care must be taken in selecting a cover crop that will not negatively affect the crop. A general rule of thumb is that small seeded crops are more likely to be affected by the allelochemicals than large-seeded crops (Finney and Creamer 2008).

Quick growing smother crops can be used to shade out weeds in fields not presently in production to maintain low weed seed levels (Grubinger 1999). In this way, cover crops may be used to reduce the overall seed bank in long-term weed control efforts. Another technique that uses cover crops is intercropping, also known as living mulches. This uses cover crops between cash crop rows throughout the growing season to reduce weed growth. The key is to select a cover crop that will not compete with the cash crop (Grubinger 1999).

5.1.2 Compost Management

Ensuring on-farm compost production is done correctly will limit the number of weed seeds that remain viable after composting. Allowing compost to reach 57–65 °C will kill most weed seeds (Montoya et al. 2013; Seal et al. 2012; Dahlquist et al. 2007).

Fig. 4 Strip-tillage pumpkin production (**a**) rolled mat of cereal rye suppressing weeds, (**b**) control treatment (no rye) allows weeds to grow

5.1.3 Banded Nutrient and Water Application

When applying nutrients and water to a crop, limiting general or overhead application will help limit the number of weeds that are fertilized and watered. Drip-irrigation can help with both of these issues, as water is directed only directly to plant roots and dissolved fertilizer can be sent through the irrigation system.

5.1.4 Mulches

By limiting light penetration, mulches can suppress weed growth either between rows or within rows, depending on where the mulch is situated. In one experiment, mulching (both biodegradable film and straw mulch) increased tomato yield by 35 % compared to mechanical-thermal weed control without mulch (Fontanelli et al. 2013). There is, therefore, incentive to find environmentally and economically feasible mulching systems.

Biological mulches can include straw, paper, or other organic material situated around crops. The more durable and secured the material, the longer it will persist throughout the growing season and the more weed protection the material will provide. A straw blower can help apply straw mulch between crop rows on a large scale.

Biological mulches also include using a cover crop mat formed by flattening a mature stand of cover crop with a roller crimper, thus avoiding cutting the cover crop from its roots. Strip-tillage or no-tillage can be employed after the mat is formed (See Sect. 7: Tillage). Using a cover crop mat has been shown to reduce weeds when compared to bare ground (Fig. 4) (Canali et al. 2013). When growing warm season crops in a location with a short growing season, however, the decrease of weed pressure may not translate into an increase in crop yield due to the decreased soil temperature under the rolled rye (Leavitt et al. 2011).

Polyethylene film provides weed suppression within the row, right around the plants (Fig. 5). Since weeds are able to emerge from the hole the crop is planted in,

Fig. 5 Weed suppression using black polyethylene plastic mulch for musk melon production

care should be taken to make the smallest hole possible when planting into plastic while allowing enough growing room for the crop. There are other advantages to plastic mulching, such as increased soil temperature, increased soil moisture retention and increased nutrient retention by the root zone (Kasirajan and Ngouajio 2012).

Many colors of plastic mulch are being manufactured, and the effects of the colors on the soil and plants are still being studied. Clear plastic warms the soil greatly (8–14 °F higher than bare soil at a 2-inch depth during the daytime), but since light is allowed to penetrate, weeds can grow under the plastic (University of Massachusetts Extension 2012). Black plastic is often touted as having the most warming effect on soil compared to other non-clear plastics, providing a daytime temperature increase of 5 °F at a 2-inch depth compared to bare soil (Ibarra-Jimenez et al. 2011; University of Massachusetts Extension 2012). However, brown infrared transmitting plastic may warm soil better than black plastic early in the spring. Red and blue plastic may encourage weed growth under the plastic (Rajablariani et al. 2012). However, red plastic may change the reflected light ratio of red:far-red wavelengths to help produce earlier yields in peppers, and blue mulch was found to increase cucurbit yields compared to black mulch. White and silver mulches slightly decrease the soil temperature compared to bare ground (2 °F cooler at a 1-inch depth). Choosing the best color of plastic to use depends on the crop, field conditions, and budget.

Plastic mulch can vary in effectiveness based on how well it is laid. The plastic must be pulled tight (but not stretched too severely or it will be prone to tears), and the edges should be buried deeply and uniformly to prevent the wind from loosening it. The closer the plastic is to the soil, the more warming effect it has on the soil. The biggest drawback about using plastic mulch is removing it from the field after the season. It is labor intensive, generates a lot of plastic waste, and often leaves bits of plastic in the soil that could interfere with root development in future seasons (Kasirajan and Ngouajio 2012).

The material waste and labor cost associated with plastic mulch systems has led to various biodegradable and photodegradable alternatives to polyethylene film. These tend to cost significantly more than polyethylene, but can provide the same weed control, soil warming, and crop yields as polyethylene film without the time and material waste associated with removing the plastic from the field (Cirujeda et al. 2012; Lopez-Martin et al. 2012). The degree to which the biodegradable alternatives persist through the season varies, as well as the degree to which they fully break down after being tilled into the soil. Degradable films can be made from many materials; a starch-based biodegradabale polymer blend was shown to be visually undetectable 2 weeks after being ploughed under, and it produced a similar yield of peppers compared to polyethylene film rows (Halley et al. 2001).

5.2 Post-emergence

5.2.1 Herbicide

There are fewer herbicides available for vegetable crops than for other agronomic crops; however, there is research being done to establish herbicides that can be used with specific vegetable crops (Robinson et al. 2008; Wallace et al. 2007). A pre-emergent herbicide spray can substantially knock down weed pressure. As herbicide resistant weeds become increasingly common, there is a need to focus on integrated weed management systems as opposed to relying solely on chemicals (Mortensen et al. 2012; Beckie 2006).

5.2.2 Cultivation

Mechanical suppression of weeds can be an alternative to chemical suppression if used properly. Cultivation tools must be suitable for the weed, crop, timing of cultivation, and soil conditions. For instance, a perennial grass weed may not be managed well with disk harrows as the rhizomes will simply spread more when cut into smaller pieces. In this situation, a tool that provides a lifting action, like a field cultivator with S-tines and shovels, will help expose the rhizomes so they will dry out and die. Cultivating before weeds become mature is important as weeds are easily managed when small, but limiting disruption to the crop is also important. In a weed-free sweet corn experiment, a rotary hoe was found to reduce the number of cobs and delay maturity when used three or more times per season (Leblanc et al. 2006).

Cultivating before the main crop emerges can give the crop an advantage over the weeds. A shallow cultivation with a tine weeder can disturb small weeds enough to cause them to dry out and die without harming deeply-sown seeds or stout transplants.

After crops have emerged, more care must be taken to minimize harm to the crop while cultivating. High clearance tractors allow for cultivation later in the season as crops grow taller, and tools should be mounted so as to allow the operator to have a clear view of the tool and crop row. Various tools can be attached to cultivation shanks for different purposes. Sweeps can be used on large weeds and to throw soil into the row, while shovels and knives are narrower and throw less soil (Grubinger 1999). Unneeded aggressive cultivation should be avoided, as it can bring buried weed seeds closer to the soil surface and enable them to germinate.

5.2.3 Flame Weeding

Flame weeding can kill weeds while limiting soil disruption and chemical use. The scale of this technique can range from backpack flame weeders in which a person carries a propane cylinder, to a tractor-mounted flame weeder that can uniformly flame a whole bed (Grubinger 1999). The flame can be broadcast over a whole bed prior to crop emergence, or directed between rows of crops once they have emerged. The purpose is not the burn the weeds, but to expose weeds to such extreme heat that their cell walls rupture, causing the plants to wither and die shortly after the flame weeder passes over them. In a trial with onions, flame weeding controlled broadleaf weeds better than it controlled grasses, and onions tolerated six flame weedings without a yield loss (Sivesind et al. 2012).

6 Irrigation

Conservation of water resource is a critical component of the overarching goal of sustainability. Irrigation is the dominant form of water use in a production system and measures that improve the efficiency of water application are of paramount importance (NRC 2010). To improve the overall sustainability of irrigation within a production system, one should focus on two main aspects: irrigation scheduling and delivery method. On-farm measurements of water are useful to irrigation scheduling. A rainfall gauge in combination with an evaporation pan provides an input/output estimate of soil water. By measuring how much water evaporates into the air each day, growers can estimate crop water use (Grubinger 1999). Most crops use only 10–15 % of pan evaporation early in the season. This consumption will gradually increase to 40–50 % of pan evaporation in the middle of their growth phase and finally reach 60–85 % of pan evaporation during crop maturity. A more technical way of scheduling irrigation would be to irrigate based on daily evapotranspiration (ET) value. Evapotranspiration is the sum of water lost from the soil surface and through transpiration from the crop leaf surface. Growers can obtain daily ET values by accessing agriculture weather stations located in their county or region. Obtaining ET values could be challenging, so growers may choose to use instruments such as tensiometers to accurately schedule irrigation in their fields.

The delivery method employed to irrigate crops carries equal weightage when it comes to sustainability. Improving the efficiency of delivery mechanism is one

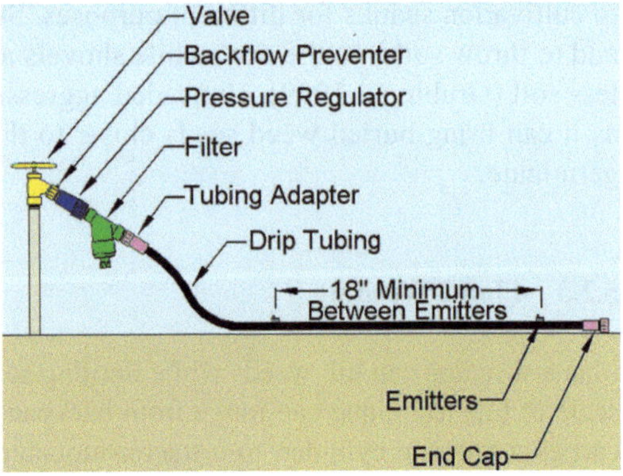

Fig. 6 Illustration and parts needed in the construction of a simple drip irrigation system. (Source: http://www.irrigationtutorials.com/drip-guide.htm.)

of the most effective means by which a farming enterprise can save water. Crops can be irrigated using sprinkler system, and trickle or drip irrigation. The choice of system mainly depends on type of crop grown and available resources. Vegetables that are grown for processing for example carrots, squash, broccoli, potato, etc. are grown on an extensive scale and require irrigation systems that are large and can carry huge volumes of water. An example of such a system would be a center pivot irrigation system. Such systems, if properly designed and operated have an efficiency range of 50–75 %. Disadvantages of using sprinkler irrigation systems include (1) inefficient use of water, (2) uneven distribution, (3) spread of diseases, and (4) higher weed populations. A sustainable and efficient alternative to sprinkler system would be trickle or drip irrigation. Shifting to drip irrigation system has been the greatest strategic improvement in water use efficiency and energy savings over the past three decades. In this system water is directly applied to the root zone of plants using a network of plastic pipes to carry a low flow of water under low pressure (Fig. 6). Schaible and Aillery (2006) reported that in 2003, low flow systems including drip, trickle, and micro-sprinklers were used on 3 million acres in the United States. Most of it was used for vegetables and perennial crops such as orchards and vineyards. Trickle irrigation has been shown to increase yields of tomatoes, peppers, onions, muskmelons, and watermelons by 15–25 % and water savings of more than 50 % compared to conventional furrow irrigation (Swiader and Ware 2002).

Advantages of drip irrigation are many:

- Delivers water directly to the root zone and minimizes water loss due to runoff, and evaporation
- Operating costs reduces due to low pressure and volume
- Allows for application of fertilizer or pest control materials through the system
- Has lower maintenance costs

7 Conservation Tillage Systems in Vegetable Production

7.1 Tillage Use in Vegetable Production

Tillage is an important tool used by vegetable growers for soil preparation and weed control. Primary tillage loosens the soil, incorporates vegetable and cover crop residues and enhances soil warming in the spring (Licht and Al-Kaisi 2005), while secondary tillage forms a seedbed. Periodic subsoil tillage can reduce compaction (Copas et al. 2009) and improve water infiltration (Pikul and Aase 2003). After planting, mechanical cultivation for weed control provides an alternative to herbicide use. However, conventional tillage has been shown to destroy soil aggregates, decrease soil organic matter, facilitate soil erosion, (Moebius-Clune et al. 2008; Magdoff and van Es 2009), decrease microbial biomass in the surface layer of soil (Doran 1987; Mathew et al. 2012), reduce water-holding capacity (Zibilske and Bradford 2007), and increase CO_2 emissions (Calderon et al. 2001). Reduced tillage systems—dubbed *conservation* tillage systems because they help conserve soil—provide alternatives to conventional tillage in vegetable production.

Conservation tillage practices—such as no-till, strip-till, and ridge till—are typically defined as those that leave more than 30 % of the soil surface covered with residue. Figure 7 shows the clear correlation between increased residue cover and decreased soil erosion in corn and soybean systems. While vegetable production has historically relied, at least in part, on mechanical cultivation for weed control, there has been an increasing amount of research looking into conservation tillage systems for vegetable production (Rogers et al. 2002; Peachey et al. 2004; Delate et al. 2012; Brainard and Noyes 2012; Haramoto and Brainard 2012). For organic producers and others interested in limiting herbicide use, options exist for using cover crops, along with a roller-crimper or flail mower, to suppress weeds in these conservation tillage systems. Two such systems—no-till and strip-till will be discussed in this section. They both rely on the same principles of cover crop utilization for weed control and enhanced ecological benefits.

7.2 Cover Crop Management

The basic idea of a cover crop-based, conservation tillage system is to grow a cover crop, mechanically kill it by mowing or rolling it flat, and then plant directly through the resulting mulch. Many different cover crops species can be used, depending on the planting date and type of the vegetable crop to follow. The most common cover crops used in the Upper Midwest are cereal rye (*Secale cereal* L.) and hairy vetch (*Vicia villosa*), often planted as a biculture. These cover crops are seeded in the fall and terminated in the late spring of the following year. In order for the cover crop to effectively suppress weeds, it must be seeded at an appropriate rate. Figure 8 shows that higher cover crop seeding rates are associated with

Fig. 7 Soil erosion decreases as surface cover increases. (Note: *FP* fall plow, *FC* fall chisel, *NT* no-till; *circles* corn, *no circles* soybeans)

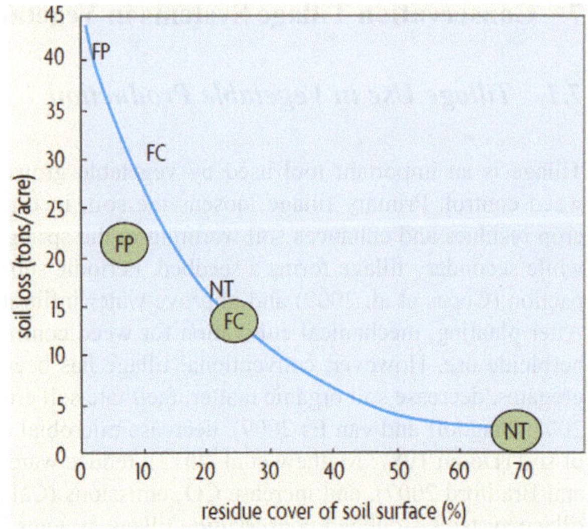

improved weed suppression (Ryan et al. 2011). Cover crops can be terminated using a number of different tools, but the roller-crimper holds advantages over alternatives, such as the flail mower, because it uses less power, can be operated at higher speeds, and provides longer lasting weed suppression because residue is not chopped (Smith et al. 2011). The roller-crimper design most commonly used now was popularized by the Rodale Institute (Kutztown, PA, USA) (Fig. 9). It is a steel cylinder (41–51 cm diameter) with blunt blades welded in a chevron pattern across the length of the cylinder. It can be filled with water for increased mass, and either front- or rear-mounted on a three-point hitch of a tractor. To achieve consistent cover crop kill, the cover crop must be at an appropriate stage of maturity. Cereal rye must be at anthesis, or Zodaks growth stage 61 (Mirsky et al. 2009); hairy vetch should be terminated at early pod stage (Mischler et al. 2010). These rigid timeframes put restrictions on the vegetable crops that can be grown using cover crop-based, conservation tillage systems. In the Upper Midwest, cereal rye and hairy vetch typically don't reach the required growth stage until late-May or early-June, making many early spring-planted crops incompatible with the system; however, breeding efforts to select for earlier maturity are underway (Mirsky et al. 2012). There is the potential to grow fall-sown, frost-tender cover crops ahead of these early spring crops, though there may not be sufficient cover crop residue remaining in the spring to suppress weeds; more research is needed to determine feasibility of such a system. Spring-sown summer annual cover crops can also be used for midsummer-planted vegetables, such as fall cabbage (Haramoto and Brainard 2012).

It should be mentioned that these systems would be better thought of as *rotational* no-till or strip-till, as full-width tillage is often used once per year after cash crop harvest to prepare the soil for cover crop establishment and to control perennial

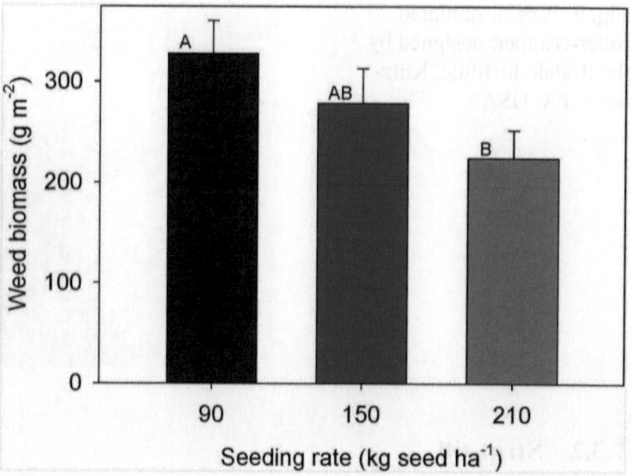

Fig. 8 Mean weed biomass at 10 weeks after rye termination at three rye seeding rates in a cover crop-based, no-till system. (Ryan et al. 2011.)

weeds, most of which tend to be difficult to control in conservation tillage systems (Buhler et al. 1994) and are largely unsuppressed by cover crop mulches (Mirsky et al. 2012).

7.3 *Reduced Tillage Systems*

7.3.1 No-till

The system that causes the least soil disturbance is no-till. In no-till systems, large-seeded crops, such as sweet corn, peas, beans, and cucurbits, can be seeded directly through the rolled-crimped mulch using a no-till planter. Vegetables which are typically transplanted can be planted through the mulch using a modified transplanter, usually with the addition of a large coulter and a subsurface shank ahead of the normal transplanter components. One of the first designs was the Subsurface Tiller-Transplanter, developed by Ron Morse in the early 1990s (Morse et al. 1993), though other no-till transplanters have been developed more recently, and are manufactured mostly on a custom basis. The goal is to create a narrow band of loosened soil into which the plugs can be planted. Small-seeded vegetable crops which must be direct seeded, such as carrots and other root crops, are not suited for high residue, no-till systems.

Because the soil remains largely undisturbed, no-till systems maximize the ecological benefits associated with conservation tillage. Some of the potential drawbacks of no-till include delayed soil warming in the spring, initially reduced inorganic nitrogen availability (Haramoto and Brainard 2012), and the limitations on direct seeding of small-seeded vegetable crops.

Fig. 9 A back-mounted roller-crimper, designed by the Rodale Institute, Kutztown, PA, USA

7.3.2 Strip-till

In strip-till systems, a wider band of soil (usually 15–30 cm) is tilled in the cover crop to create a seedbed for the vegetable crop establishment, leaving the cover crop in between rows to mature and be terminated mechanically, as in no-till (Fig. 9). The ideal date for strip-tilling depends on soil type, cover crop species, vegetable crop planting date, and other factors. In a system using a fall-seeded winter annual, such as cereal rye, strip-tillage can be done in the fall, allowing the residue in the strips plenty of time to break down before cash crop planting the following spring. strip-tilling can also be done in the spring, though there is an increased risk that prolonged moisture might prevent timely field operations. In either case, precautions should be taken to assure that the strip-tiller can adequately incorporate the cover crop. Most commercially-available strip-tillers are designed for use in grain cropping systems, where residue is dried down and root mass is much less than what would be found in a dense, vigorous cover crop stand.

Because more soil is tilled and exposed, strip-till systems have benefits and drawbacks compared to no-till systems. The wider strip enhances soil warming, allowing for earlier crop maturity, increased nitrogen mineralization of soil organic matter (Al-Kaisi et al. 2005), and greater vigor of warm-season crops. It also allows for the direct seeding of small-seeded crops. However, the increased tillage in strip-till systems makes the soil in that tilled zone vulnerable to many of the same factors as conventional tillage, such as increased soil erosion, decreased aggregation, and reduced soil organic matter. Additional weed control may be required in the tilled zone because of absence of mulch (Fig. 10).

Conclusion

There are different ways in which a crop production enterprise can contribute to various aspects of sustainability. A systems approach to production is necessary to identify and understand the significance of the linkages between grower practices and their implications on crop growth, productivity and the environment.

Fig. 10 Tilled strips in author's rye/vetch cover crop plot in mid-November. (Photo credit: Dana Jokela.)

In vegetable production systems, the diversity of the enterprise, size and scale of the farm, market and labor demands, and climatic conditions provide unique opportunities or barriers to improving the overall sustainability of the production system. More research and extension programs are needed to help growers develop and adopt production practices and strategies that increase the productivity of their cropping systems without compromising economic, social, and environmental sustainability of their communities.

Acknowledgements This work was supported by Iowa State University College of Agriculture and Life Sciences and the Leopold Center for Sustainable Agriculture.

References

Abdul-Baki AA, Teasdale JR (1993) A no-tillage tomato production system using hairy vetch and subterranean clover mulches. HortScience 28:106–108

Abdul-Baki AA, Teasdale JR, Lorcak R (1996) Fresh-market tomato production in a low-input alternative system using cover crop mulch. HortScience 31:65–69

Adkins JI, Stall WM, Santos BM (2010) Critical period of interference between American black nightshade and triploid watermelon. Weed Technol 24(3):397–400. doi:10.1614/wt-d-10-00014.1

Al-Kaisi MM, Yin X, Licht MA (2005) Soil carbon and nitrogen changes as influenced by tillage and cropping systems in some Iowa soils. Agric Ecosyst Environ 105:635–647

Amarasiri SL (1990) Phosphorus management in intensive vegetable cultivation. In: Phosphorus requirements for sustainable agriculture in Asia and Oceania. Proceeding of a symposium, 6–10 March 1989, pp 453–460

Andow DA (1991) Vegetational diversity and arthropod response. Annu Rev Entomol 36:561–586

Beckie HJ (2006) Herbicide-resistant weeds: management tactics and practices. Weed Technol 20(3):793–814

Brainard DC, Noyes DC (2012) Strip-tillage and compost influence carrot quality, yield, and net returns. HortScience 47:1073–1079

Buhler DD, Stoltenberg DE, Becker RL et al (1994) Perennial weed populations after 14 years of variable tillage and cropping practices. Weed Sci 42:205–209

Bulluck LR III, Brosius M, Evanylo GK et al (2002) Organic and synthetic fertility amendments influence soil microbial, physical and chemical properties on organic and conventional farms. Appl Soil Ecol 19:147–160

Burgos NR, Talbert RE (1996) Weed control and sweet corn (*Zea mays* var. *rugosa*) response in a no-till system with cover crops. Weed Sci 44:355–361

Buyer JS, Teasdale JR, Roberts DP (2010) Factors affecting soil microbial community structure in tomato cropping system. Soil Biol Biochem 42:831–841

Calderon FJ, Jackson LE, Scow KM (2001) Short-term dynamics of nitrogen, microbial activity, and phospholipid fatty acids after tillage. Soil Sci Soc Am J 65:118–126

Campiglia E, Paolini R, Colla G (2009) The effects of cover cropping on yield and weed control of potato in a transitional system. Field Crops Res 112:16–23

Canali S, Campanelli G, Ciaccia C (2013) Conservation tillage strategy based on the roller crimper technology for weed control in Mediterranean vegetable organic cropping systems. Eur J Agron 50:11–18

Carrera LM, Buyer JS, Vinyard B (2007) Effects of cover crops, compost, and manure amendments on soil microbial community structure in tomato production systems. Appl Soil Ecol 37:247–255

Cirujeda A, Aibar J, Anzalone A (2012) Biodegradable mulch instead of polyethylene for weed control of processing tomato production. Agron Sustain Dev 32(4):889–897

Clark A (2007) Managing cover crops profitably, 3rd edn. Sustainable Agriculture Network, Beltsville

Copas ME, Bussan AJ, Drilias MJ (2009) Potato yield and quality response to subsoil tillage and compaction. Agron J 101:82–90

Costello MJ, Daane KM (1998) Influence of ground cover on spider populations in a table grape vineyard. Ecol Entomol 23:33–40

Creamer NG, Bennett MA, Stinner BR (1996) A comparison of four processing tomato production systems differing in cover crop and chemical inputs. J Am Soc Hortic Sci 121:559–568

Dabney SM, Delgado JA, Reeves DW (2001) Using winter cover crops to improve soil and water quality. Commun Soil Sci Plant Anal 32:1221–1250

Dahlquist RM, Prather TS, Stapleton JJ (2007) Time and temperature requirements for weed seed thermal death. Weed Sci 55(6):619–625

Dean JE, Weil RR (2009) Brassica cover crops for nitrogen retention in the mid-Atlantic coastal plain. J Environ Qual 38:520–528

Delate K, Cambardella C, McKern A (2008) Effects of organic fertilization and cover crops on an organic pepper system. HortTechnol 18(2):215–226

Delate K, Cwach D, Chase C (2012) Organic no-tillage system effects on soybean, corn and irrigated tomato production and economic performance in Iowa, USA. Renew Agric Food Syst 27:49–59

Doran JW (1987) Microbial biomass and mineralizable nitrogen distributions in no-tillage and plowed soils. Biol Fertil Soils 5:68–75

Drinkwater LE, Wagoner P, Sarrantonio M (1998) Legume-based cropping systems have reduced carbon and nitrogen losses. Nature 396:262–265

Dyck E, Liebman M (1994) Soil fertility management as a factor in weed-control—the effect of crimson clover residue, synthetic nitrogen-fertilizer, and their interaction on emergence and early growth of lambsquarters and sweet corn. Plant Soil 167(2):227–237

Evanylo G, Sherony C, Spargo J (2008) Soil and water environmental effects of fertilizer-, manure-, and compost-based fertility practices in an organic vegetable cropping system. Agric Ecosyst Environ 127(1–2):50–58

Finney D, Creamer N (2008) Weed management on organic farms. North Carolina Cooperative Extension Service, North Carolina

Fontanelli M, Raffaelli M, Martelloni L (2013) The influence of non-living mulch, mechanical and thermal treatments on weed population and yield of rainfed fresh-market tomato (*Solanum lycopersicum* L.). Span J Agric Res 11(3):593–602

Friedel JK, Gabel D, Stahr K (2001) Nitrogen pools and turnover in arable soils under different durations of organic farming: II: Source- and -sink function of the soil microbial biomass or competition with growing plants? J Plant Nutr Soil Sci 164:421–429

Garvey PV Jr, Meyers SL, Monks DW (2013) Influence of Palmer amaranth (*Amaranthus palmeri*) on the critical period for weed control in plasticulture-grown tomato. Weed Technol 27(1):165–170

Griffin TS, Larkin RP, Honeycutt CW (2009) Delayed tillage and cover crop effects in potato systems. Am J Potato Res 86:79–87

Grubinger VP (1999) Sustainable vegetable production from start-up to market. Natural resource, agriculture, and engineering services. Cooperative Extension, Ithaca

Hallberg GR (1989) Nitrate in groundwater in the United States. In: Follett RF (ed) Nitrogen management and groundwater protection. Elsevier, Amsterdam, pp 35–74

Halley P, Rutgers R, Coombs S (2001) Developing biodegradable mulch films from starch-based polymers. Starch-Starke 53(8):362–367

Haramoto ER, Brainard DC (2012) Strip-tillage and oat cover crops increase soil moisture and influence n mineralization patterns in cabbage. HortScience 47:1596–1602

Haramoto ER, Gallandt ER (2004) Brassica cover cropping for weed management: a review. Renew Agric Food Syst 19:187–198

Ibarra-Jimenez L, Lira-Saldivar RH, Valdez-Aguilar LA (2011) Colored plastic mulches affect soil temperature and tuber production of potato. Acta Agric Scand Sect B-Soil Plant Sci 61(4):365–371

Ilbery B, Maye D (2005) Food supply chains and sustainability: Evidence from specialist food producers in the Scottish/English borders. Land Use Policy 22:331–344

Kasirajan S, Ngouajio M (2012) Polyethylene and biodegradable mulches for agricultural applications: a review. Agron Sustain Dev 32(2):501–529

Kimbrell A (ed) (2002) Fatal harvest: the tragedy of industrial agriculture. Island, Washington, DC

Krishnan G, Holshouser DL, Nissen SJ (1998) Weed control in soybean (*Glycine max*) with green manure crops. Weed Technol 12:97–102

Kristensen HL, Thorup-Kristensen K (2007) Effects of vertical distribution of soil inorganic nitrogen on root growth and subsequent nitrogen uptake by field vegetable crops. Soil Use Manage 23:338–347

Kruse R, Nair A (2013) Summer cover crop influence on fall vegetable production. http://www.ag.iastate.edu/farms/2013/hort/SummerCoverCrop.pdf. Accessed 17 Mar 2014

Kuo S, Sainju UM, Jellum EJ (1997) Winter cover cropping influence on nitrogen in soil. Soil Sci Soc Am J 61:1392–1399

Langdale GW, Blevins RL, Karlen DL (1991) Cover crop effects on soil erosion by wind and water. In: Hargrove WL (ed) Cover crops for clean water. Soil and Water Conservation Society, Ankeny, pp 15–22

Leavitt MJ, Sheaffer CC, Wyse DL, Allan DL (2011) Rolled winter rye and hairy vetch cover crops lower weed density but reduce vegetable yields in no-tillage organic production. HortScience 46(3):387–395

Leblanc ML, Cloutier DC, Stewart KA (2006) Rotary hoe cultivation in sweet corn. HortTechnology 16(4):583–589

Licht MA, Al-Kaisi M (2005) Strip-tillage effect on seedbed soil temperature and other soil physical properties. Soil Till Res 80:233–249

Lopez-Marin J, Gonzalez A, Fernandez JA (2012) Biodegradable mulch film in a broccoli production system. In: Mourao I, Aksoy U (eds) Xxviii international horticultural congress on science and horticulture for people. Int Soc Hortic Sci, Leuven 1, Acta Hortic 933:439–444

Lundquist EJ, Jackson LE, Scow KM (1999) Changes in microbial biomass and community composition, and soil carbon and nitrogen pools after incorporation of rye into three California agricultural soils. Soil Biol Biochem 31:221–236

Magdoff F, van Es HM (2009) Building soils for better crops, 3rd edn. Sustainable Agriculture Research and Education, Washington, DC

Martinez S, Hand M, Da Pra M, Pollack S (2010) Local food systems concepts, impacts, and issues. http://www.ers.usda.gov/media/122868/err97_1_pdf. Accessed 17 Mar 2014

Mathew RP, Feng Y, Githinji L (2012) Impact of no-tillage and conventional tillage systems on soil microbial communities. Appl Environ Soil Sci 2012:1–10

McCracken DV, Smith MS, Grove JH (1994) Nitrate leaching as influenced by cover cropping and nitrogen source. Soil Sci Soc Am J 58:1476–1483

Mirsky SB, Curran WS, Mortensen DA (2009) Control of cereal rye with a roller/crimper as influenced by cover crop phenology. Agron J 101:1589–1596

Mirsky SB, Ryan MR, Curran WS (2012) Conservation tillage issues: cover crop-based organic rotational no-till grain production in the mid-Atlantic region, USA. Renew Agric Food Syst 27:31–40

Mischler R, Duiker SW, Curran WS (2010) Hairy vetch management for no-till organic corn production. Agron J 102:355

Mithen R (2001) Glucosinolates and the degradation products. In: Callow J (ed) Advances in botanical research, vol 35. New York Academic Press, New York, pp 214–262

Moebius-Clune BN, van Es HM, Idowu OJ (2008) Long-term effects of harvesting maize stover and tillage on soil quality. Soil Sci Soc Am J 72:960

Mohler CL, Johnson SE (2009) Crop rotation on organic farms: a planning manual. NRAES 177, Ithaca

Montoya JE, Waliczek TM, Abbott ML (2013) Large scale composting as a means of managing water hyacinth (*Eichhornia crassipes*). Invasive Plant Sci Manag 6(2):243–249

Morse RD, Vaughan DH, Belcher LW (1993) Evolution of conservation tillage systems for transplanted crops; potential role of the Sub-surface Tiller Transplanter (SST-T). Proc South Conserv Conf Sustain Monroe, 145–151

Mortensen DA, Egan JF, Maxwell BD (2012) Navigating a critical juncture for sustainable weed management. Bioscience 62(1):75–84

Nair A, Ngouajio M (2012) Soil microbial biomass, functional microbial diversity, and nematode community structure as affected by cover crops and compost in an organic vegetable production system. Appl Soil Ecol 58:45–55

Nair A, Shaw A, Lawson V (2012) Cover cropping and strip-tillage to improve crop performance and food safety in muskmelon production. http://www.ag.iastate.edu/farms/2012%20Farm%20Reports/Muscatine/CoverCropping.pdf. Accessed 17 Mar 2014

National Research Council (2010) Towards sustainable agriculture in the 21st century. National Academies Press, Washington, DC

Ngouajio M, Mutch DR (2004) Oilseed radish: a new cover crop for Michigan. Extension Bulletin E 2907

Ngouajio M, McGiffen ME Jr, Hutchinson CM (2003) Effect of cover crop and management system on weed populations in lettuce. Crop Prot 22:57–64

Peachey RE, William RD, Mallory-Smith C (2004) Effect of no-till or conventional planting and cover crops residues on weed emergence in vegetable row crop. Weed Technol 18:1023–1030

Pikul JL, Aase JK (2003) Water infiltration and storage affected by subsoiling and subsequent tillage. Soil Sci Soc Am J 67:859–866

Pimentel D, Harvey C, Resosudarmo P (1995) Environmental and economic costs of soil erosion and conservation benefits. Science 267:1117–1123

Putnam AR, DeFrank J, Barnes JP (1983) Exploitation of allelopathy for weed control in annual and perennial cropping systems. J Chem Ecol 9:1001–1010

Rajablariani HR, Hassankhan F, Rafezi R (2012) Effect of colored plastic mulches on yield of tomato and weed biomass. Int J Environ Sci Dev 3(6):590–593

Robinson DE, McNaughton K, Soltani N (2008) Weed management in transplanted bell pepper (*Capsicum annuum*) with pretransplant tank mixes of sulfentrazone, s-metolachlor, and dimethenamid-p. HortScience 43(5):1492–1494

Rogers GS, Little SA, Silcock SJ (2002) No-till vegetable production using organic mulches. XXVI Int. Hort Cong Sustain Hort Syst 21st Century 638:215–223

Russo VM, Webber CL III (2007) Organic agricultural production in the United States: an old wheel being reinvented. The Americas J Plant Sci Biotechnol 1:29–35

Ryan MR, Curran WS, Grantham AM (2011) Effects of seeding rate and poultry litter on weed suppression from a rolled cereal rye cover crop. Weed Sci 59:438–444

Sainju UM, Singh BP, Whitehead WF (1998) Cover crop root distribution and its effects on soil nitrogen cycling. Agron J 90:511–518

Sainju UM, Singh BP, Whitehead WF (2002) Long-term effects of tillage, cover crops, and nitrogen fertilization on organic carbon and nitrogen concentrations in sandy loam soils in Georgia, USA. Soil Till Res 63:167–179

Sanchez E, Lamont WJ, Orzolele MD (2008) Newspaper mulches for suppressing weeds for organic high-tunnel cucumber production. HortTechnology 18(1):154–157

Sainju UM, Singh BP, Whitehead WF (2001) Comparison of the effects of cover crops and nitrogen fertilization on tomato yield, root growth, and soil properties. Scientia Hortic 91:201–214

Sarwar M, Kirkegaard JA, Wong PTW, Desmarchelier JM (1998) Biofumigation potential of brassics. Plant and Soil 201:103–112

Schaible G, Aillery M (2006) Irrigation water management. In: Agricultural resources and environmental indicators, 2006 Edition/EIB–16. U.S. Department of Agriculture Economic Research Service, Washington, DC

Seal A, Bera R, Chatterjee AK (2012) Evaluation of a new composting method in terms of its biodegradation pathway and assessment of compost quality, maturity and stability. Arch Agron Soil Sci 58(9):995–1012

Shennan C (1992) Cover crops, nitrogen cycling, and soil properties in semi-irrigated vegetable production systems. HortScience 27:749–754

Sitarz D (1998) Sustainable America: America's environment, economy and society in the 21st century. Earth Press, Carbondale, p 312

Sivesind EC, Leblanc ML, Cloutier DC (2012) Impact of selective flame weeding on onion yield, pungency, flavonoid concentration, and weeds. Crop Prot 39:45–51

Skinner EM, Diaz-Perez JC, Phatak SC (2012) Allelopathic effects of sunnhemp (*Crotalaria juncea* L.) on germination of vegetables and weeds. HortScience 47(1):138–142

Smith AN, Reberg-Horton SC, Place GT (2011) Rolled rye mulch for weed suppression in organic no-tillage soybeans. Weed Sci 59:224–231

Snapp SS, Fortuna AM (2003) Predicting nitrogen availability in irrigated potato systems. HortTechnology 13:598–604

Snapp S, Swinton S, Labarta R (2005) Evaluating cover crops for benefits, costs and performance within cropping system niches. Agron J 97:322–332

Stirzaker RJ, White I (1995) Amelioration of soil compaction by a cover crop for no-tillage lettuce production. Aust J Agric Res 46:553–568

Swanton CJ, O'Sullivan J, Robinson DE (2010) The critical weed-free period in carrot. Weed Sci 58(3):229–233

Swiader JM, Ware GW (2002) Producing vegetable crops, 5th edn. Interstate, Danville

Teasdale JR (1996) Contribution of cover crops to weed management in sustainable agricultural systems. J Prod Agric 9:475–479

Thurston HD (1992) Sustainable practices for plant disease management in traditional farming systems. Westview, Boulder

Tillman A, Batzer J, Nair A (2013) Strip-tillage and row cover use in organically and conventionally grown summer squash. http://www.ag.iastate.edu/farms/2013/muscatine/StripTillSquash.pdf. Accessed 17 Mar 2014

Treonis M, Austin EE, Buyer JS (2010) Effects of organic amendment and tillage on soil microorganisms and microfauna. Appl Soil Ecol 46:103–110

University of Massachusetts Extension (2012) Using colored plastic mulches in the vegetable garden. http://ag.umass.edu/sites/ag.umass.edu/files/fact-sheets/pdf/mulch_colored_plastic.pdf. Accessed 17 Mar 2014

University of Minnesota Extension (2005) Nutrient cycling and maintaining soil fertility in fruit and vegetable crop systems. http://www.extension.umn.edu/garden/fruit-vegetable/nutrient-cycling-and-fertility. Accessed 19 Mar 2014

Wallace RW, Phillips AL, Hodges JC (2007) Processing spinach response to selected herbicides for weed control, crop injury, and yield. Weed Technol 21(3):714–718

Walters TW, Eckenrode CJ (1996) Integrated management of the onion maggot (Diptera: Anthomyiidae). J Econ Entomol 89:1582–1586
Wilson DO, Hargrove WL (1986) Release of nitrogen from crimson clover residue under two tillage systems. Soil Sci Soc Am J 50:1251–1254
Wyland LJ, Jackson LE, Chaney WE (1996) Winter cover crops in a vegetable cropping system: impacts on nitrate leaching, soil water, crop yield, pests and management costs. Agric Ecosyst Environ 59:1–17
Zadoks JC, Schein RD (1979) Epidemiology and plant disease management. Oxford University Press, Oxford, p 427
Zasada IA, Ferris H (2004) Nematode suppression with brassicaceous amendments: application based upon glucosinolate profile. Soil Biol Biochem 3697:1017–1024
Zibilske LM, Bradford JM (2007) Soil aggregation, aggregate carbon and nitrogen, and moisture retention induced by conservation tillage. Soil Sci Soc Am J 71:793

A Small-Scale Integrated Farm System in a Tropical Insular Environment of Guam: A Case Study

Mari Marutani and John Brown

1 Introduction

Guam, located at 13° 28″ N and 144° 47″ E, is a tropical island in the western Pacific region having the total area of 549 km^2. Agriculture on Guam has drastically changed after the end of WWII by shifting from traditional subsistence agriculture to economy-based cash-crop production. Guam now imports more than 90 % agricultural produce due to the rapid military build-up and development of tourism in recent years (Marutani et al. 1997). Today, agriculture on Guam is characterized as a small multi-cropping farm operation in insular urban environment with farm size ranging from a half-acre (0.2 ha) truck-crop production to up to 20 acres (8 ha). With the decreasing availability of arable land and small labor pool, Guam's agriculture can only be sustained by each local grower's effort to advance a farm management system by innovative and efficient use of available natural resources, production of high-valued marketable commodities, and linking their market system closely with local consumers.

Construction of an integrated model farm at the University of Guam was initiated in 2001 with the support of US Department of Agriculture/Initiative for Future Agriculture and Food Systems Program as a 4-year project with the total budget of $ 297,759. The main goal of the project was to establish an integrated model farm, appropriate as a small agricultural enterprise operated by Asian-Pacific Islanders. The whole farm system emphasized having diversification of agricultural products and an integration of farming practices in a total area of less than 5 acres (2 ha). There were three main sections in the farm: (1) aquaponics where the system consists of aquaculture and hydroponics to produce both fish and plants in re-circulating water, (2) horticultural crop production in soils, and (3) animal production. The

M. Marutani (✉) · J. Brown
Agriculture and Life Science Division,
College of Natural and Applied Sciences, University of Guam,
UOG Station, Mangilao, GU 96923, USA
e-mail: marutani@uguam.uog.edu

D. Nandwani (ed.), *Sustainable Horticultural Systems,* Sustainable Development and Biodiversity 2, DOI 10.1007/978-3-319-06904-3_4

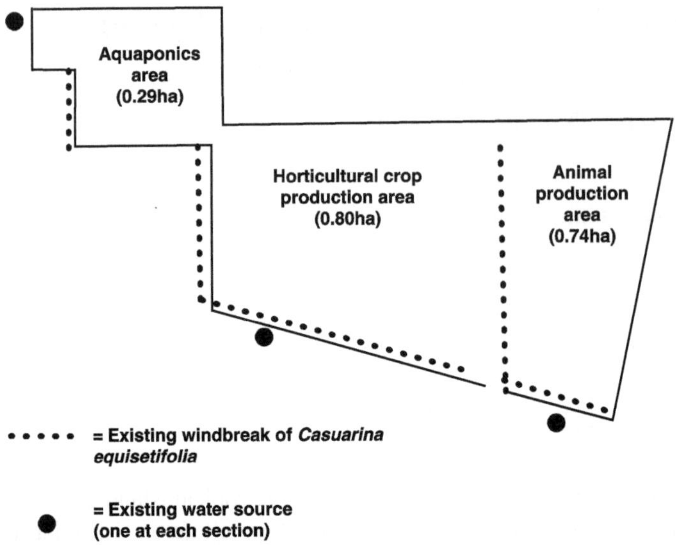

Fig. 1 A schematic map of the model farm at the University of Guam

biggest challenge was finding profitable crops and developing sustainable farming practices suited to small farm agriculture in a tropical environment.

2 Establishment of the Farm

2.1 Design

The first step to construct the model farm was to conduct a survey of the site. The survey included the measurement of the field size, shape, slope and soil type. The soil was classified as Guam cobbly clay soil (calcareous soil) which is a typical soil type found in the central and northern Guam and is characterized as a very shallow, well drained soil on limestone plateaus (Young 1988). Three sections of the farm were divided with existing windbreaks of *Casuarina equisetifolia* L. (Fig. 1) and identified as "aquaponics area" in the smallest site of 0.72 acre (0.29 ha), "horticulture area" in the middle section of the farm with 1.98 acre (0.80 ha), and "animal production area" with the field of 1.83 acre (0.74 ha). After surveying the land, the main activities during the first stage of the farm development involved installation of water lines and planting windbreaks/hedgerows and perennial crops such as fruits and ornamental foliage. Windbreaks were essential plants for protecting crops and farm structures from typhoons and strong trade winds on the island of Guam. Field irrigation systems were installed utilizing existing water pipes near the horticultural crop production and animal production areas (Fig. 1). To construct an aquaponics

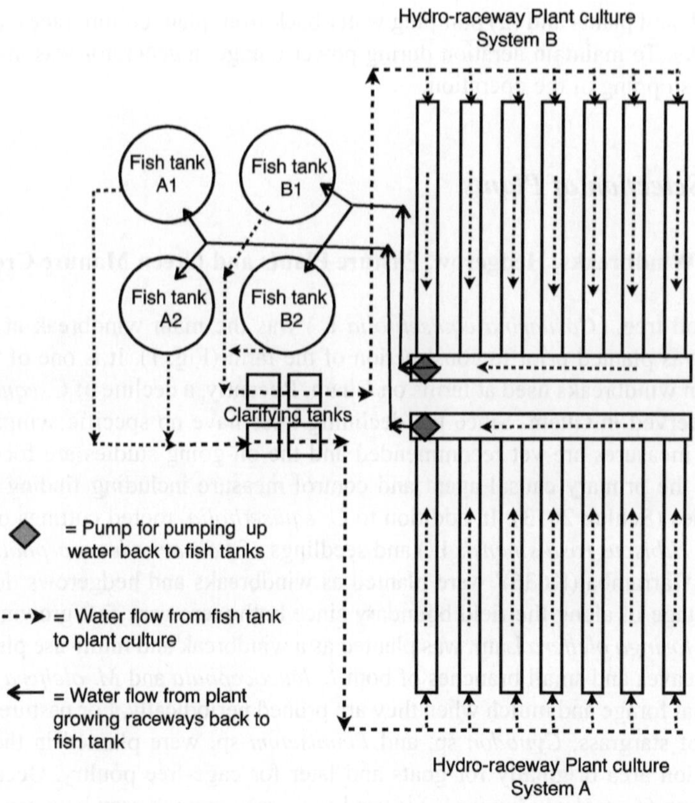

Fig. 2 A schematic diagram of the aquaponics system at the model farm at the University of Guam

system we followed the basic design of Rakocy-type aquaponics system developed at the University of Virgin Islands (Rakocy 1999a, b). The aquaponics has two main components, fish culture with tilapia (*Oreochromis niloticus*) and hydro-culture to grow plants. Our aquaponics has two independent water re-circulating systems, A and B (Fig. 2). Each system consists of two fish tanks, two clarifying tanks and six hydro-culture raceways. The size of each raceway is 15 m × 1.5 m × 25 cm for System A and 14 m × 1.3 m × 25 cm for System B. The total plant growing area in this aquaponics system is 244 m^2. Recirculating water flows down by gravity force (1.2 m ground drop) from the fish tank area and clarifying tank to the site of the plant culture raceways. After plants absorb water-soluble nutrients, the water is pumped back to fish tanks. Plants are a bio-filter of the water. Guam's municipal water is characterized as alkaline water with a high content of calcium and magnesium, which is the ideal water for tilapia culture. Besides fish feed, Fe chelate is added to water to maintain the plant culture. Shade covers were placed on tops of canopy structure for fish tanks and plant growing raceways. There is a water source near the aquaponics system and also electricity to power: (1) aeration for both fish and plants and (2) pumping water back from plant culture raceways to the fish tanks. To maintain aeration during power outage, a generator was installed to prevent stopping of the operation.

2.2 Selection of Plants

2.2.1 Windbreaks, Hedgerow, Pasture Plants and Green Manure Crops

Ironwood tree, (*Casuarina equisetifolia* L.) was the main windbreak at the site, which was planted prior to construction of the farm (Fig. 1). It is one of the most common windbreaks used at farms on Guam. Recently, a decline of *C. equisetifolia* was observed in Guam. Since the declining trees have no specific symptoms, no control measures are yet recommended and the on-going studies are focusing on finding the primary causal agent and control measure including finding resistant germlines (Schlub 2013). In addition to *C. equisetifolia*, rooted cuttings of red hibiscus (*Hibiscus rosa-sinensis* L.) and seedlings of *Leucaena leucocephala* (Lam.) de Wit 'Tarramba (K636)' were planted as windbreaks and hedgerows during the initial stage of along the field boundary since both were very fast growing plants. Later, *Moringa oleifera* Lam. was planted as a windbreak and multi-use plant in the farm. Leaves and small branches of both *L. leucocephala* and *M. oleifera* are used as animal forage and mulch when they are pruned periodically. For pasture grasses, sprigs of stargrass, *Cynodon* sp. and *Pennisetum* sp. were planted in the animal production area originally for goats and later for cage-free poultry. Occasionally Sunnhemp (*Crotalaria juncea* L.) is used as a green manure crop between growing vegetable crops when seeds are available.

2.2.2 Fruit Crops

Calamansi (× *Citrofortunella microcarpa*) and eating banana (*Musa* spp. 'Macau') were originally selected for fruit production mainly due to their great marketability, year-round production and availability of planting materials. Rooted stem cuttings of calamansi and offshoots of banana were planted in the field. Several other fruits crops were considered as potential commercial crops. However, the limited size of the model farm has eliminated many fruit crops growing commercially at the model farm (Table 1).The schematic diagram of the horticulture production section is shown in Fig. 3. Calamansi plants were planted in 2003 in four rows (60–70 m long) having that the distance of two adjacent trees was 8 ft (2.4 m) and the distance between rows was 20 ft (6 m). Calamansi is the only fruit crop currently grown at the farm that is sold regularly to the market. There are four rows with 28 plants in a 60 m row with a total of 112 trees in the field and 39 plants in 15–20 gal (57–75 L) pots. Figure 4 shows the site of field preparation for calamansi planting in 2003 and fruit harvest in 2010. Water stress appears to trigger flowering; especially with

Table 1 Fruit crops considered to grow at the model farm

Crops	Advantages	Disadvantages
1. Acerola or Barbados cherry (*Malpiphia punicifolia*)	a. Fruits with high in vitamin C b. Processing juice and jelly c. Easy propagation	a. Not common commercial crop b. Cultivation method is unknown in Guam
2. Banana (*Musa* spp.)	a. Locally marketable b. Cultivation method established in Guam c. Fruit ripening facility available d. Planting materials available	a. Need a large area b. Many pests and diseases
3. Calamansi (× *Citrofortunella microcarpa*)	a. Locally marketable b. Adapted to Guam c. Established market available d. Processing possible e. Planting materials available	a. Need pruning for easy harvest b. Harvest may be tedious due to small sized fruits
4. Mulberry (*Morus nigra*)	a. Grow well in Guam b. Can make jelly and wine	a. Not commercial fruit crop in Guam
5. Passion fruit (*Passiflora* spp.)	a. Good processing fruit	a. Not commonly grown in Guam b. Not popular fruits in Guam
6. Papaya (*Carica papaya*)	a. Locally marketable b. Both yellow and green fruits are eaten	a. Need a large area for crop rotation as disease control
7. Pineapple (*Ananas comosus*)	a. Tropical fruit	a. Need a large area b. Slow growing
8. Star fruit (*Averrhoa carambola*)	a. Marketable (flea market) b. Processing possible	a. Need a large area b. Delicate fruits c. Need good harvest handling Seasonal
9. Mango (*Mangifera indica*)	a. Marketable	a. Need a large area b. Not Adapted to soil type of the farm c. Seasonal

plants growing in pots with the limited volume of medium. This plant response to water deficit is shown in the record of the 4-year monthly gross income data compared with rainfall data (NOAA 2009, 2010, 2011, 2012) (Fig. 5), indicating that the amount of calamansi fruit harvest (recorded in the $ value) was the greatest during the dry months of March (2010), March (2011) and April (2012), and a drastic decrease in rainfall observed on November of 2009, 2010 and 2011, which were 4–5 months before the peak month of fruit harvest.

To increase the variety of fruit crops at the farm, local avocado (*Persea americana* Mill.), soursop (*Annona muricata* L.) and more recently breadfruit (*Artocarpus* altilis 'Ma`afala') were planted at the farm as a living fence or specimen plants to study potential use of fruits to develop value added products. Papaya (*Carica papaya* L.) plants are grown from seeds scattered in the pasture areas of the chicken

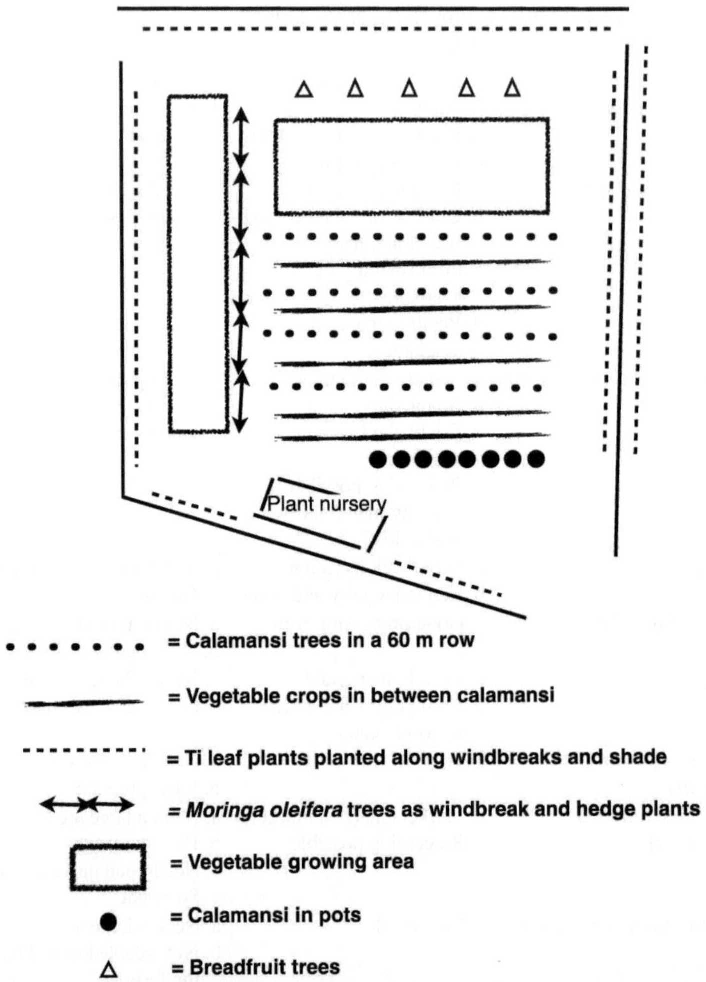

Fig. 3 A schematic diagram of horticultural crop production area at the model farm

layers as shade plants and additional feedstock, for which both leaves and fruits of papaya are used. All fruit crops except calamansi cannot be produced commercially in this small-scale farm. The harvest volume is not large enough and they are best grown as supplemental chicken feed or to distribute specialty buyers when the fruits are available.

2.2.3 Ornamental Plants

Since Guam's main industry is tourism, lei flowers and foliage plants were thought as important crops at the model farm. Three lei flower plants were considered as

Fig. 4 Field preparation of calamansi plants in 2003 (*above*) and harvesting fruits in 2010 (*below*) at the model farm

potential ornamental crops grown at the farm (Table 2). However since a fresh lei flower industry has not developed on Guam during last decade, the plan to grow lei flower plants was eliminated. A couple of flower shops buy Ti leaf (*C. fruticosa*) from the model farm to reduce or substitute their importation of Ti leaves from Hawaii. Although it is a small portion of the farm operation, a weekly supplying Ti leaves to a local flower shop has contributed a steady income to satisfy revenue from the ornamental production area. The supply record of Ti leaf from 2009 to 2012 indicates that the month of May was the peak month of demand for leaves on Mother's Day, Memorial Day and graduation day.

2.2.4 Field-grown Vegetable Crops

Seven fruiting vegetable crops were considered initially as field grown commercial crops at the model farm (Table 3). Cucumber (*Cucumis sativus* L.), eggplant

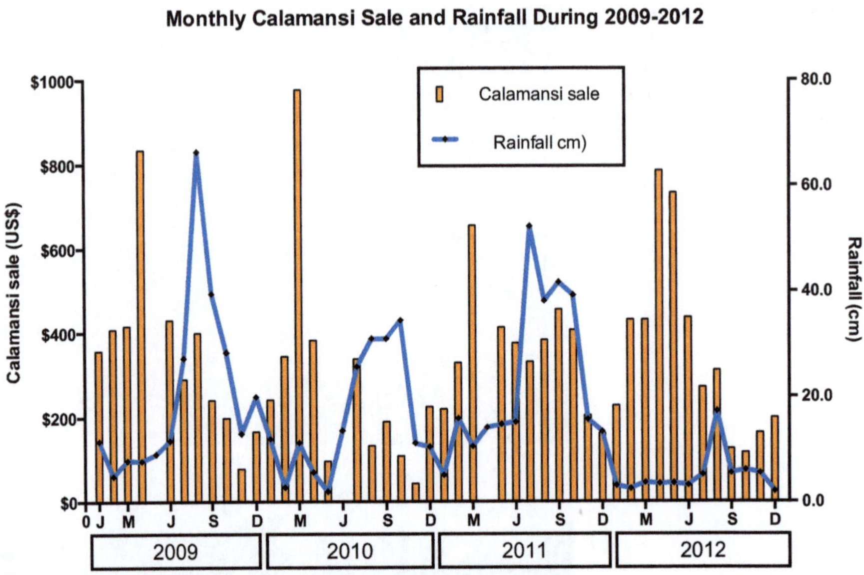

Fig. 5 Monthly calamansi sale (US$) from the model farm and rainfall record (NOAA) during 2009–2012 ($ 100 calamansi fruit sale = 32 kg)

Table 2 Ornamental plants considered as lei flower and foliage production at the model farm

Crop	Advantages	Disadvantages
1. Foliage of Ti leaf (*Cordyline fruticosa*)	a. Good for flower arrangement b. Use to make a lei c. Grow well in Guam	a. Need windbreak and shade b. Commercial cultivation method unknown in Guam
2. Bracts of bougainvillea (*Bougainvillea* spp.)	a. Common plants b. Various colors available	a. Seasonal b. No information as lei flowers in Guam
3. Flowers of Vanda orchids (Vanda Miss Jacuin)	a. Common orchid in Guam	a. No information as lei flowers in Guam b. Commercial cultivation method and marketability unknown
4. White flowers of Plumeria (*Plumeria obtuse*)	a. Common lei flowers with a very nice scent	a. Easy to damage flowers b. Commercial cultivation methods and market—unknown c. Need a large area for produciton

(*Solanum melongena* L.), tomato (*Solanum lycopersicum* L.), watermelon (*Citrullus lanatus* (Thunb.) Matsum and Nakai) and hot peppers (*Capsicum annuum* L. and *C. frutescens* L.) were planted at different time period as one or two rows (60 m per row) between the calamansi planting rows in an intercropping system or areas designated as vegetable growing areas (Fig. 3) or as pot culture. Among potential vegetable crops for the model farm, it appears that hot peppers (*Capsicum* spp.)

Table 3 Vegetable crops considered to grow at the model farm

Crops	Advantages	Disadvantages
1. Cucumber (*Cucumis sativus*)	a. High demand in Guam b. Cultivation method established	a. Very short harvest period b. Need netting support c. Sometimes market is flooded
2. Eggplant (*Solanum melongena*)	a. Marketable b. Cultivation method established	a. Need a large area b. Many pests and diseases
3. Hot peppers (*Capsicum annuum* and *C. frutescens*)	a. High demand b. Adapted to Guam c. Established market d. Processing products e. Planting materials available	a. Need pruning b. Harvest tedious and labor intensive c. Diseases
4. Pumpkin (*Cucrubita maxima* or *C. moschata*)	a. Both fruits and leaf tips are edible b. Local lines available c. Use as animal feedstuffs	a. Need a large area
5. Tomato (*Solanum lycopersicum*)	a. High demand	a. Cracking especially during wet months b. No control for new virus diseases c. High temperature and high humidity affect fruit quality
6. Watermelon (*Citrullus lanatus*)	a. Grow well in Guam b. Marketability-Excellent	a. Need a large area
7. Yard long beans (*Vigna unguiculata* subs. *sesquipendalis*)	a. Grow well in Guam b. Marketability-Excellent c. Direct seeded d. Local lines available	a. Many pests on legumes in Guam

is the best crop to be grown in our system because of the high demand and is less affected by rainy season. Development of local value-added products using chili peppers is also a factor increasing demand for hot peppers.

Vegetable crops listed in Table 3 were planted several times at different growing seasons during the last decade of farm operation as field and as pot culture trials. However, it has been a great challenge to produce labor-intensive vegetable crops profitably in the limited land space and with the limited labor of the model farm. Only hot peppers are consistently grown at the farm except during 2012 when a severe disease problem with anthracnose resulted in no marketable yields. Hot pepper is still grown in pots or in shaded area of the farm as this crop requires less care and this is a high demand crop in Guam for local cuisine. Further studies will be focused on how to increase yield by selecting disease resistant lines.

Traditional roots/tubers crops were also considered for mainly feedstock. Examples are taro, *Colocasia esculenta* (L.) Schott and *Xanthosoma sagittifolium* (L.) Schott; dagu, *Dioscorea alata* L.; mendioka or cassava, *Manihot esculenta* Crantz; and kamuti or sweetpotato, *Ipomoea batatas* (L.) Lam.

Table 4 Crops considered to grow in the aquaponics system at the model farm

Crops	Advantages	Disadvantages
1. Leafy lettuce (*Lactica sativa*)	a. Popular salad green b. Need more local lettuce c. Easy propagation by seeds d. Common crop in hydroculture	a. Insect pests unknown b. Very perishable
2. Kangkong (*Ipomoea aquatica)*	a. Popular leafy green b. Easy propagation by stem cutting c. Planting materials available	a. Short-day promote flowering b. Iron chelate needed a large amount
3. Petsai (*Brassica rapa* subsp. *chinensis*)	a. Very popular vegetable b. Fast growing c. Established market	a. Many insect pests in brassicas in Guam b. Retailed price is low
4. Green onion (*Allium fistulosum*)	a. Very popular vegetable b. Established market	a. Retailed price may be too low for aquaponics
5. Basil (*Ocimum basilicum*)	a. Niche market available b. Grow well in hydroculture	a. Diseases and pests are unknown b. Perishable
6. Mint (*Mentha spicata*)	a. Niche market available b. Grow well in hydroculture	a. Diseases and pests are unknown b. Perishable
7. Fruiting vegetable crops such as tomato and cucumbers	a. Popular vegetable crops	a. Require more nutrients than leafy greens b. Longer growing period than leafy greens c. Feeding required nutrients is difficult with the aquaponics

2.2.5 Aquaponics Crops

Commercial scale aquaponics is a new plant growing system in Guam. Among the potential aquaponics plants listed in Table 4, leaf lettuce (*Lacutica sativa* L.), basil (*Ocimum basilicum* L.), mint (*Mentha* sp.), kangkong (*Ipomoea aquatica* Forssk.), cherry tomato, cucumber, and Ti leaf were examined for adaptability of crops grown in the system. It appears fast-growing leafy vegetable or herbs were more suited to our aquaponics system. Sources of plant nutrients in the system come from; fish feed, fish waste, minerals of water, addition of iron chelate. Nutrient levels are not sufficient to support plant growth and development of some crops, especially fruiting vegetable like tomato and cucumbers. Among leafy greens tested, *I. aquatica* required a much greater amount of Fe and possibly N to have lush green leaves. It was removed from our cultivation system. Similarly Ti leaf plants needed more Fe and N to maintain and produce marketable dark green leaves. Typically kangkong and Ti leaf plants had chlorotic leaves in our aquaponics system. Additionally, the pest control is potentially a big problem with long growing plants since we have limited use of pesticides for plants growing in the re-circulating water with symbiotic relation with fish. A variety of leafy lettuce is grown in the aquaponics system

Fig. 6 Growing lettuce in hydro raceways (*above*) and tilapia (*below*) of the aquaponics system

at the model farm now. Selection criteria of cultivars are heat tolerance, slow bolting and multi-disease resistance. Figure 6 shows lettuce growing in hydro-raceways and tilapia in the holding tack ready to sell in the aquaponics.

2.3 *Incorporation of Animal Production in Sustainable Agriculture*

Livestock production is an important component in sustainable agriculture. The animal production section of the integrated model farm generates not only revenue but also waste materials used as bio-fertilizer or soil amendment at the farm. Originally goats (*Capra aegagrus hircus*), chickens (*Gallus gallus domesticus*) and pigs (*Sus* spp.) were considered as farm animals. Currently three cage-free chicken coops with extension fenced pasture areas support the average of 150 layers/coop generate the highest revenue by egg production for the model farm. Besides commercial feeds, any farm waste (unmarketable vegetables and fruits) and *L. leucocephala* and papaya leaves are used as supplemental feeds. The size of the fenced pasture area (paddock) and rotation of paddocks for chickens need to be further investigated in order to prevent the paddock from becoming bare by over-grazing of stargrass (*Cynodon* spp.), which was planted at the beginning stage of the farm construction. Figure 7 shows that chickens are feeding on grasses and weeds in pasture. Chicken manure is used as organic fertilizer in calamansi and vegetable production at the farm. Composting is in progress using chicken manure with wood chips from pruning farm trees to promote a recycling system of organic matter and natural resources conservation within the farm. Breadfruit, avocado, papaya, starfruit and soursop are grown in the animal production area as shade plants. Leaves of *C. equisetifolia* and woodchips of *L. leucocephala* are used as bedding materials of chicken coops in addition to shredded papers generated from offices of University of Guam.

3 Biodiversity and Integrated Farm Management

The goal of the project was to develop a model farm with a profitable integrated farming system in less than 2 ha of land. The model farm has presented both the importance and limit of biodiversity that is defined as "genetic variation, species variation, or ecosystem variation within a small area" (United Nation Environmental Program http://www.unep-wcmc.org/what-is-biodiversity_50.html. Accessed 7 Dec 2013). Diversification with a wide range of species (plants, fish and farm animals) and agricultural practices at the farm has contributed to develop a robust farm management system by supplying several produces to a local market. The farm was able to become sustainable by using a good strategy of selection of species and cultural practices to achieve diversification. Produce sales were recorded from the beginning of the project and the first gross income was recorded at the end of 2003. The first annual gross income of $ 4348 was recorded in 2004 (Fig. 8) by mainly selling fish, eggs and any short-term vegetable crops mainly on university campus. The farm started to increase gross income yearly by incorporating feedback from local market studies, consumer responses to improve species selections and cultural practices. In 2008, by increasing the number of egg-layers in a cage-free operation, a big increase of egg production was recorded to reach the gross income of $ 45,431. A steady increase in gross income from 2008 to 2012 was due to increased production of leafy lettuce in the aquaponics. Improvement of the system maintenance and finding a niche market for the aquaponics grown lettuce, the increase of lettuce sale would be expected more in future years. Vegetable crop production section remains the toughest area in the farm to grow crops profitably. Each crop has different cultural requirements. Labor-intensive crops with less profit would be eliminated from this farm operation. On the other hand, gourmet varieties of vegetable crops can be included when a niche market is found, and linking directly with consumers would make the farm operation feasible and sustainable.

Fig. 7 Chickens are grazing weeds and grasses under *L. leucocephala* and Ti plants (*above*); Chickens made almost the soil bare after 7 days of grazing; at 0 day (*middle*) and seventh day (*bottom* photo). Remaining pieces of grass in photos are stargrass (*Cynodon* sp.)

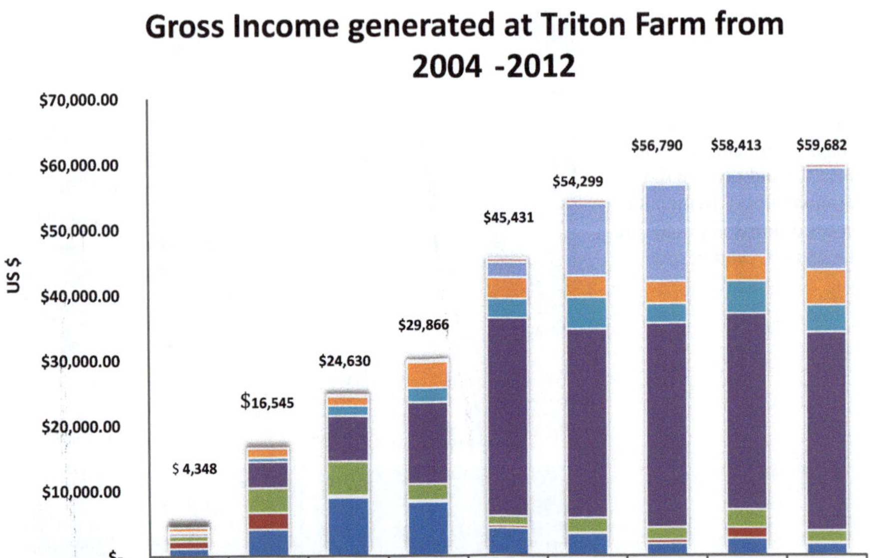

	Y2004	Y2005	Y2006	Y2007	Y2008	Y2009	Y2010	Y2011	Y2012
Basil	$-	$-	$-	$-	$594.50	$590.00	$272.00	$284.00	$548.00
Lettuce	$-	$-	$291.35	$291.80	$2,328.2	$11,046.	$14,666.	$12,411.	$15,560.
Ti leaf	$561.00	$1,421.5	$1,388.8	$3,917.0	$3,224.0	$3,229.0	$3,372.0	$3,820.0	$5,338.0
Calamonsi	$278.25	$686.50	$1,542.2	$2,190.1	$2,994.3	$4,911.4	$3,076.8	$5,008.9	$4,232.8
Egg	$360.50	$4,039.7	$7,013.1	$12,535.	$30,381.	$28,821.	$31,222.	$29,963.	$30,323.
Tilapia	$841.50	$3,785.5	$5,168.0	$2,502.0	$1,370.0	$2,297.5	$1,822.0	$2,759.5	$1,800.0
Hot pepper	$1,157.0	$2,560.7	$345.00	$239.25	$430.95	$54.00	$615.50	$1,612.0	$180.75
Others	$1,149.9	$4,051.4	$8,881.7	$8,190.2	$4,108.4	$3,348.7	$1,744.3	$2,554.7	$1,699.7

Fig. 8 Annual gross income of products at the model farm from 2004 to 2012. Egg and lettuce were two main products since 2009 while basil, Ti-leaf, calamansi and hot peppers generated additional income

The two most important factors in determining suitable crops grown in the integrated farm were (1) marketability and (2) plant suitability to the farm system. The farm system includes not only the ecosystem of the field, but also availability of constant supply of farm inputs including availability of planting materials, fertilizers, animal feeds and labor. The two biggest expenses of the model farm operation are animal feed cost and labor cost. Finally, in addition to main profitable species, it is valuable to include farm plants as farmers can consume produce by themselves to reduce cost of food expenses and also for creation of future commercial crops or value added products.

4 Summary

This case study on the whole farm development suggests that sustainable agriculture can be achieved by utilizing the diverse farm species that are screened on the basis of their marketability and adaptability to a specific farm's agroecosystem. The climatic, ecological, economical and social factors all influence the selection of species and the degree of biodiversity that results in the farm becoming sustainable and profitable. Understanding these factors is essential to make a strategy to advance innovative farming practices that lead to achieving the construction of a successful integrated farm.

Acknowledgements The work was supported by grants from US Department of Agriculture/Initiative for Future Agriculture and Food Systems Program, and USDA/HATCH project. Authors thank Mr. Joe Tuquero Mr. Ken Perez, Mr. Jerome Andrew, Mr. Glenn Takai, Mr. Eric Esteves, Ms. Leanne Obra, Ms. Lauren Gutierrez, Ms. Ashley Randall, Mr. Donavan Leon Guerrero, Mr. Gabriel Camacho, Ms. Sheeka Afaisen, Mr. Manny Bermudas, Mr. Chris Posadas, Mr. Craig Borja, Mr. Dennis Velasco, Mr. Art Moon, Ms. Anne Yalung, Ms. Kateri Cruz, Mr. Gary Baxter, Mr. Cameron Santos, Mr. Norman Santos, Dr. Aubrey Moore, students of UOG Americorp, Interns from the Okayama University, Interns from the Ryukyu University of Okinawa, and Guam Agricultural Experiment Station farm crew for their technical support and assistance in data collection. Our appreciation goes to Dr. James McConnell, Dr. Bob Barber and Dr. Manny Duguies for technical advice, and a special thank goes to Mr. Karl (Von) Nelson for his dedicated work managing the farm.

References

Marutani M, Brown J, Wall G et al (1997) Agricultural crop production on Guam during 20th century. Micronesia 30*(2):*389–415

National Oceanic and Atmospheric Administration (2009) Local climatological data, annual summary with comparative data, Agana, Pacific. http://www.ncdc.noaa.gov/IPS/lcd/lcd.html. Accessed 3 Dec 2013

National Oceanic and Atmospheric Administration (2010) Local climatological data, annual summary with comparative data, Agana, Pacific. http://www.ncdc.noaa.gov/IPS/lcd/lcd.html. Accessed 3 Dec 2013

National Oceanic and Atmospheric Administration (2011) Local climatological data, annual summary with comparative data, Agana, Pacific. http://www.ncdc.noaa.gov/IPS/lcd/lcd.html. Accessed 3 Dec 2013

National Oceanic and Atmospheric Administration (2012) Local climatological data, annual summary with comparative data, Agana, Pacific. http://www.ncdc.noaa.gov/IPS/lcd/lcd.html. Accessed 3 Dec 2013

Rakocy J (1999a) The status of aquaponics, part 1. Aquacult Mag 25(4):83–88

Rakocy J (1999b) Aquaculture engineering—the status of aquaponics, part 2. Aquacult Mag 25(5):64–70

Schlub R (2013) Gago, Guam ironwood tree, *Casuarina equisetifolia* past, present, future. University of Guam Cooperative Extension Service, Mangilao

Young F (1988) Soil survey of territory of Guam. USDA Soil Conservation Service, Washington, DC

Impacts of Vulnerabilities and Climate Change on Sustainable Agriculture for Caribbean Small Island Developing States (CSIDS)

Rohanie Maharaj and Dimple Singh-Ackbarali

1 Introduction

Small Island Developing States (SIDS) are small islands and low-lying coastal countries with varied geography, climate, social, cultural, political and economic development. Currently, fifty-one (51) SIDS are included in the list used by the United Nations Department of Economic and Social Affairs (UNDESA) in monitoring the sustainable development of SIDS (UNEP 2008). These countries are often categorized by their three regions; the Caribbean, the Pacific, and the AIMS (Africa, Indian Ocean, Mediterranean and South China Sea). These states and territories often work together in the United Nations (UN) through the Alliance of Small Island States (AOSIS)[1]. These countries share a number of environmental and socio-economic characteristics which highlight their vulnerability to emerging challenges and climate change, which undermine efforts to achieve sustainable development.

Most SIDS recognize the importance of sustainable development[2] as a key component in their developmental path due to a number of structural problems that they may have (UNEP 2005; UNFCCC 2007b):

[1] The Alliance of Small Island States (AOSIS) is a coalition of Small Island and low-lying coastal countries that share similar development challenges and concerns about the environment, especially their vulnerability to the adverse effects of global climate change. It functions primarily as an ad hoc lobby and negotiating voice for SIDS within the United Nations system. AOSIS has a membership of 43 States and observers, drawn from all regions of the world: Africa, Caribbean, Indian Ocean, Mediterranean, Pacific and South China Sea. Thirty-seven are members of the United Nations, close to 28 % of developing countries, and 20 % of the total membership of the United Nations. Together, SIDS communities constitute some 5 % of the global population. http://www.un.org/esa/sustdev/sids/sids.htm.

[2] According to the International Institute for Sustainable Development (IISD), sustainable development has been defined in many ways, but the most frequently quoted definition is from Our

R. Maharaj (✉) · D. Singh-Ackbarali
Biosciences, Agriculture and Food Technologies (BAFT), Eastern Caribbean Institute of Agriculture and Forestry (ECIAF) Campus, The University of Trinidad and Tobago, Caroni North Bank Road, Centeno, Piarco, Trinidad and Tobago, W. I.
e-mail: rohanie.maharaj@utt.edu.tt

D. Nandwani (ed.), *Sustainable Horticultural Systems*, Sustainable Development and Biodiversity 2, DOI 10.1007/978-3-319-06904-3_5

- Their populations, and markets, are small; and depend on international trade
- Their resource base is usually narrow (e.g. limited land space and finite natural resources), fragile (e.g. their ecosystems) and can be prone to disruption by natural disasters especially among low-lying islands;
- Serious vulnerability to extreme climate events and other environmental disasters;
- Isolated geographically from the global market and vulnerable to global developments
- Typically SIDS depend on foreign exchange on a small range of primary product exports; as well as high import food bill and
- Local capital for productive investment is limited.

Climate change has been identified globally as the single environmental issue of the twenty-first century that poses unprecedented threats to mankind. The accelerated rate of climate change as a result of human activity, can be attributed to the increases in concentration of greenhouse gases in the atmosphere as a result of deforestation, fossil fuel combustion, industrial processes and waste management (IPCC 2007a). Climate change is anticipated to have far reaching effects on the sustainable development of SIDS including their ability to attain the United Nations Millennium Development Goals by 2015 (UN 2007). Caribbean Small Island Developing States (CSDIS) are particularly vulnerable to severe consequences of climate change, which results in sea level rise, increased flooding, an increased frequency and intensity of hurricanes, hillside erosion, and loss of coastal habitats. These environmental concerns, combined with their particular socio-economic situations, make CSIDS, some of the most vulnerable countries in the world to climate change, despite the fact that they produce very low levels of greenhouse gas emissions. The main sectors in CSIDS that are likely to be impacted are agriculture, human health and settlements, coastal zones, and water resources as well as cross sectoral socio-economic systems (UNEP 2005).

Agriculture has been the backbone of many CSIDS economies; however the impact of climate change on the agriculture sector and its effect on food and nutrition security is of great concern. Figure 1 shows the different components of food security that are affected by climate change. In the Caribbean, agricultural productivity has declined and this has been exacerbated as a result of environmental issues such as depletion of marine and coastal zone resources, contamination of fresh water resources, deforestation, decline in biodiversity, extended periods of the dry season, hurricanes, flooding, lack of proper water supply, land degradation and soil erosion, loss of soil fertility and shortening of the growing season, all of which lead to major economic losses and seriously affect food security. In the Caribbean region, the 2004 hurricane season caused damages estimated at US$ 2.2 billion in four (4) countries alone: Bahamas, Grenada, Jamaica and Dominican Republic. The relative magnitude of economic losses due to climate change is likely to differ among

Common Future, also known as the Brundtland Report (1987), which states that: "Sustainable development is development that meets the needs of the present without compromising the ability of future generations to meet their own needs."

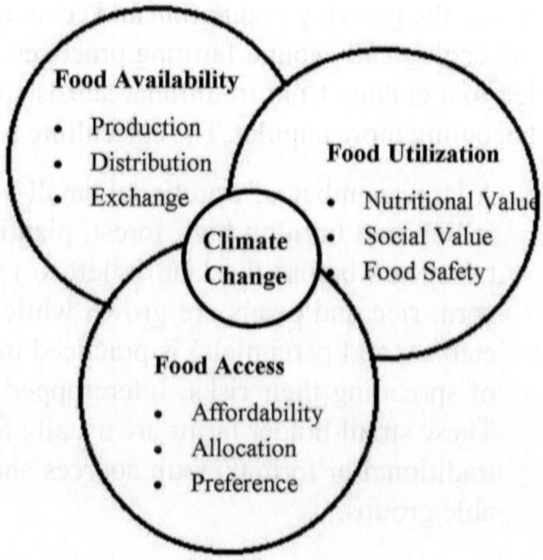

Fig. 1 Climate change linked to components of food security. (Reproduced from Ericksen et al. 2011)

islands (UNFCCC 2007b). Food and nutrition security is now an important item on the political agenda in this region.

The term sustainable agriculture has been defined (U.S. Code Title 7, Sect. 3103) as an integrated system of plant and animal production practices which will have a site-specific application that will, over the long term:

- Satisfy human food and fiber needs
- Enhance environmental quality and the natural resource base upon which the agricultural economy depends
- Make the most efficient use of nonrenewable resources and on-farm resources and integrate, where appropriate, natural biological cycles and controls
- Sustain the economic viability of farm operations and
- Enhance the quality of life for farmers and society as a whole.

Adaptation is a process through which societies make themselves better able to cope with an uncertain future. As a result, CSIDS require a diversity of adaptation measures very much depending on individual circumstances in order to propel economic development. This chapter will highlight and discuss the challenges and constraints of sustainable agriculture including the impacts of climate change most commonly faced by the agricultural sector in CSIDS and also present the opportunities and options that can allow for sustainable farming systems and sustainable agriculture.

2 Agricultural Production Systems

Traditionally CSIDS practice subsistence agriculture including livestock farming where family labour and few purchased inputs are employed. This practice has the advantage of being environmentally friendly but the output is low productivity.

Today the growing population and economic pressures faced by CSIDS are disrupting ecologically sound farming practices, such as fallow rotation systems. This has led to a change from traditional subsistence agriculture to moderate input systems becoming more popular. The agriculture sector in many CSIDS generally comprises:

- A large number of traditional small-holder farming systems which consists of felling and burning high forest, planting food crops for subsistence for one or two years before the land is left to recover under fallow. Staple food such as corn, rice and beans are grown while mixed cropping (root crops, pulse, vegetables, and perennials) is practiced mostly for home-consumption with a view of spreading their risks. Intercropped are plantains, yams and sweet potatoes. These small-holder farms are usually found in rural areas, have limited access to traditional or formal credit sources and are the most disadvantaged and vulnerable groups.

- A small number of small commercial farms. These are a transition between subsistence and mechanized production and primarily use family labour and on occasion, hire seasonal labour. Land as well as all other production practices are done manually by family and usually the women in the family are responsible for marketing. Technological inputs such as improved seeds and irrigation are either by hand or simple systems involving a pump for farms located near ponds and rivers. Output is staggered to provide a small but steady stream of income. Farms are located near roads that increase access to transportation, have high capital base and easy access to traditional credit source. Subsystems are:

1. Domestic crops—plantains, rice, corn, beans, peanuts
2. Vegetables—potatoes, onion, cabbage, tomatoes, sweet peppers, carrots
3. Fruit Trees—mangoes, coconuts, soursop, guavas, avocado, cashew
4. Livestock—beef, pigs, poultry, eggs, dairy, sheep (local and improved breeds reared on mostly natural pastures)

- A small number of large commercial farms with high resource base and access to commercial credit. These farm systems dominate the agricultural sector and account for a large portion of production of traditional export crops (banana, sugar, coconuts, cocoa, coffee etc.). Such farms are family owned with hired labour during harvest and the land is normally cleared by hand and then planted. Most farms are owned by persons having other employment. The large farm units with mechanization can be found in Guyana, Belize, Trinidad, Jamaica and Suriname. There is also use of improved technology such as protected agriculture and inputs. Two subsystems are identified:

1. Traditional crops—such as sugarcane, bananas and citrus primarily grown with other food crops on a smaller scale.
2. Non-traditional crops—such as papayas, hot peppers, cacao

- A small number of large farms and estates which are currently idle, even though there is severe competition for limited agricultural land space. This group is beginning to consider investment in traditional commodities. These large acreage farms which are not idle, are highly mechanized and have large investments that include technological inputs as well as secure land tenure. Such farms grow grains and other non-traditional export crops. The principal characteristic is that they have a well-established marketing system.

The agricultural sector in most CSIDS have either declined or stagnated in recent years. This is primarily because of the contraction in traditional exports and a reduction in the human resource invested in agriculture. The erosion of the European Union (EU) trade preferences have led to price volatility of sugar, bananas, cocoa and rice in commodity markets. As expected, declining output levels have not only had an impact on employment in the sector but have also led to the growing tendency to import poor quality food displacing nutritionally valuable local crops. The declining contribution of domestic agriculture and traditional food crops and increase in dependence on imported food, often results in unbalanced diets, an increase in incidence of under-nutrition, malnutrition, and diet-related disorders such as diabetes, obesity and cardiovascular diseases. It is estimated that the Caribbean region has a food import bill between US$ 3–5 billion which represents the potential for billions worth of business opportunities for Caribbean producers including jobs, profitable Caribbean business enterprises, and billions of dollars of avoided imports every year, therefore improving external balances (FAO 2013).

CSIDS response to the decline in traditional agriculture, which has been driven by the changing world markets, trade imbalances, the quest for food security and growing human population, have been to implement programmes to raise productivity, differentiate the mix within traditional agriculture and introduce new crops such as papaya, mango, guava, sorrel, spices, herbs and root vegetables. Unfortunately, there are no official data on the performance of non-traditional exports, but countries such as Guyana have benefited greatly from increased trade with its centres of its diaspora for example New York and Toronto in products such as fish, fruits and vegetables.

Today, there is international recognition that CSIDS face special challenges with respect to agriculture sector development and management, and with attaining United Nations Millennium Development Goals (UN 2007). This recognition is increasingly reflected in recent international instruments which are applicable to CSIDS and have a direct bearing on food security such as the 1995 Kyoto Declaration and Plan of Action on the Sustainable Contribution of Fisheries to Food Security, the 1996 Rome Declaration on World Food Security and the World Food Summit Plan of Action (1996). With this recognition and the assistance from key players such as World Bank (WB), International Fund for Agricultural Development (IFAD), World Food Programme (WFP) and Food and Agricultural Organization (FAO) to name a few, CSIDS now have opportunities for increasing agricultural production and growth, and improving farmer's income, provided policies support advantages and address constraints. Before the constraints can be addressed; land, labour and capital markets have to be operating efficiently.

Three essential goals for sustainable agriculture and development in CSIDS include:

1. Food security through an appropriate and sustainable balance between self-sufficiency and self-reliance;

2. Employment and income generation in rural areas, particularly to eradicate poverty; and
3. Natural resource conservation and environmental protection.

Achieving these goals implies a lengthy process requiring a comprehensive approach and significant investments of labour, capital, technology and research, all of which require political intervention in CSIDS. The FAO has offered its expertise in order to help revert the tendency of CSIDS to monoculture and high dependency on fertilizers and pesticides. However it must be noted that a more integrated approach which takes into account the traditional production systems and builds on modern tools and technologies is required with the participation and full collaboration of both environment and agriculture line institutions. Integrated production systems, agroforestry, agricultural tourism, diversified and certified organic products for export markets, as well as increased horticultural production, are some of the emerging trends for agriculture. These require a better knowledge of past and present production systems adapted to local conditions and require coordination and concerted actions amongst all stakeholders.

The key characteristics of successful activities are:

- The return to farm labour where average wage rates are above what was acceptable in the past; avoiding labour intensive activities, unless returns are high;
- Export of fresh produce (e.g. seasonal supply of fruits and vegetables), or with inexpensive processing that can either be done on the farm, or centrally, without the need for substantial economies of scale (e.g. specialist products);
- Improving post-harvest storage so that quality loss of stored produce is reduced within the product life cycle;
- Produce and products must be high value in relation to transport cost;
- Improving produce quality according to consumer and phytosanitary requirements, including presentation and packaging.

3 Macroeconomic Issues and Constraints in Integrating CSIDS into the Global Community

Economic development in the Caribbean started with agriculture and despite it being a key economic driver in many Caribbean countries such as Belize, Cuba, Dominica, Guyana, Haiti, Jamaica and Suriname, the scale of this sector cannot be compared to that in developed countries. In the case of rice, CPEC (2005) reported that none of the major exporters in the region were likely to be competitive on the world market, and that Guyana had the greatest opportunities for selling its rice on the Caribbean Community and Common Market (CARICOM). In 2000, contribution from agriculture sector to Gross Domestic Product (GDP) ranged from 7.7 % in St. Lucia, and 9.2 % in Grenada to over 21 % in Dominica. In Guyana, agriculture contributed to 25 % GDP compared to Jamaica, where it contributed 7.3 % (Government of Guyana 2002; Government of Jamaica 2000). Agricultural contributions to GDP in Trinidad and Tobago fell from 4.2 to 1.02 % between 1984 and 2004. However, during the past five (5) years, Caribbean agriculture has experienced declining production and hence declining contributions to GDP, mainly due to liberalized trading practice, loss from natural disasters and experiences of climate change. The exceptions are Belize, Dominican Republic, Trinidad and Tobago, and Jamaica, where agricultural output has been growing at around 2–5 % per year. The agricultural sector continues to be a major employer of labour in the region, however the percentage of the labour force employed in agriculture range from < 10 % for Trinidad and Tobago, Antigua and Barbuda, Bahamas and Barbados; to 30.5 % in Belize and 60 % in Haiti and between 20 and 40 % in Dominica, Jamaica, St Vincent and the Grenadines, Guyana, Suriname, Dominican Republic and St Lucia (Pemberton 2006; IICA, ECLAC, FAO 2011) . On average Agriculture directly accounts for about 21 % of CARICOM's labour force and 6.5 % of its total GDP (CARICOM 2012).

As agriculture has declined, the role of the services sector has grown in importance, especially in the areas of tourism, construction, telecommunications and financial services. Many economists suggest that the services sector, in CSIDS economies is not achieving the type of economic effect that resulted from high levels of agricultural production and export. As the factors of production in the agricultural sector are locally owned, profits are passed on for consumption, investment and savings in the same local economy. The opposite is true for the service sector, where a significant portion of the capital invested is from foreign investments and hence, the multiplier and distribution effects are less than those from agricultural earnings.

A characteristic of CSIDS that differentiates them from other countries is their high dependence on the international trading systems for their livelihood, which makes them vulnerable to any unexpected changes or "shocks" in the world market. This makes achieving sustainable agriculture more complex and increases their economic vulnerability[3].

The issues and constraints which contribute to the difficulties on successfully integrating CSIDS into the global economy are:

- Size—the smallness of CSIDS usually means limited or poor natural resources and a narrow resource base, which can result in relatively high imports of capital and consumer goods in relation to GDP. The need for a large amount of foreign exchange to pay high import bills, leads to a high dependence on exports. The small size also restricts the ability of many SIDS to diversify exports, rendering them dependent on a very narrow range of goods (primary products from agriculture, forestry, fisheries and mining) and services and hence intensifying the problems associated with dependence on international trade and on foreign exchange earnings (Briguglio 2004).

[3] One of the central themes that informed the deliberations at the United Nations Global Conference on Small Island Developing States was the proposition that the "sustainable development capacity of SIDS was severely undermined by a number of characteristics that were unique to such entities and which trans-late into specific development problems that impede their achievement of such development". (ECLAC 2000, p 2)

- Vulnerability to exogenous environmental shocks—CSIDS appears to be disproportionally vulnerable to natural disasters as compared to large nations. The effects of a disaster are expected to be greater, that is the damage per unit area and costs per capita are higher due to their small size.
- Geographic dispersion or remoteness—since islands are separated by sea, they are constrained to using air and sea transport, for their imports and exports. This places many CSIDS at a disadvantage economically, leading to high freight costs, inability to respond efficiently to unexpected changes in demand and reduced competitiveness (Pelling and Uitto 2001).
- Price competitiveness on exports—Many small-sized farms use low levels of technology and coupled with low levels of investment, producers in the CSIDS are unable to achieve economies of scale and are high-cost producers. This negatively affects the price competitiveness of exports. Comparative studies of banana exporters have shown that the Caribbean islands with smaller farms are higher cost producers than most of their competitors. Companies charged with marketing produce from Caribbean islands, usually have no control over production costs or supply quantities. In addition to the negative effects of weather, farmers in the Caribbean islands enter and exit the industry regularly depending on available prices.
- Vulnerability to exogenous economic shocks—CSIDS are more exposed to trade related shocks, including slumps in demand, instability in world commodity prices and international fluctuations on interest rates (Guillaumont 2010). The per capita income of many CSIDS are higher than those of other developing countries as a group and this leads to CSIDS having limited access to concessionary resources. Current incomes of CSIDS are often facilitated by migrant remittances and preferential market access for some major exports. The EU and the US, grant preferences to a number of SIDS under the Generalized System of Preferences. The EU also provides additional preferences to 26 SIDS under the Lomé Convention for African, Caribbean and Pacific (ACP) countries.
- Limited internal market hinders CSIDS ability to reap the benefits of economies of scale—a relatively low population can mean that domestic demand is below the minimum efficient scale of production. This leads to an inability to capture the benefits of economies of scale and also implies that typical import substitution possibilities are limited.
- Small population and migration of population that are "highly skilled"—CSIDS have small populations in absolute terms, insufficient to generate economies of scale in several areas. They therefore have limited scope for the full utilization of certain types of highly specialized experts. This results in unpredictable shortages of specific skills and difficulty in adjusting to shortfalls or surpluses in labour market segments. Thus they experience high levels of migration, particularly with respect to skilled human resources, which not only places a burden on training facilities but also forces the import of high-cost foreign expertise. A small labour pool also implies that CSIDS are unable to compete with larger nations that have large endowments of low-skilled and skilled labour.

Table 1 Climate projections for the insular Caribbean, based on global predictions from IPCC (2007b)

Climate parameter	Predicted change for the insular Caribbean
Air temperature	Increase of 1.8–4.0 °C by 2099
Sea surface temperature	~ 1.7 °C by the end of the century
Sea level rise	Rise of 0.18–0.59 m by 2099
Carbon dioxide	Reduction in pH of the oceans by 0.14–0.35 units by 2099
Hurricanes	More intense with larger peak wind speeds and heavier precipitation

- Limited commodities can lead to a dependence on consumer imports and capital goods. A lack of opportunities for achieving economies of scale, coupled with a narrow resource base, limits the total production of CSIDS to a narrow range of crops, value added products and services. CSIDS attempts to diversify at low cost can be problematic when considering the economies of scale and higher per unit costs of production, infrastructure development and training manpower. CSIDS that adopt protectionist policies to counter difficulties in diversification can run the risk of encouraging inefficient growth, or growth in uncompetitive sectors.

4 Climatic Changes Impacts on Agriculture Sector

In the Caribbean region, climate change involves a symbiosis of threats, risks, challenges and opportunities. The region faces impacts of climate change, a problem to whose making it has had little to contribute (Table 1).

Some expected impacts on CSIDS as a result of vulnerability to the effects of climate change, sea level rise and extreme events are listed below (IPCC 2007b; UNFCCC 2007a):

- Deteriorating coastal conditions, e.g. through beach erosion and coral bleaching, are expected to adversely affect local resources, e.g. fisheries, and reduce their value as tourist destinations.
- Floods, storm surge, erosion and other coastal hazards, exacerbated by sea-level rise that threaten vital infrastructure, settlements and facilities that support the livelihood of island communities.
- Reduction in freshwater resources by mid-century, to the point where they cannot meet demand during low rainfall periods.
- Increased invasion by non-native species as a result of higher temperatures is also expected, particularly on middle and high-latitude islands.
- Economic losses from reduced agricultural yields as a result of shortening of the growing season and drought.
- Loss of mangrove forests and coral reefs due to sea level rise.
- Bleaching and acidification of the ocean.

Table 2 Factsheet on impact of climate change on Latin America and the Caribbean. (Source: International Food Policy Research Institute, Climate Change: Impact on Agriculture and Costs of Adaptation 2009)

Climate change will have a negative effect on crop yields in Latin America and the Caribbean in 2050 with average yield declines of up to 6.4 % for rice, 3 % for maize, 3 % for soybean, and up to 6 % for wheat
With climate change, average food availability in the region will decrease by more than 300 kcal per person, a decline of about 12 %
Climate change will result in 6.4 million malnourished children in 2050, 1.4 million more than in a no-climate change scenario
Latin America and the Caribbean require additional annual investments of about US$ 1.3 billion, the majority of which should be used for agricultural research, irrigation efficiency and rural road improvements in order to counteract the effects of climate change on nutrition

- Damage to terrestrial forest caused by extreme events.
- Reduction in the size of freshwater lenses and of general water resource availability due to decreased rainfall and saltwater intrusion.
- Inundation of coastal settlements and arable land on the coast.
- Reduction in tourism due to increased frequency and severity of extreme weather.

While the successes in the agricultural sector over the last decade are heralded, the inequitable distribution of benefits and unsustainable impacts on the natural resources in CSIDS are becoming more evident and has direct effects on agricultural production and food security. FAO's assessments have indicated that the target of the World Food Summit to reduce the number of the world's food insecure persons is not being met and that, despite the signing of agreements; carbon emissions continue to rise, and species extinction and climate change effects continue. Paragraph 28, of the Basis for Action of the World Food Summit Plan of Action (1996) notes that, "Small Island Developing States face the threat of land loss and erosion due to climate change and sea level rise and have particular needs for their overall sustainable development." The increasing frequency of storms, flooding and drought have implications on viability of agricultural production and food availability. In many CSIDS, the small holder or lower income farms are forced to move in marginal areas (floodplains and exposed hillsides), putting them at risk from the negative impacts of climate variability and change (Table 2).

5 Production Constraints

5.1 Constraints in Farm Labour

The stigma attached to agriculture from the colonial period has unfortunately branded itself into the Caribbean psyche, where it is percieved that only slaves, paupers and uneducated persons plant crops and actively fish. The out-migration of a large part of the young male population in many CSIDS, the alternative higher-income employment opportunities in commercial farming, urban areas and the tourist sector have regrettably reduced the human resource invested in agriculture, painting a somewhat uncertain future for the sector. The agriculture sector is also plagued with an aging farming population and a disinterested youth population in farming.

5.2 Constraints in Water Available for Irrigation

In order to feed the increasing population, food and feed production will also need to increase substantially including water to produce the required crops. Even if it is assumed that groundwater extractions can be continued at present rates, approximately 30 % of the global irrigation water demand in the year 2100 cannot be met, resulting in a loss of global (irrigated) crop production of around 400 metric t (Biemans et al. 2012). With groundwater already depleting in many regions, due to an uncertain supply source, will only exacerbate crop production losses. Small islands affected by climate change impacts including reduced rainfall, will not have their water sheds and aquifers recharge as effectively. Rural farms may not have the resources to invest in wells to tap into ground water sources or may not have access to pipe borne water or proper road facilities to deliver water to the farms, and thus will rely solely on water from rainfall to irrigate their crops. Unless many efforts are made to increase water harvesting and storage capacity via retention ponds with mobile pumps and to increase water use efficiency in both rainfed and irrigated agriculture, water availability will put a serious constraint on food security by the year 2100 (Biemans et al. 2012). There is a serious challenge in CSIDS in ensuring that water resources for agriculture can be sustainable with respect to competition from other industries for this limited resource. The poor infrastructure for delivery of water to users and the uncertain potential effects of continued climate change make this a high priority issue for CSIDS.

5.3 Constraints in Genetic Resource

Planting material availability is generally inadequate to meet demand particularly if rapid expansion of production is to be undertaken. This leads to a strong dependence of CSIDS on imported seeds for many crops. Although FAO has implemented various interventions in the field of germplasm, information management and seed programmes, availability of quality planting material is still generally inadequate to meet demand. There is also a fundamental problem related to quality control due to the continued and extensive use of planting material from preceding crops rather than certified and clean source of material.

5.4 Constraints in Plant Protection

CSIDS must cope with the stringent quality demands of agricultural produce from importers in industrialised countries and also with the changing requirements of local food consumption patterns especially in the tourist industry. This together with the favourable climatic conditions for pest and diseases, and the high cost of labour, has led to the widespread abuse of pesticide in crop production systems. There is a lack of efficient pest management and monitoring programs in CSIDS. Only a small number of countries in this region are participating in the Convention on the Prior Informed Consent (PIC) arrangement and so only a few are exposed to information between exporting and importing countries about potentially hazardous chemicals.

5.5 Constraints in Livestock and Fisheries Production

Commercially oriented livestock systems in CSIDS are relatively new, because of the high cost and land requirement associated with livestock farming. However, most smallholder farms keep a few domestic pigs, cattle and poultry. Another challenge for commercial livestock farming in CSIDS is their geographic isolation and the associated high transportation costs which can seriously affect the financial viability of any livestock enterprise dependent on imported inputs. Major constraints include inadequate nutrition, ineffective animal health services, high cost of purchased feed, problems in procurement of commercial feed, poor genetic stock, labour shortages and inadequate management of breeding farms and hatcheries. Due to these constraints, the scope for any rapid development of livestock production appears to be limited, but potential exists for import substitution since high transport costs counteract imports.

Fishing resources are being depleted in the oceans as a result of environmental and climate change issues which also puts at risk food and nutrition security. The fisheries sector in CSIDS have been neglected over the years with poor infrastructure for fish landing sites, fish repair nets, boat repair sheds and jetties which do not meet international standards. Global research done by the FAO have concluded that offshore seismic survey activity has negative impacts on fish stocks and even stated that there was an "urgent need for the development of a strongly regulated regime for the mitigation of seismic surveys and for further research so as to minimise the impact on coastal communities and fishers" in their 2010 dossier (FAO 2013). However to this day, no research has been done regionally on the impact of seismic surveys on marine life in the Caribbean. Invasive species such as lionfish have also negatively impacted Caribbean marine ecosystems and fish stocks. Traditional inland aquaculture farms and hatcheries in CSIDS, with the exception of Belize and Jamaica, cannot compete in the regional and global market because it is almost impossible for them to competitively produce and distribute large and uniform fish in mass production. Technical personnel with training in aquaculture both at university and lower level are lacking within the region with the exception of Jamaica, Dominican Republic, and Trinidad and Tobago. With respect to higher education, various universities are active in aquaculture, both at the regional level, such as the University of West Indies (UWI), and at the national level in the Dominican Republic such as the Universidad Autonoma Central de Santo Domingo (UASD) and the Universidad Central del Este (UCE) (FAO 1993). Aquaculture training at the technician or skilled level is not easily available and is a common constraint for the development of the industry throughout the region. Many CSIDS are net importers of fishery products which are for use both by island populations and, in some cases, for the tourist industry. In view of this situation, and the need to ensure that fisheries continue to contribute to the maximum extent possible to food security in CSIDS, every effort should be made to facilitate long-term sustainable resource use.

6 Food Losses

One-third of all food produced in the world is lost or wasted from farm to fork, according to estimates calculated by FAO (2011). This wastage not only has an enormous negative impact on the global economy and food availability, but it also has major environmental impacts. The direct economic cost of global food wastage of agricultural products (excluding fish and seafood), based on producer prices only, is about US$ 750 billion, equivalent to the GDP of Switzerland, (FAO 2013).

Since 2005, the cost of food and its imports have been spiraling upwards. There is a more heightened awareness of food loss in the Caribbean amid concerns about nutrition and hunger, resource conservation, and economic costs associated with food security and waste management. Currently no public or private efforts have been made to improve available food supplies or recover safe and nutritious food that would otherwise be wasted. Food security problems will persist if the Government's action plan for the sector looks only inwardly to identify growth poles for development and not consider measuring and reducing food wastage. There have been quantifications of consumption of six commodity groups, namely staples, vegetables, legumes and pulses, fruits, livestock and aquaculture but to this date no public or private efforts have been made to quantify food that is wasted or lost, or to better use available food supplies, or recover safe and nutritious food that would otherwise be wasted (Figure 2).

It is expected that not all food that is lost would be suitable for consumption nor be economically recoverable. However, large quantities of wholesome, edible foods, such as blemished or over-ripe produce, which may be unmarketable for cosmetic reasons, but are otherwise nutritious and safe, are lost at every stage of the food value chain. In the developing world, losses are mainly attributable to the absence of food-chain infrastructure and the lack of knowledge or investment in storage technologies both on and off the farm. Food losses in the agriculture sector of CSIDS occur at both the farm (post-harvest) and farm-to-retail level (processing and wholesaling):

1. Farm and post-harvest level
   - Pre-harvest losses due to severe weather, disease, and predation.

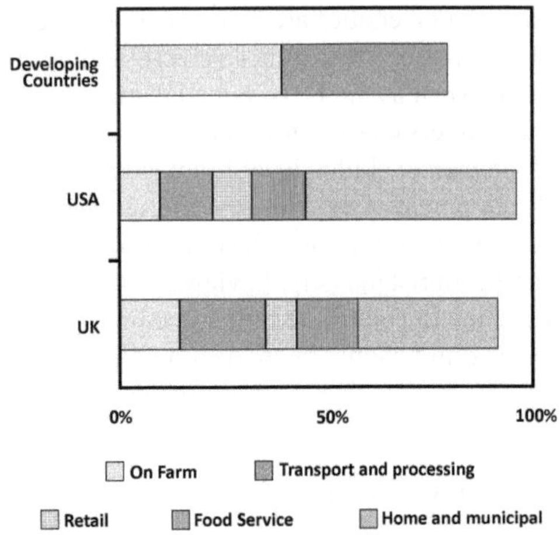

Fig. 2 The makeup of total food waste in developed and developing countries. Retail, food service, and home and municipal categories are lumped together for developing countries. (Reproduced from Nellemann et al. 2009)

- Harvest losses attributed to manual or mechanization, production practices, and decisions.
- Storage losses due to insects, mold, deterioration, shrinkage, and spoilage.

2. Processing and wholesaling level
   - Removal of inedible portions bones, blood, peels, pits, etc.
   - Discard of substandard products (bruised and diseased fruit, etc.)
   - Deterioration and shrinkage in storage
   - Poor handling or package failure
   - Transportation losses

A lack of knowledge and appropriate post-harvest technologies have led to improper selection of produce for harvest and post-harvest handling, packaging and storage techniques, which in turn contribute to the substantial post-harvest losses in many CSIDS. Exports of smallholder-grown root crops, fruits and vegetables are constrained by the irregular supply and frequent failure to meet the quality and quarantine standards of importing countries.

Even a modest recovery of such wholesome foods could improve food security by supplementing food-assistance efforts and lessen the environmental impact of waste disposal. In order to drive this reduction in food wastage, there must be an understanding of where and how much food is lost along the food value chain. This information should be used to plan and increase the efficiency of food recovery efforts and measures adapted to reduce the amount of food lost in different areas of the food value chain.

7 Market Constraints

Agriculture has traditionally benefited from special arrangements which sheltered it from the full impact of General Agreement on Tariffs and Trade (GATT)[4] disciplines. The World Trade Organization (WTO) framework is a legal term indicating the government-imposed conditions under which a product may enter a country and be released for free circulation within that country under normal conditions. WTO Members agreed upon a set of principles and disciplines that were designed to help liberalize international trade in agricultural products.

The Agreement on Agriculture (AoA) seeks to reduce restrictions on trade in agricultural products by introducing disciplines to:

- increase market access;
- reduce domestic support measures;
- reduce subsidized exports.

The domestic marketing system in CSIDS is usually based on non-institutional channels formed by private intermediaries and is generally poorly developed. Establishing an efficient marketing system is difficult due to several reasons: farms are remote, infrastructure is poor which can lead to high transport costs, a lack of investment in market research and rudimentary market information systems. Small-volume producers and exporters have no control over production costs or supply quantities since these farms enter and exit the industry regularly depending on available prices. Thus marketing economies of scale, or the bargaining power possessed by larger entrepreneurs, are not available to smaller farmers. There are provisions for CSIDS domestic support of their agriculture products however not all the Governments of CSIDS have the fiscal resources to provide such support to the extent that it will result in lower prices of key foods and contribute to overall food security.

The situation for export subsidies is similar to that for domestic support. CSIDS are net food importing countries and thus benefit from cheap, subsidized exports from developed countries, except, where such products are in competition with domestic products. The modalities of export competition are not likely to have any significant impact on any of the targeted commodities in CSIDS.

The critical areas of market access for the CSIDS, in terms of the AoA and the on-going negotiations, are in the areas of tariff binding reductions, tariff peaks and escalation, tariff rate quotas (TRQs), special agricultural safeguards (SAGs) and non-trade concerns. The latter issues have implications for key exports that traditionally benefited from preferences, as well as products targeted for food security and rural development.

---

[4] GATT (1994); the Agreement on Safeguards or the Safeguards Agreement; the Agreement on Import Licensing Procedures or the Import Licensing Agreement; the Agreement on the Application of Sanitary and Phytosanitary Measures or the SPS Agreement; the Agreement on Technical Barriers to Trade or the TBT Agreement and, the Agreement on Trade Related Aspects of Intellectual Property Rights or the TRIPs Agreement.

A tariff is a trade barrier that takes the form of a government tax imposed on goods (usually imports and occasionally on exports) when they cross borders. Like internal taxes, a tariff generates revenue for the government of the importing country.
A tariff quota is a quantity of imports or exports within which a lower tariff applies. A higher tariff applies within the border (UN 2003)

CSIDS that lack the fiscal resources to provide most types of domestic support, have used common external tariffs (CETs) as a measure to safeguard domestic producers from declines in world prices, protect them from cheap imports and stimulate domestic production. Tariffs are the only mechanism allowable under AoA that can be placed to safeguard domestic production. Without flexibility in the use of tariffs as a safeguard, Caribbean producers of vegetables, eggs, poultry, pigs and vegetable oils, in particular, will be at a significant disadvantage unlike their competitors in the developed world. These commodities are key elements of food security strategies in the CSIDS. The reverse is true for bananas, where erosion of tariff preferences places banana producers at a significant disadvantage.

Trade liberalization has disadvantaged CSIDS from earning foreign exchange, employment and economic growth from exports, especially banana export. The inherent characteristics of CSIDS have placed them at a disadvantage in banana exports and will plague them in other economic endeavours as well. Thus tariff preferences must be tied to assistance to improve competitiveness as well as market-based options such as branding, organic production and fair trade.

Given their high dependence on banana exports, the Windward CSIDS would benefit if historical allocations were maintained in TRQs. The AoA can explicitly recognize the need for full compensation (by developed countries) for the loss of preferences as a condition for developing countries agreeing to give up country-specific quotas, from which they have historically benefited. Such compensation would reflect the losses incurred by the country's economy in terms of foreign exchange, employment and linkages to consumption and investment. For the estimated 4000 farmers that have left the industry, the Government of St. Lucia, with assistance from the EU, has initiated a US$ 44 million programme of diversification and social revival (CARDI 2011).

8 Institutional Constraints

Currently, in most of the Caribbean, resource allocation to Research and Development (R&D) is much less than the benchmark figure of 1 % of the GDP. In fact, between 2000 and 2005, Caribbean countries such as Trinidad and Tobago and Jamaica spent approximately 0.1 % of its GDP on R&D (UNDP 2008). Most of the knowledge creation in this region is done in Cuba, Puerto Rico, Jamaica, Barbados and Trinidad & Tobago universities, where research capacity is related to academic knowledge creation.

Another institution that faces constraints is farmers organizations that generally suffer from weak management, low level and quality of services, low participation of members and inadequate financial resources. However, there are some farmers organizations where private exporters are the backbone of the export trade in non-traditional crops. These small and medium size operators tend to avoid formal credit institutions because of fear of debt and the associated collateral requirements.

9 Agricultural Policy Constraints

The policy goals for agricultural development in CSIDS have been pursued through a mix of national, regional and international policies and have focused on key areas of export promotion, food security, agricultural diversification, biodiversity and environmental concerns.

The main trading partners for food are the United States (US), the United Kingdom (UK) and CARICOM, especially Trinidad and Tobago and Barbados. With the exception of phytosanitary considerations, OECS countries can trade freely with each other in agricultural goods. At the regional level, CARICOM countries have agreed a common external tariff (CET), and goods from other CARICOM States are allowed tariff-free access to the Caribbean Windward Islands.

Some CSIDS have import-substitution policies that use quantitative restrictions as a means of protection of local agricultural production. Pigs, poultry, vegetables and food crops are the major beneficiaries of such restrictions. These are key commodities for domestic food security; as limited land space does not allow any significant ruminant, pig and poultry production. These commodities face significant competition from US production, especially chicken, as domestic support policies in that country result in low prices, which allows chicken and its subsidized cuts to be readily imported into the Caribbean. Import substitution for some commodities such as egg, fish, vegetables and pig production have increased in some CSIDS however chicken and tilapia production continue to experience difficulties due to lack of scale in production and processing and makes it difficult for local producers to compete with giants such as US and China. In other sectors such as fruit production, endeavours at import substitution and diversification achieved less-than-anticipated success for several reasons. These include the lack of a holistic approach, an inadequate focus on shipping and transport, the low involvement of industry partners and insufficient emphasis on commercial production and marketing.

Food security and agriculture are tightly linked to overall rural development in CSIDS. The potential of many farmers and farm workers to participate in areas such as financial services and information technology are limited by low levels of education. In the foreseeable future, agriculture is the sector with the largest employment possibilities and multiplier potential for rural communities. In Dominica, it is estimated that there are three persons dependent on each person employed in bananas. Employment statistics for St. Lucia show that a number of rural communities experienced unemployment that was significantly higher than the national average of 17.5 %, and increases in rural unemployment and poverty are linked to the decline in banana exports (CDB 1999). This statistic has since increased to 20.6% unemployment in 2010 (CARICOM 2010).

The effective design and implementation of diversification initiatives has many challenges including the difficulties that techniciáns and administrators will encounter in sorting out the complex rules of any new trade environment and developing policies that can realistically achieve their objectives.

The main environmental issues with respect to agriculture in the CSIDS include:

- Deforestation,
- Solid and liquid waste management,
- Unplanned development,
- Natural disasters, and
- Squatting.

Until the removal of quantitative restrictions on key commodities, including fruits, vegetables and meat products, the impact of liberalization on domestic agriculture is difficult to judge. The only real changes in trade measures in agriculture since the inception of the WTO, has been the implementation of the common external tariff (WTO 2001).

10 Strategies for Addressing Issues and Constraints

The reality is that there remain significant impediments to agricultural sustainability and food security in the region, and there are many pre-requisites that need to be met in order to realise this objective. Figure 3 shows the key elements that should be considered when governments of CSIDS are implementing or adapting strategies to encourage sustainable agriculture. There must be political, entrepreneurial, and popular will to achieve sustainable agriculture and food security in the Caribbean region. A key enabler must be a binding political undertaking by decisions makers at the highest level, both regionally and within individual national jurisdictions, that regional agricultural sustainability is an objective to be pursued with a unified approach. The region's private sector must also be recruited to the view that food security is good for regional business and for the other indirect economic advantages that it will bring to the regional economy. The declining role of agriculture in the region, the continuing loss of preferential markets for the region's traditional products and the rapidly increasing extra-regional food import bill are among the serious and challenging issues highlighted in the Jagdeo Initiative (Private Sector Commission 2007). Efforts should be made to:-

- Ensure all appropriate steps taken to establish the physical infrastructure conducive to private sector investment in agricultural and food production in the region. This includes investment of public resources in food processing infrastructure areas including in farm to market roads, drainage, irrigation, in affordable energy including renewable sources for food processing and packaging and downstream production.
- Large scale private investment in agricultural enterprise must be promoted and facilitated. Also small scale farmers must be supported with extension services,

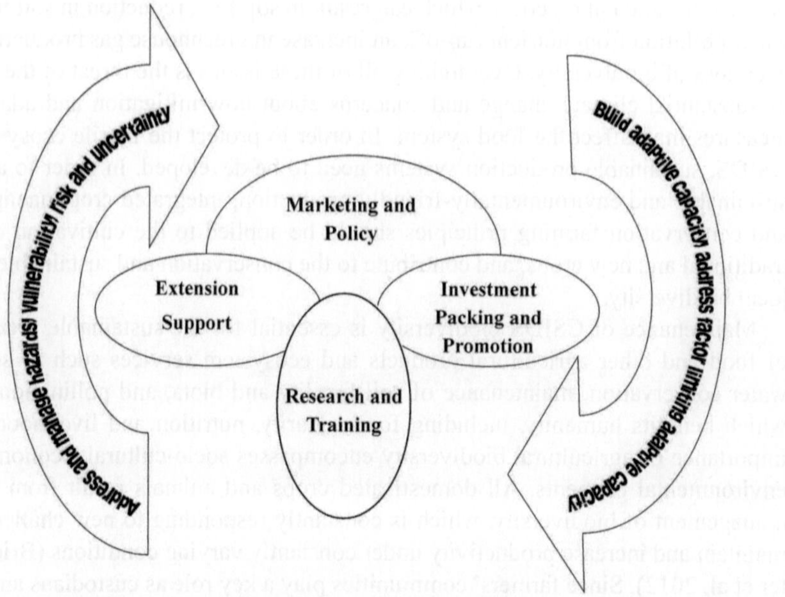

Fig. 3 Elements that should be considered to achieve sustainable agriculture

supportive services to access markets, technology and technological advice including disease resistant crops and improving productivity.

- Affordable and efficient intra-regional transport so that if we produce the food for example in the southern Caribbean for example Guyana, transportation costs should make it affordable for the purchaser in the northern Caribbean to source their food products from within the Caribbean, thereby promoting regional food security.
- Establishing a unified system of sanitary and phytosanitary standards to facilitating trade in agri-food products which facilitate intra-regional trade and not perceived to be used to protect inefficient domestic producers against more competitive extra-national regional producers.
- A competitive agriculture sector in the region undoubtedly requires scientific research and development, in such areas as crop development and husbandry, soil management and enhancement, disease diagnosis and response, plant and animal productivity, to name just a few.

10.1 Sustainable Farming Systems

Domestic market needs will increase with an increasing population and this coupled with a growing competition for land use, water and energy normally results in intensification of agriculture production in CSIDS. However intensification can lead to poor environmental practices which can result in soil loss, reduction in soil fertility, water pollution from nutrient run-off, an increase in greenhouse gas production and even loss of biodiversity. Overarching all of these issues is the threat of the effects of substantial climate change and concerns about how mitigation and adaptation measures may affect the food system. In order to protect the fragile ecosystem of CSIDS, sustainable production systems need to be developed. In order to achieve sustainable and environmentally-friendly production, integrated crop management and conservation farming principles should be applied to the cultivation of both traditional and new crops, and contribute to the conservation and sustainable use of local biodiversity.

Maintenance of CSIDS biodiversity is essential for the sustainable production of food and other agricultural products and ecosystem services such as soil and water conservation, maintenance of soil fertility and biota, and pollination, all of which benefits humanity, including food security, nutrition and livelihoods. The importance of agricultural biodiversity encompasses socio-cultural, economic and environmental elements. All domesticated crops and animals result from human management of biodiversity, which is constantly responding to new challenges to maintain and increase productivity under constantly varying conditions (Bridgewater et al. 2012). Since farmers' communities play a key role as custodians and managers of agricultural biodiversity local and traditional knowledge and culture are integral parts of agricultural biodiversity management. Much agricultural biodiversity is now conserved ex situ in gene banks or breeders' materials. The interaction between the environment, genetic resources and management practices that occurs in situ within agro-ecosystems often contributes to maintaining a dynamic portfolio of agricultural biodiversity.

Conservation agriculture (CA) is recognized as essential for sustainable agriculture and covers three core principles: minimal soil disturbance, retention of soil cover and a rational use of crop rotations. Land and water use in the Caribbean region are not only being threatened by the impacts of climate change but by improper human practices. Contamination of water courses by agro-chemicals and improper disposal of household waste as well as the high sedimentation rates as a result of soil erosion during the rainy season threaten aquatic ecosystems and food security due to intense flooding of rural homes and agricultural lands. These challenges require changes in the agricultural production system that currently exists in CSIDS.

The following actions merit consideration in trying to achieve sustainable agriculture by using technology and sustainable practices while reducing negative externalities:

- Develop sustainable *agro-forestry systems* to raise and diversify production, improve soil fertility, prevent soil loss and environmental degradation, and reduce dependence on external inputs. Leveraging endemic species (flora and fauna) should be an important feature of our development strategy for sustainable agriculture development. The Caribbean is a unique place in that many plant species including those for herbal and medicinal purposes can only be found in this part of the world.

- Applying *precision agriculture* by implementing technologies that allow the application of water, nutrients, and pesticides only to the places and at the times they are required, thereby optimizing the use of inputs. However it should be noted that maintaining and increasing, productivity depends on continued innovation to control weeds, diseases, insects, and other pests as they evolve resistance to different control measures, or as new species emerge or are dispersed to new regions.
- Introduce short-duration cover crops and legumes to improve soil fertility and structure, conserve moisture within *intensive high-input agricultural systems on lowlands*, to reduce build-up of weeds and pests, reduce reliance on imported chemicals and fertiliser, minimise environmental degradation, and increase green fodder availability.
- Using *farming systems research* techniques to appraise socio-economic issues and feed this information into cropping trials and extend technology to the farming community using a farmer-to-farmer approach (e.g. Farmer Field Schools).
- *Focus research* on tree crops, introduction of new crops, mixed perennial cropping systems, the multi-crop smallholders that utilise agroforestry systems, and livestock and fisheries rearing to optimise production.

10.2 Plant Genetic Resources (PGR)

Modern genetic techniques and a better understanding of crop physiology allow for a more directed approach to selection across multiple traits (Charles et al. 2010). Currently, the major commercialized genetically modified (GM) crops involve relatively simple manipulations, such as the insertion of a gene for herbicide resistance or another for a pest-insect toxin. Future advances in this technology can eventually produce crops with a combination of desirable traits and the introduction of new traits such as drought tolerance. Biotechnology could also produce plants for animal feed with modified composition that increase the efficiency of meat production and lower methane emissions rather than direct plant food resources that can be used for human consumption to feed animals. CSIDS will need the expert assistance from the developed world and their research into unexploited genetic material from land races, rare breeds, and wild relatives and the adaptation of this locally so that breeders can respond to new challenges.

Intensified and diversified crop production requires a strong seed programme, either at regional or sub-regional level. Elements of such a strategy include: (i) the development of regional technical capability for seed supply; (ii) the development of appropriate seed policies to enhance national and regional efforts in PGR utilisation and seed supply; and (iii) development of a germplasm information network to link the separate islands and to bridge information gaps in the region. However it is important that countries who have access to seeds from international gene banks ensure that locally adapted crop and livestock germplasm is not lost in the process of their displacement by modern, improved varieties and breeds.

10.3 Plant Protection

There is a need to review the current pest and disease control strategies in CSIDS as these are based mainly on the use of pesticides. Integrated Pest Management (IPM) should be considered as an alternative strategy for effective, efficient, balanced and environmentally safe pest control. Due to the small size of CSIDS and the fact that pests cannot easily spread to the neighbouring territories that are separated by oceans biological control can be effective and relatively easy to implement in CSIDS. Production could be undertaken in official Pest Free Areas (PFA) so as to meet the special needs of niche export markets.

At the special Ministerial Conference on Agriculture in Small Island Developing States held in Rome on March 1999, the issue of Plant Protection by Integrated Pest Management (IPM) as an alternative strategy for effective, efficient, balanced and environmentally safe pest control was discussed as follows (Sustainable production 1999).

> Pest eradication programmes should be seriously considered when dealing with important crops and a small number of pests with a simple life cycle and infection/infestation process. Other areas of intervention are: (i) establishment of harmonised pesticide legislation and registration requirements to establish a common legal framework for the importation and distribution of pesticides; (ii) monitoring of pesticide residues in the environment (for e.g. drinking water) and in agricultural produce; (iii) active participation and joint decision making within the PIC procedure; (iv) verification of the existence of obsolete pesticides and their improved storage.

10.4 New Crops

CSIDS in general have a strong comparative advantage in the production of certain tropical crops (e.g. papaya, banana, plantain, mango, pineapple, watermelon, hot peppers etc.), tuber/root crops (e.g. dasheen, tania, taro, yams, sweet potato, cassava), nuts and spices (e.g. coconut, nutmeg) vegetables and cut flowers. Production of these indigenous crops for domestic consumption and export can be increased as these would be consumed in either the domestic market (e.g. local tourist market) or as substitute products for the export markets. In the latter case, the potential of these crops—possibly under organic farming conditions—for niche markets should be explored with the assistance of the global experts in this area.

Developing sector niches that offer opportunities to create value and develop businesses that have strong long term growth prospects can increase export earnings. Cocoa, hot peppers and herbs are potential crops that could be cultivated for niche markets. It must be noted that fresh produce exports must comply with sanitary and phytosanitary regulations while processed foods must comply with the regulations of the importing countries. Shipping foods to the USA requires an understanding of the Food Safety and Modernisation Act (FSMA), and in Canada an understanding of the regulations set out by the Canadian Food Inspection Agency (CFIA). Products entering the EU markets are also subjected to European

Commission Marketing Standards. The EU also monitors pesticide levels in produce sold in their markets. Capacity building in terms of research and development and quality infrastructure which encompasses accredited laboratories, regulatory and institutional frameworks must be enhanced in Caribbean region if we are serious about having a competitive advantage from niche products.

10.5 Irrigation

Poor management of watershed catchment areas, over-exploitation of aquifer resource and of water flowing through agro-ecosystems, particularly irrigation supplies, can lead to water scarcity, pollution of downstream supplies, salt water intrusion in lands near the shore and salinisation of groundwater. All of these translate into less water being available for agricultural purposes. Irrigation is an opportunity to reduce the unit cost of growing a crop by increasing yields and improving quality, and diversifying into other crops. However, technologies for new irrigation schemes should be thoroughly evaluated not only from an efficiency and ecomonic point of view but also with respect to their potential conflicts with other land uses. In islands where water is scarce, localised irrigation systems, such as high tech commercial drip systems and micro-aspersion, or low-tech simple drip irrigation systems appropriate for gardening, can be introduced for high-value crops. Drip irrigation when combined with training has been proven to increase productivity, increase yield, positively impact individual farmers and address water challenges, arable land reduction and soil erosion in India and Kenya. These technologies and training should be easily adapted to CSIDS. In the highlands of semi-arid islands, various types of water harvesting and runoff capture techniques for drinking water, water for animals and gardening can be introduced. All new irrigation schemes should be accompanied by an environmental impact assessment, depending upon the scale of the scheme, in case significant negative environmental impacts are expected. Technologies for new irrigation schemes should be thoroughly evaluated, including their potential conflicts with other land uses. The active involvement of water-users groups is a supporting objective. It is not only important that all the different water user groups be consulted and actively involved when water from fresh ground water resources and water sheds is extracted. Policies should also be put in place that can control potential environmentally unfriendly facilities from extracting and contaminating these water sources. Introducing irrigation technology and training farmers on these will require a significant investment. For this reason it is important to do studies on how much additional water could be saved through water use efficiency improvements so that the best technology can be applied to each individual farm that may have unique constraints. For example, it may be best to improve irrigation scheduling using local climate and soil information to more precisely meet crop water needs during the "rainy season" or use a regulated deficit irrigation system where crops are watered less during any drought-tolerant growth stage in their life cycle which can lead to an improved crop quality or yield.

10.6 Agro-processing

The government of CSIDS should mount programmes to introduce or extend appropriate technology and to develop expertise in agro-processing. Farms in CSIDS should be encouraged to develop small-scale, low-cost agro-processing ventures to increase the value and market potential of products. The market for chilled fresh-cut produce has witnessed dramatic growth in recent years, stimulated largely by consumer demand for fresh, healthy, convenient and additive-free foods which are safe and nutritious. Apart from presenting the consumer with a range of options in a single package, fresh cut produce reduce wastage at the household level, in that they allow the consumer to procure only the quantities of fresh produce required, while allowing the opportunity to readily assess the quality of the produce being purchased. In the US and Europe, shifting population demographics have also indicated a growth trend in the consumption of ethnic foods. Growing consumer interest in international markets in new or exotic tastes has promulgated growth in the international trade of fresh-cut products, exotic spices, flavours, oils and food colourings. CSIDS need to respond to this growing demand by producing fruit and vegetable fresh-cuts, spices and other "exotic" or "ethnic" food products for export. Rather than just look at exporting new crops, CSIDS farms should be encouraged and be given assistance with making the investment in technology, equipment and training on food safety principles and practices so that they can transform their crops, into pre-packaged ready to use goods.

10.7 Animal and Fisheries Production and Health

Small scale poultry and pig farms should consider using locally available feeds from high energy crops—coconuts, yams and cassava, to reduce their costs. The CSIDS that have a fishing industry have an added advantage where the fish waste from processing can be used to make silage that can provide a high quality protein supplement which can complement the "energy" feed available locally. The development of feed mills that use these local ingredients could encourage semi-intensive indigenous livestock industries that can meet the demands of urban markets. Farms that have integrated livestock into their mixed farming systems should be educated and trained to practice recycling and reusing wastes. The wastes from the livestock pens can be diverted to biogas digesters which can produce nutrient rich irrigation water, fertilizers and methane gas for any agro-processing or energy consumption activities. Farmers can also optimise use of crop residues and vegetation on uncultivated land, as well as to assist the recycling of soil fertility. Farm animals such as goats and sheep can be used as an environmentally friendly and cost effective method of clearing land in preparation for planting rather than use fire or machinery. With the incorporation of livestock into mixed farming or increasing the intensity of livestock farming management skills should be upgraded and feeds and pastures improved in order to realise the full potential of the selected breeds. Rather than just having fresh milk or meat as products from livestock, processing should be done to supply more versatile products where niche markets for yoghurt, cheese, ready to eat or pre-seasoned vacuum packed steaks and burgers could be developed.

Governments of CSIDS should promote aquaculture development and intensify and diversify existing production systems. Cultivation of tilapias should be encouraged as they are suited to low technology farming systems and can be cultivated in ponds, concrete tanks intensively at both the large and small scale level. Technology for combating problems of overcrowding and market demand for large and uniform tilapia sizes is well established and tested in some of the Caribbean countries as well as Nigeria, Thailand and Japan. Strategic partnerships with regional global leaders in aquaculture should be fostered so that aquaculture training at the technological and skill level can be increased and proven technologies and policies can be adapted. CSIDS should become members of Council overseeing the Caribbean Regional Fisheries Mechanism (CRFM) where they can benefit from partnerships and strategies to increase the supply of fish, adopt an ecosystem approach to aquaculture and guide on best management practices and standards.

Farm workers should be trained on animal health and food safety, farms should readily have access to animal health services and specific disease eradication programmes should also be promoted (e.g. Tropical Bont Tick in the Caribbean).

10.8 Reducing Climate Change Shocks

There exist a multitude of proposals for measures that effectively mitigate climate change, including from the Committee for Development Policy (CDP). Many SIDS have called for more ambitious international climate change negotiations however to this day attempts to find effective international solutions have been elusive. In a 2013 background report submitted by the CDP secretariat, Bruckner (2013) highlighted that, "Climate change mitigation requires massive investments in technology and infrastructure, in particular in the generation of low-carbon emission energy but also in transport and buildings. Such investments simultaneously contribute to other sustainable development objectives such as expanding access to electricity or increased eco-efficiency in consumption patterns". However why should CSIDS have to make these heavy investments when they are not producers of very high levels of greenhouse gas emissions? The larger emitters of greenhouse gases in the developed world should bear the majority of responsibilities and costs of not only reducing their carbon emissions but in the mitigation of the effects faced by CSIDS with respect to reducing land degradation, biodiversity preservation and food production. Agreements on this should be done at the international level and expressed in either the Kyoto Protocol, the Durban Platform for Enhanced Action or in other forms.

Effective climate change mitigation should be integrated into the broader sustainable development agenda and not treated as a stand-alone environmental concern. Transformation of socio-economic paradigms should also be a focus rather than simple technological fixes. This should all be done at both an international and Caribbean Regional level since uncoordinated actions by individual or small groups of countries cannot be expected to lead to sufficient emission reductions.

10.9 Recovering Lost Food or Reducing Food Losses

Post-harvest food losses can occur on the farm, between the farm and retail levels or even when a commodity moves into the marketing system, to the point of consumption. This can be due to several reasons such as severe weather, including droughts and floods, pest and disease infestations, selective harvesting leaving small, misshapen, or otherwise blemished produce in the field, or discarding blemished products in the packing shed or processing plant. Farmers can mitigate post-harvest losses by using leftover crops for secondary processing or as fertilizer or animal feed.

During processing and marketing, food is subject to additional loss as it leaves the farm and enters the food marketing system. Some loss occurs in storage, due to insect or mold infestations, deterioration, or improper transportation and handling. The post-harvest losses experienced at this stage can be mitigated through improved farm management, marketing practices and through use of technologies. Food waste from agro-processing can be recovered by diverting it for use in animal feed or as ingredients in other food products. For example, cassava lose significant weight when they are processed into frozen wedges or fries. Although this appears to represent a "loss" of edible fresh cassava, most of the "loss" can be recovered and used by processors for other products, such as animal feeds.

10.10 Marketing

Food production in CSIDS can be severely affected by market interventions in the developed world, such as subsidies or price supports. These need to be carefully designed and implemented so that their effects on global commodity prices do not act as disincentives to production in other countries (Anderson 2009).

The marketing strategy for entry of agricultural exports into mainstream markets and building strategic alliances with market participants should be a joint venture between marketing services of the private sector and farmer organisations (FOs). Both entities should be knowledgeable concerning available markets, set standards, provide information and co-ordinate decisions on buying raw materials, transportation, packaging, training, quality, etc. in order to be more competitive. The agencies that are best placed to provide market intelligence and information; exporter facilitation, and develop FOs are the existing marketing boards, cooperatives and private exporters in CSIDS. A proactive, project oriented and multi-disciplinary approach to develop the production and marketing system of non-traditional agricultural crops is required for agricultural diversification.

10.11 Institutional Policies

Governments need to commit to funding research, creating government-industry-academia alliances and collaboration networks to enhance competitive capacity of the region's knowledge societies and develop innovative and sustainable responses to Caribbean issues.

FOs should have access to low interest credit, entrepreneurial development and marketing support to develop strategies that include; infrastructure development in the form of feeder roads and marketing facilities, input supply, setting of grades and standards, and agro-processing of targeted commodities. In order to further commercialise agricultural operations, there is a need to change negative attitude to credit and a need for the credit institutions to devise financing schemes for small to medium commercial operators.

10.12 Agricultural Policies

Generally, agricultural policies should incorporate the following elements:

1. Protection for local production as a means of import substitution, foreign exchange savings and food security;
2. Agricultural diversification to reduce overdependence on a few exports;
3. Removal of barriers to trade at both the sub-regional (OECS) and regional (CARICOM) levels to stimulate trade in the regional market; and finally
4. Tax concessions and subsidies to local farmers to stimulate increased local production and exports.

However successful implementation of revised and newly drafted national agricultural policies/strategies to address constraints that affect agricultural production and trade can only be achieved through programmes and projects in technology transfer (e.g. irrigation, germplasm and product quality), fiscal incentives, trade policy development, market development and entrepreneurial development.

Import-substitution policies are only successful when a non-governmental entity e.g. farmers' marketing cooperative carries out the processing and marketing functions and does checks on the evasive measures commonly employed by importers. Quantitative restrictions have been used by government as a means of protection of local agricultural production. These measures can create a balance between: (i) protection for food security and overall import substitution; and (ii) revenue generation and provision of food for the populace. Protection measures include non-automatic licensing, local-content requirements and import quotas. Governments can grant concessions to farmers, including waivers on income tax payments for incomes below a set level and duty free concessions for farm vehicles and certain agricultural inputs. Governments should be fully cognizant of the need to convert quantitative restrictions to tariffs, as part of their Uruguay Round commitments, and must commit to the process.

The Governments of CSIDS, in collaboration with the FAO and CARICOM, are developing a regional Food Security Programme targeting the development of production and marketing of a group of select commodities for national and regional food security. These commodities include, hot peppers, small ruminants, poultry, sweet potatoes and herbs and spices.

Diversification policies can be revamped to include low interest credit, entrepreneurial development and marketing support, infrastructure development in the form of feeder roads and marketing facilities, input supply, setting of grades and standards, and agro-processing of targeted commodities. A major aspect of these policies should be the redirecting of marketing boards from a buying and selling operation to that of providing market intelligence and other facilitation support.

10.13 Human Resource Development

Education at all levels and research institutions should begin the process of sensitization to vulnerabilities among children, as well as to educate and train technical persons especially at the tertiary level. The acquisition of appropriate skills and technologies will constitute an essential part of the process of building the resilience needed to reduce vulnerability to global threats, whether natural, economic, social or political.

Caribbean agriculture needs a continuous flow of competent professionals trained in agriculture and the conduct of relevant research to support its transformation from an agriculture based on protected markets to an agriculture that is technology-driven, competitive and market oriented. Continued modernization of the sector requires agricultural workers with far greater levels of skills and competencies than currently obtains. Agricultural research is mainly government-funded and in CSIDS is constrained by limited human and financial resources. Governments also provide technical assistance in the form of extension research, testing and other services but the links from extension services to the farming community, and to other key sectors such as agro-processing and agri-tourism are weak. Few CSIDS have clearly identified research priorities, which are relevant to the smallholder sector. There is a considerable need to train local manpower in research and technology development at the post-graduate level.

At the national and regional level, both Governments and donors need to significantly increase the resources allocated for human resource development consistent with the goal of the Social Summit and the in the spirit of United Nations Agenda 21. Incorporation of ICT should be considered in the formulation of strategies in this area. In the Caribbean there is concern over the rural-urban drift accelerated by income disparity and producing a fall in employment, wages and productivity in rural areas as skilled workers move to higher-earning, urban opportunities. Youth involvement in agriculture has been a major target in the revitalization thrust of agriculture for many Governments in the Caribbean.

At the global level the international community, should play a role in assisting governments and private sector actors in CSIDS to better understand their opportunities and obligations within the international trading system. The skills of farmers, exporters, administrators, extension personnel and other players need to be upgraded through training. In particular, extension staff will require a reorientation to market-oriented farm management, with attention paid to both increased productivity and better business.

10.14 Integrated Land Resource Planning and Management

Land availability for large scale farming may bring major benefits to CSIDS, especially where investors bring considerable improvements to crop production and processing, but only if the rights and welfare of the tenants and existing resource users are properly addressed (Cotula et al. 2009).

In the past government's solution to producing more food involved bringing more land into agriculture. The competition for land from other human activities makes this an increasingly unlikely and costly solution, particularly if protecting biodiversity and the public goods provided by natural ecosystems (e.g., carbon storage in rainforest) are given higher priority (Balmford et al. 2005). In recent decades, agricultural land that was formerly productive has been lost to urbanization and other human uses, or are no longer arable as a consequence of unsustainable land management.

The lack of secure land rights can be problematic for small holder farms in many poor communities, and can be a disincentive for small farm holders to invest in managing the land productively. This creates an additional challenge by making it difficult to raise investment capital. Modern information and communication technologies (ICTs) need to be applied to the relevant government bodies to assist small holder farms with title definition, land protection and address the rights and welfare of agriculture tenants on leased government land.

CSIDS that do not have this in place can adapt the Antigua and Barbuda Agricultural Land Use Policy. This policy can foster and facilitate rapid development of the island's agricultural land resource, while ensuring the continued productive capacity of its agricultural land resource and guarantee economic profitability of all producers. Any agricultural land use policy that is developed/adapted and implemented should consider the location, size, function and growth of existing and new settlements and their spatial and functional relationships as well as provide a framework for the provision of physical and social infrastructure and opportunities for economic activity, in accordance with a comprehensive settlement strategy. The policy must also articulate a sustainable development strategy that allocates the most appropriate land for various activities and in so doing considers the capacity of such land to sustain development in the long term without degrading or damaging the scarce and fragile land resource base of CSIDS. Improved land management is a key strategy to achieve food security. The use of protected agriculture systems, soilless culture, multi-layered cropping and fully integrated aquaponics all offer adjunct solutions to reduced land availability (Maximay 2013).

An agroforestry strategy that integrates agriculture and forestry including flora and fauna as mutually compatible and complementary can provide a scope for joint development that can bring about mutual benefits. This approach will allow a wider range of agroforestry enterprise mix, optimize land resource utilization, and enhance the income generating potential of agroforestry investments. These approaches can sustain and enhance the growth of the agricultural sector and allow it to become more globally competitive.

Conclusion

This chapter highlights the major agricultural issues and constraints faced by CSIDS and suggest strategies to address them. There is no simple solution to sustainable agriculture in CSIDS, a broad range of options, including those that are highlighted in this chapter, should be pursued simultaneously. A lot of hope is placed on scientific and technological innovation. Any optimism must be tempered by the enormous challenges of making food production sustainable while addressing the wide-ranging effects of climate change, including managing our biodiversity, conserving dwindling water supplies, as well as meeting the Millennium Development Goal of ending hunger. The temptation to further sacrifice CSIDS biodiversity for short term gains in food production must be avoided, not only because biodiversity provides many of the public goods on which mankind relies but also because we do not have the right to deprive future generations of its economic and cultural benefits. Together, these challenges amount to a perfect storm. Reducing the effects of this storm will require further research in the social and natural sciences and partnerships between CSIDS and the developed world should be encouraged. In CSIDS, a multi-sectorial policy is needed to deal with the issues of climate change and sustainable agriculture which can be affected through regional and local strategic development plans. However in order to achieve this, strong decision making in terms of direction and sustained action is required from our leaders in the Caribbean. Institutional mechanisms at the national and regional levels must be put in place, to undertake the regular dissemination of user-friendly information on such technologies as well as to assist with the training of nationals in use of such technology, and introduction of incentives to encourage the use of appropriate technology. The goal of CSIDS in achieving sustainable agriculture should not be to just maximize productivity, but to optimize across a far more complex landscape of production, environmental, and social justice outcomes. Developing a sustainable agricultural sector for the Caribbean region requires collaboration among all stakeholders—economists, agrologists, teachers, researchers, farmers, extension officers, consumers and public and private sector representatives.

References

Anderson K (ed) (2009) Distortions to agricultural incentives, a global perspective 1955–2007. Palgrave Macmillan, London

Balmford A, Green RE, Scharlemann JPW (2005) Global change. Biology 11:1594–1605

Biemans H, Kabat P, Ludwig F (2012) Water constraints to future food supply and an assessment of potential adaptation measures. Stockholm International Water Institute, Stockholm

Bridgewater P, Régnier M, Zhen W (2012) Healthy planet, healthy people—a guide to human health and biodiversity. Secretariat of the Convention on Biological Diversity, Montreal

Briguglio L (2004) Economic vulnerability and resilience: concepts and measurements. Paper presented at vulnerability and resilience of small states, University Gozo Centre, Malta

Bruckner M (2013) Effectively addressing the vulnerabilities and development needs of small island developing States. CDP Background Paper No. 17

Brundtland Report (1987) Our Common Future, Chapter 2: Towards Sustainable Development

CARDI (2011) St Lucia Country Highlights 2010 CARDI's contribution to agricultural research and development, food production and the reduction of poverty and hunger PSC# SL/003/10

CARICOM (2010) Highlights to CARICOM'S selected economic indicators: 2002, 2006–2010. Statisitics programme CARICOM Secretariat. http://www.caricomstats.org/Files/Publications/Economic%20Handbook/ExecutiveSummary.pdf. Accessed 10 Aug 2014

CARICOM (2012) CARICOM's Selected Economic Indicators: 1997, 2002–2006 and 2001, 2004–2008. http://www.caricomstats.org/Files/Publications/Economic%20Handbook/Draft%20Selected%20Economic%20Indicators%202001%202004%20-%202008.pdf>. Accessed 24 June 2013

CDB (1999) Caribbean development bank poverty assessment reports

Charles H, Godfray J, Beddington JR, Crute IR, Haddad L, Lawrence D, Muir JF, Pretty J, Robinson S, Thomas SM, Toulmin C (2010) Food security: the challenge of feeding 9 billion people, review. Sciencexpress 327:812, Washington, DC

Cotula L, Vermeulen S, Leonard L, Keeley J (2009) Land grab or development opportunity? Agricultural investment and land deals in Africa. International Institute for Environment and Development, London

(CPEC) (2005) Agriculture and Agro-Industry. http://www.cpechrd.org/regagri.htm. Accessed 24 June 2013

ECLAC (2000) The vulnerability of small island developing states of the Caribbean. http://www.eclac.cl/publicaciones/xml/8/8118/G0588.html. Accessed 24 June 2013

Ericksen P, Thornton P, Notenbaert A, Cramer L, Jones P, Herrero M (2011) Mapping hotspots of climate change and food insecurity in the global tropics. CCAFS Report no. 5. CGIAR Research Program on Climate Change, Agriculture and Food Security (CCAFS). Copenhagen, Denmark. http://www.ccafs.cgiar.org. Accessed 24 June 2013

FAO (1993) Status report on Caribbean Aquaculture. GCP/RLA/102/ITA

FAO (2011) Global food losses and waste—extent, causes and prevention, by J. Gustavsson, C. Cederberg, U. Sonesson, R. van Otterdijk & A. Meybeck. Rome

FAO (2013) CARICOM food import bill, food security and nutrition. http://www.fao.org/fsnforum/caribbean/sites/caribbean/files/files/Briefs/Food%20Import%20brief%20.pdf. Accessed 10 Aug 2014

Food and Agriculture Organization of the United Nations (FAO). (2001). What is the code of conduct for responsible fisheries? FAO, Rome, Italy Available: http://www.fao.org/docrep/003/x9066e/x9066e00.html. Accessed 16 Dec 2013

GATT (1994) Article XIX Agreement on Safeguards

Government of Guyana (2002) First national communication to the conference of the parties of the United Nations framework convention on climate change, July 2002

Government of Jamaica (2000) Initial National Communication of Jamaica. November 2000. http://unfccc.int/resource/docs/natc/jamnc1.pdf. Accessed 24 June 2013

Guillaumont P (2010) Assessing the economic vulnerability of small island developing states and the least developed countries. J Dev Stud 46:828–854

IICA, ECLAC, FAO (2011) The outlook for agriculture and rural development in the Americas: a perspective on Latin America and the Caribbean

International Food Policy Research Institute (2009) Climate change: Impact on agriculture and costs of adaptation

IPCC (2007a) Climate change 2007: synthesis report. IPCC AR4 SYR 2007, p 37. http://www.ipcc.ch/pdf/assessment-report/ar4/syr/ar4_syr.pdf. Accessed 24 June 2013

IPCC (2007b) Summary for policymakers. In: Parry ML, Canziani OF, Palutikof JP, Van der Linden PJ, Hanson CE (eds) Climate change 2007: impacts, adaptation and vulnerability. Contribution of Working Group II to the Fourth Assessment Report of the Intergovernmental Panel on Climate Change. Cambridge University, Cambridge, pp 7–22

Maximay S (2013) Land for farming. Contact (Trinidad and Tobago Chamber of Industry and Commerce) 13(1):6–8

Nellemann C, MacDevette M, Manders T, Eickhout B, Svihus B, Prins AG, Kaltenborn BP (eds) (2009) The environmental food crisis-the environment's role in averting future food crises. A UNEP rapid response assessment. United Nations Environment Programme, GRID-Arendal. http://www.grida.no. Accessed 16 Dec 2013

Pellin M, Uitto JI (2001) Small island developing states: natural disaster vulnerability and global change. Global Environ Change B: Environ Hazards 3(2):49–62

Pemberton C (2006) Agricultural development and employment in the Caribbean: challenges and future prospects. International Labour Office, Geneva

Private Sector Commission (2007) The Jagdeo Initiative. Technical Information Bulletin No. 8; The Private Sector Commission of Guyana Ltd, Georgetown

Sustainable production, intensification and diversification of agriculture, forestry and fisheries in small island developing states (1999) Special Ministerial Conference on Agriculture in small island developing states, Rome

UN (2003) Conference on trade and development. Dispute settlement module 3.15. New York and Geneva

UN (2007) The Millennium Development Goals Report, United Nations. New York. http://www.un.org/millenniumgoals/pdf/mdg2007.pdf. Accessed 24 June 2013

UNDP (2008) Human Development Report 2007/2008—fighting climate change: human solidarity in a divided world. http://hdr.undp.org/en/media/HDR_20072008_EN_Complete.pdf. Accessed 24 June 2013

UNEP (2005) Caribbean Environment Outlook. United Nations Environment Programme

UNEP (2008) Climate change in the Caribbean and the challenge of adaptation. Panama City, Panama

UNFCCC (2007a) Report on the expert meeting on adaptation for Small Island Developing States FCCC/SBI/2007/11. Nairobi, Kenya

UNFCCC (2007b) Vulnerability and adaptation to climate change in small island developing State. Background paper for the UNFCCC Expert Meeting on Adaptation for Small Island Developing States (SIDS) Part I—Caribbean and Atlantic Ocean SIDS, Jamaica, February 2007

World Food Summit Plan of Action (1996) Basis of action: commitment three, Paragraph 28. http://www.fao.org/docrep/003/w3613e/w3613e00.HTML. Accessed 24 June 2013

World Trade Organization (2001) Trade policy review of the OECS. WTO, Geneva

Part II
Section B: Biodiversity in Sustainable Horticultural Systems

Agroforestry in the Caribbean, Traditional Systems, both Sustainable and Biodiverse

Michael Morgan and Thomas W. Zimmerman

1 Introduction

This chapter will discuss various agroforestry practices as they are practiced on the islands of the Caribbean Sea. Agroforestry is the association of crops and trees. It has been practiced since the dawn of agriculture, some 11,000 years ago ever since crops and trees grew in proximity to each other. Since agroforestry is such an ancient practice, we would like to show the historical continuity of these practices in the Caribbean from the Pre -Columbian times to the present. Agroforestry is sustainable because it is low-input and biodiverse because several different plant species grow in association. That association leads to a diversity of animals and insects. The authors live and work on the island of St Croix, US Virgin Islands, so there is going to be a certain emphasis on agroforestry as it is practiced on St Croix. We discuss forest gardens, home gardens of trees, shade-grown crops, shifting agriculture and fallows, alley cropping and living hedges, and finally, silvopastoral systems.

2 Background

The islands of the Caribbean can be divided into three groups: the Greater Antilles, the Lesser Antilles, and the Bahamas. The Greater Antilles consist of the big islands of Cuba, Hispaniola, Jamaica, and Puerto Rico with some associated smaller islands such as the US and British Virgin Islands. The Lesser Antilles consist of the small islands that extend in an arc from the continent of South America, starting at Trinidad and ending at the island of Anguilla. The islands of the Bahamas, some-

M. Morgan (✉) · T. W. Zimmerman
Agricultural Experiment Station, University of the Virgin Islands,
Kingshill, VI 00851 USA
e-mail: mmorgan@uvi.edu

T. W. Zimmerman
e-mail: tzimmer@uvi.edu

D. Nandwani (ed.), *Sustainable Horticultural Systems,* Sustainable Development and Biodiversity 2, DOI 10.1007/978-3-319-06904-3_6

what to the north of the Antilles, are also considered part of the Caribbean Basin, but not covered in this paper.

People have been living on the various Caribbean Islands from roughly 4000 BC onwards. The first inhabitants were Amerindians. The earliest group or groups to arrive were the Guanajatabey and/or Ciboney, hunter-gathers who came out of the west from Central America, north from South America, or even south from Florida. About 500 BC, the Arawak speaking Taino and Igneri peoples island hopped north from the South American mainland. They were followed by the more warlike Caribs (Rouse 1992). In 1492 the first Europeans and Africans arrived to these already inhabited islands with Christopher Columbus, who claimed all the islands of the Caribbean and the rest of the New World for Spain. Within 100 years of discovery by Europeans, the Amerindians were almost exterminated by war, enslavement, and disease. Africans were imported as slaves to replace the now almost extinct Amerindians in the fields and mines. At the same time, once gold was discovered on the mainland, large numbers of Spaniards left the islands to conquer and settle Mexico in 1519, and the Incan empire, in 1532, once gold was discovered on the mainland. Other European powers, most noticeably England and France, started prying islands from Spain and setting up their own New World colonies. The current population of the Caribbean Islands is predominantly African with admixtures of European and Native American genes.

3 Forest Gardens and Home Gardens

Forest gardens are probably the oldest agroforestry system in the world. They are forests with useful plants growing in them. People favoured and protected the plants they found useful and eliminated the ones they found not useful. They also introduced to the forest, plants from other places that they found useful.

The islands of the Caribbean share many plant species from South America and Central America. Some species were brought by birds and bats but others were brought by people. For example, Columbus came across people cultivating pineapple (*Ananas comosus* (L.) Merr.) on the island of Guadeloupe in 1493, although it is believed that the species originated somewhere in southern Brazil and Paraguay. It is easy to imagine the Guanajatabey and the Taino who followed them, bringing seeds of fruit trees in their canoes to their new island homes.

Three speculative examples of human dispersed forest garden trees on the island of St Croix are: Mamey Apple (*Mammea americana* L), Stinking Toe (*Hymenaea courbaril* L.) and Genip (*Melicoccus bijugatus* Jacq.). But, first let us provide some background. St Croix is isolated from the other Caribbean Islands because of a deep marine trenches between it and Puerto Rico and another marine trench between it and the other islands that compose the US and British Virgin Islands. St. Thomas, USVI is 68 km away and Puerto Rico is 116 km away. Furthermore there are no large mammals native to St Croix. White-tailed deer (*Odocoileus virginianus* Zimmermann) were introduced in the 1790s and the mongoose (*Herpestus auropunctatus*) in 1884 (Weaver 2006). Weaver goes on to writes of three genera of forest rodent *Capromys, Isolobodon*, and *Dasyprocta* on the island, probably going extinct during the destruction of the islands original forests. Bones of agouti (*Dasyprocta* spp), which are large and edible forest rodents have also been found in excavations of Amerindian settlements on the island of St Croix. They did not swim here. Agouti can be somewhat domesticated, so it is easy to imagine them being brought to St Croix and released into the forest as a future food source. Now let us go back to discussing the trees.

Mamey apple (*Mammea americana*) has a very large fruit that ranges from 10 to 20 cm in diameter (Little and Wadsworth 1964) and one or two very large seeds inside. The average weight of 54 seeds collected on St Croix was 55 g (Morgan, unpublished data). It would have to be a very large bird or bat that could swallow, and then defecate a seed this big. Bats, birds, and supposedly on St Croix, mongoose, eat the fruits but not eat the seed, which is poisonous (Little and Wadsworth 1964).

Stinking Toe or West Indian Locust (*Hymenaea courbaril* L.) is another tree that is considered "native" to St Croix but probably has an Amerindian origin. It is a large tree that occurs in mature forests in the north-western part of the island. The fruit is a very hard seed pod 5–10 cm long and 4–5 cm wide. Inside are several seeds surrounded by a dry yellow pulp; powdery would be a better description of its texture because it is not sticky. The pulp is edible and has been somewhat in jest described as a primitive energy bar. However, its pod is so hard that it can only be opened by smashing it with a rock, or a hammer, or with chisel-like rodent teeth. How do the seeds get dispersed if there are no wild dispersers left other than a few people who like the taste? Birds, bats, nor wind brought the tree to St Croix. A hypothesis is that Amerindians brought the tree seeds to St Croix, and planted them. Later on, people and their, now, wild agouti proceeded to disperse the tree seeds.

Genip, *Meliococus bijugatus*, native to north western South America is the most common roadside tree on the island of St Croix (Geographic Consulting 2010). It is variously considered an invasive exotic and a tree introduced some 1,000 years ago to the Caribbean islands by Amerindians for food and spiritual reasons. Fruit-eating bats and people disperse these seeds (Francis 1992). The Genip trees' four-lobed leaf is said to look like a man. The Africans who replaced the Amerindians also considered it a jumbey tree, that is to say, a tree imbued with magical powers or home of a spirit (pers. com, Olasee Davis 2010).

After the demise or assimilation of the Taino, people of African descent, whether Maroons or not, continued to plant useful fruit trees in the forest for future use: "Along the stream beds, various species of fruit trees were planted historically, especially mangos trees. The trees held the banks of the streambeds together, improve underground water supply, and control soil erosion. Traditionally and culturally, fruits were gathered during season." (Olasee Davis, UVI Cooperative Extension Service Forester, St Croix 2014). Many of these trees were brought from the Old World, species such as mango (*Mangifera indica* L.), coconut (*Cocos nucifera* L.) and Sea or Indian almond (*Terminalia catappa* L.). It is also worth mentioning that on the islands of Dominica and St John, wild trees of Bayrum (*Pimenta racemosa* (Mill) J.W. Moore) had their leaves and berries regularly harvested to make perfume and cosmetics. Branches were pruned and allowed to grow back. This was a big industry on the island of St. John, USVI from the 1880 to 1950 (Weaver 2009).

Home gardens are a step removed from forest gardens. They are simply trees and bushes planted around houses with crops along. In the Caribbean, these gardens have an origin in slavery, where slaves were granted the use of a small plot of land to grow their own food and supplement rations received from the slave owner. Later, when slavery was abolished, excess crops were sold or traded in markets to supplement wages. Geilfus (1994) describes Grenadian home gardens with an average of 2000 m^2 as being composed of three, maybe four layers. The ground layer has root crops such as "sweet potatoes (*Ipomea batatas* (L). Lam.) and Cassava (*Manihot esuculenta* Crantz.), grains, beans, and other vegetables for consumption by the family or sale. The next layer consists of bananas and plantains (*Musa* spp). On Grenada, there is a commercial shrub layer of nutmeg, (*Myristica fragrans* Gronov). In other islands, the shrub layer is provided by cacao (*Theobroma cacao* L.) in lower areas, and coffee (*Coffea arabica* L.) in higher areas. The tree layer mainly consists of fruit trees such as bread fruit (*Artocarpus altitis* (Parkinson) Fosberg), avocado (*Persea americana* Mill.) and citrics (*Citrus* spp). The animal component is provided by free-ranging chickens and a pig or two tied to a tree. On the island of St Croix, what remaing elements of a home garden seems to be a few bread fruit and avocado trees planted around a home along with some bananas and papaya plants (*Carica papaya* L).

4 Plantations of Shade-Grown Crops

Another variation upon the forest garden, plantations of shade grown crops tend to be bigger than the home garden just discussed and instead of being designed by hunter-gatherers, it was designed by farmers with the goal of selling the production of the shade-grown crop in the market place, whether that market is domestic or international. Coffee and Cocoa/Cacao are traditionally grown under the shade of a tree overstory. At first, the original forest was used for shade, but later as the original forest trees were cut down, they were replaced by fruit trees. Coffee tends to be grown at higher elevations than cocoa. The islands of the Greater Antilles (Cuba, the Dominican Republic and Haiti on the island of Hispaniola, Jamaica, and Puerto Rico) all produce these crops for export. The islands of the Lesser Antilles, where conditions are right, produce these products for local consumption.

Coffee is grown under the shade of Guamo (*Inga laurina* Sw. Willd) or Guaba (*Inga vera* Willd), in Puerto Rico and the Dominican Republic. Its wide crown casts a light shade, and nodules in the roots fix atmospheric nitrogen that is added to the soil. In the Dominican Republic, cacao is grown under the shade of Amapola *Erythrina poeppigina* (Walp. O.F. Cook). Other trees and palms are added to the mix to provide additional products such as fruit and lumber. On the island of St Lucia, cocoa is grown under the shade of mangos, *Mangifera indica* L. avocados, citrus, bell apples, (*Eugenia javanica* Lam) and residual forest trees (George and Krauss

2012). Often times, plantains or bananas are used to provide temporary shade and a crop in new coffee and cacao plantations until overstory trees can get established and grow above the plantains.

5 Shifting Agriculture, Fallows and Taungya Systems

One cannot write about fallows without discussing shifting agriculture. Shifting agriculture goes by many names: the action packed "slash and burn", archaic "swidden", and more neutral sounding "shifting". Presently, shifting agricultural associated with "primitive" tribespeople. Yet shifting agriculture was practiced in the "modern" societies of Europe from the Neolithic up to the middle of the twentieth century. Under whatever name it is called, farmers would cut fields out of the forest, burn the downed timber, and then plant crops. When yields declined, or weeds and pests, became too much of a problem, the field was abandoned and a new field would be cut out of the forest. In the meantime, the old plot would lie fallow and a new forest would grow up. The original farmer might move far away, or come back some years later to reclear the land, or another farmer would take his place.

Early European accounts of Caribbean agriculture describe the Tainos and Caribs practicing shifting agriculture with corn (*Zea mays* L.), tobacco (*Nicotianatabucum* L.), pineapples and such being grown in fields cut out of the forest. Tubers like sweet potato and cassava were grown in earthen mounds called *conucos*. They are circular mound of soil some 3 feet high (0.9 m) and 9 feet (2.8 m) wide. The mounds improve aeration and drainage of the soil (Rouse 1992). The Guanajatabey, still present when Columbus arrived, were hunter-gathers that did not practice agriculture; see forest gardens, mentioned previously.

However, as populations grew larger, particularly on the bigger islands, the length of fallow periods would have to decrease. The islands of the Caribbean were said to support some 750,000 Amerindians at the time of European contact, with the island of Hispaniola supporting some 500,000 people (Rouse 1992; Knight 1990). Maybe, the location of "permanent" settlements would change from time to time as soils around the original settlement became exhausted and a new forest would be cleared or cleared again. The community would just pick up and move to a new location keeping its old name. There is a precedent for this. The primary author knows of several towns in coastal Ecuador of pre-Colombian origins that changed location when environmental or political conditions became unfavorable after Spanish conquest. Also it must be mentioned, that until the arrival of Europeans, there were no large domesticated livestock in the Americas, with the exception of Llamas (*Lama glama* L.) in the Andes, whose manure could be used to fertilize agricultural fields.

Later on, shifting agriculture would have been adopted by escaped African slaves called Maroons and their descendants, Europeans too poor to receive plantations, and mestizos. The Maroons sometimes joined groups of Amerindians hiding in the mountains, and formed societies of mixed African and Amerindian ancestry.

Interestingly enough, Africans in Zambia raise crops in mounds of earth and leaves as well. The leaves provide organic matter and improve chemical and pH qualities of the soil (Rocheleau, Weber and Field-Juma 1988). It appears that this tradition has lived on, at least in folk memory, because poor slash and burn agriculturists are called "*conocuceros*" in the Dominican Republic. They will be discussed later.

In modern times, shifting agriculture is practiced by poor farmers on forested land too hilly or mountainous for large fields. They may not have title to the land they use. The real owner, being either an absent or disinterested citizen, or the government itself. Often, marijuana (*Cannabis sativa* L.) is grown as a cash crop (Honychurch 1995; Fraser 1995).

A fallow refers to agricultural fields left uncultivated to rest for varying periods of time from anywhere from a year to 20 or 30 years. Fallows can either be managed or unmanaged, dominated by herbaceous vegetation, bushes, or trees. The use of modern fertilizers has made fallows unnecessary, but in the past, soils would need to rest after several years of cultivation in order to restore their fertility. Once the field was abandoned a new forest would come up. In fact research shows that in Puerto Rican secondary forests about 40 years old resemble old growth forests (>80 years) in above ground biomass, basal area, and species richness (Aide et. al 2000). Fourty years is about the length of a tree fallow.

A field kept fallow for a year or two after abandonment is called a weed fallow because it is dominated by herbaceous vegetation that just spontaneously grow on site. If left for a longer period of time, trees and bushes can get established. Fallows that exist for five to seven years are called bush fallows and those that are allowed to grow longer are called tree fallows. These plants, especially if they are able to fix atmospheric nitrogen, will improve soil nutrient levels through the decomposition of litterfall and roots.

Long fallows are characteristic of shifting agriculture, after which a new forest would have regenerated upon the once cleared land. Often times, the new forest would be different from the original forest, especially if that tract of land had never been cleared before. It must be mentioned that the resting crop land is not considered worthless or useless in the fallow state. Fruit, fiber, fodder, fuel, wild medicines, wild game, and lumber can be and are harvested from the regenerating forest. However, population levels in the world are such that decades-long fallows are only practiced in remote areas, sparsely-populated areas.

Bush fallows are more typical these days, simply because of population pressures. In an example cited in Gelfius (1994), farmers in the south western Dominican Republic use trees of *Cassia* or *Senna spectablis* (D.C) Irwin & Barneby that spontaneously invade their parcels for firewood. On the island of St Croix, abandoned fields are dominated by trees of *Leucaena leucocephala* (Lam.) de Wit which can be put to good use as a, forage, fuel, and fertilizer.

6 Accelerated Fallows

If plant species that fix nitrogen did not colonize the site one could always plant them on the site. For example one could plant trees of *Cajanus cajan* L. Millsp. It is a small tree that fixes nitrogen, provides a pea that is very popular in Caribbean cooking, and provides a good quality firewood of kindling size, once the field is cleared for short-term crops again. If one does not have the luxury of having land in fallow for more than a year or two, one could plant a nitrogen-fixing cover crop such as Sunn Hemp, (*Crotalaria junceae* L). However, if one is deliberately planting nitrogen-fixing grasses or other herbaceous plants, weed fallows look very similar to crop rotation. On St Croix, during the era of sugar cane production pigeon pea bushes were planted in the fields as a green manure and then plowed over (Olasee Davis, pers comm 2013).

7 Taungya

This is an interesting variation upon the theme of shifting agriculture. Unfortunately, landless or impoverished farmers will often clear a patch of forest for richer landowner. In exchange for clearing bushes and trees from the land, they are allowed to make charcoal of the fallen wood, cultivate the land for a few years and just before abandonment, establish pastures. This negative aspect of shifting agriculture is practiced through the tropics and was also practiced on the North American Frontier in the eighteenth and nineteenth centuries. In the Dominican Republic, these people are called *conocuceros,* which in turn derives from the Taino word *conuco,* which were mounds of soil used to grow sweet potatoes and other tubers.

In the taungya system, instead of using the rural poor to establish pastures, they are used to establish forestry plantations. Landless famers are contracted to establish tree plantations and make sure the young trees survive. In exchange for their labour, the farmer is allowed to grow crops on the site of the young plantation. Since forestry plantations usually consist of rows of trees planted anywhere from 2 to 10 m of distance between rows, there is plenty of room between the rows for crops. The farmer tends the young trees as he weeds and fertilizes his agricultural crop. Once the forest canopy begins to close and shade affects the growth of the agricultural crop, it is time to move on. But in the meantime, the farmer has been able to harvest crops for two, three or four years. The farmer's use of the plot can sometimes be extended by planting more shade tolerant crops or by reopening the forest stand by the use of thinnings. This system was first invented or described in Burma, hence, the Burmese name, taungya, but it is called the parcelero system in Puerto Rico.

In the Dominican Republic, the NGO Plan Sierra used the taungya system to establish new pine tree and coffee plantations (Brown et al 1991). In Haiti, farmers sometimes use the taungya system to establish "jaden chabon" or charcoal gardens.

They would clear land, plant trees, and plant annual crops amongst the trees until the shade from the trees too much for the crops, cut the trees down, convert the wood to charcoal, and start over (Murray and Bannister 2004.)

8 Dispersed Trees in Fields

These are trees just left scattered about in a field either purposely planted in the field; or just quietly protected by the farm for the goods and services provided. In most farm fields, there are patches that are either too rocky or too wet or too something to farm. In the Dominican Republic, the combination of field crops with fruit trees such as mangos, avocados, citrus, and palms is very common. Often times, the growth of these trees is almost spontaneous. The products these trees produce (fruit, forage and building material) justify their presence. Trees, such as mango, which cast a dense shade and compete with the crop will have its branches pruned and turned in firewood or charcoal (Geilfus 1994). Surely, areas of scattered trees with associated crops were more common on Caribbean Islands before the introduction and cultivation of sugarcane or mechanized agriculture.

An example from the island of St. Croix relict is the frequent presence of tamarind (*Tamarindus indica* L.) and rarely, baobab trees (*Adansonia digitata* L.) in old fields and pastures. The baobabs were introduced by African slaves for spiritual purposes. The tamarind trees were introduced to the Caribbean in the 1500's from Africa. Tamarind trees provide a refreshing shade for workers in the fields besides a sour fruit used for making drinks. The ripe fruits which conserve well were exported to Denmark. Most old tamarind trees show signs of pollarding, where branches were cut off to make charcoal. Although agroforestry systems in Africa and India are beyond the scope of this paper, it is worth mentioning that in Africa and India, trees are purposely left on the field with a spacing of 10 that is to say 12 m between trees or between 100 and 50 trees/ha. Besides providing useful products such as tamarind fruit, they provide some light shade, and add nutrients to the soil, sometimes increasing crop yields. In the agroforestry literature, they are referred to as parklands.

9 Alley Cropping

This is the growing of crops between two rows of (mostly) nitrogen-fixing trees. Trees are planted at close spacings of 0.5 or 0.3 m in a row along the contour line. Each row is separated from the other row, some 4 to 6 m; 3 m appears to be the minimum. Distance between rows varies as the slope varies. Trees are pruned to chest, waist, or knee height several times a year, and the prunings are thrown on the ground between the rows. As the leaf litter and small branches decompose, nitrogen and various other elements within the leaves, such as phosphorus and carbon, are added to the soil. There are two ideas behind the use of alley cropping: providing fertilizer in areas where manufactured fertilizers are unavailable or too expensive to be used, and the prevention of erosion on slopes. Secondary products are firewood or poles from the prunings, fodder for animals, and sometimes, fruit (Murray and Bannister 2004).

An online search of the scientific literature makes it appear that Haiti is where the most work with alley cropping occurs. Species used are: *Leucaena* spp., *Delonix regia* (Bojer ex Hook, Raf.), *Gliricidia sepium* (Jacq.) Kunth. ex Walp., *Acacia angustissima* (Mill.) Britton & Rose, *Cassia siamea* Lam., *Erythrina variegata* L. *Inga* spp, and *Moringa olerifera* Lam. (Inga Foundation 2013; Isaac et al. 2000). In Jamaica, hedgerows made of *Calliandra callothyrus* Meissner are used in an alley cropping system. The hedgerows are planted along the contour lines of hillsides (McDonald, Stevens and Healey 2004). A Haitian variation of this is the "*Ban Manjé*" or food band. It uses non-woody plants to form a band some 2–3 m wide instead of the traditional hedgerow width of 1 m. An overstory of perennial food crops, especially plantain, pigeon pea, Sugar cane (*Saccharum officinarum* L.), Cassava and pine apple as the soil-holding component of the system along with a ground cover/annual crop of sweet potato or yam (*Diascorea* spp). Most of the produce is sold in the market. What was supposed to be soil conservation, or soil fertility enhancing strategy became a micro-economic income enhancing strategy (Murray and Bannister 2004).

Some experiments in alley cropping were performed at the University of the Virgin Islands Agricultural Experiment Station- St Croix in 1994 with tomato and eggplant, and again in 2004 with medical plants. The tree species tested were *Leucaena leucocephala*, *Gliricidia sepium*, *Moringa olerifera* and *Cajanus cajan*. It was found that the alley of trees competed with the planted crops for water and yields of the crops were suppressed.

However, it must be pointed out that alley cropping is usually used in association with agronomic crops like sorghum (*Sorghum bicolor*, L. Moench) and corn (*Zea mays* L.), not horticultural crops like tomato and eggplant, and that the experiment was performed on flat agricultural land and not sloping marginal lands where most poor farmers cultivate crops. Also the economic yield of the products from the alleys themselves was not accounted for: i.e., peas, firewood and fodder, reduced soil erosion, or reduced fertilizer use.

10 Living Fences, Wind Breaks, Trees in Other Places

Living fences are barriers made of plants. One way to do so is just plant some spiny species at a close spacing to form hedges impenetrable to both animals and people. An example would be the prickly bromeliad Pingüin (*Bromelia penguin* L) planted along field edges. Other (non-tree) species used in both Haiti and the Dominican Republic are: Cabuey (*Agave sisalana* Perrine), Piñon de leche (*Jatropha curcas* L.), Crotón, (*Codieuam variegatum* L.), la raqueta (*Euphorbia lacteal* Han) and the grass *Andropogon muricatus*, although it must be mentioned, not all these species are spiny. In that case, they would only serve for field delimitation. A variation found on the island of St. Croix is the planting of pigeon pea (*Cajanus cajan*) along the edge of a property. The pigeon pea trees or bushes serve as a hedge while growing food. A "useless" piece of property is put into production (personal observation, Michael Morgan).

Another method is the use of living fence posts in combination with wire. Some tree species can take root from large cuttings. When the fence is being installed large cuttings are planted as fence posts and the fencing wire attached to the posts. Examples of common Caribbean trees that take by cuttings and used for living fence posts are *Spondias mombim, Spondias purpurea*, and *Cordia alba* (Wadsworth and Little 1964). Living fence posts of Turpentine tree (*Bursera simaruba L.* Sarg.) are used in St Croix, US Virgin Islands. In the Dominican Republic, living fence posts are made from *of Gliricidia sepium* (Geilfus 1997). Trees left along field edges at establishment or those trees that get established later, such as Genip (*Melicoccus bijugatus*) allowed to grow into fence posts.

Sometimes trees are planted as windbreaks. The winds in the Caribbean tend to blow steadily from east to west. Windbreaks are barriers of trees and bushes used to reduce the force of the wind, thereby reducing the erosion of soil by the wind and evapo-transpiration of plants. There is not much in the way of scientific literature. However, in at least one case, rows of Australian Pine (*Casuarina equisetifoilia* L.) in the sugarcane fields of St Croix as windbreaks. Sugar Cane production stopped on St Croix in 1964, but a few rows of Australian pine still exist in the landscape (Personal communication, Olassee Davis 2014).

It is worth mentioning two other early developments, that are not exactly agroforestry practices but rather conservation measures that affect agricultural production. The first is the establishment of buffer zones along water courses in the US Virgin Islands back when they were still a Danish colony. The Danes established a law prohibiting the deforestation of land along water courses for some 25 feet to either side of a water course or 30 feet from the center, which ever was greater. The idea behind this was to protect the navigability of water courses, and indirectly protect agricultural lands from being flooded (Olasee Davis, pers.com 2013). This law is still operative in Virgin Islands Law: V. I. Code Annot. Title 12, sections 121–125, (Thomas and Devine 2005).

The second development is the creation of protected forests by the British Empire on some of their Caribbean possessions. Soon after colonizing the relatively flat island of Barbados in 1627 and dedicating the vast majority of the island to sugar cane cultivation, they ran out of wood to boil sugar cane, and noticed problems with soil erosion, irregularity of rainfall and sedimentation of water courses. They did not want to repeat the mistakes made on their first Caribbean colony, so the high volcanic peaks of Monserrat in as early as 1702 forest reserves were established on the high volcanic peaks of the island of Monserrat (Grove 2006). After they took over the formerly French owned islands of St Vincent and Tobago, in 1764, forest reserves were established in Tobago (Main Ridge Reserve 1776) and St Vincent (Kingshill Reserve 1790) to protect the headwaters of rivers and provide rainfall to the rest of the islands. The Main Ridge Reserve claims to be the earliest protected area in the New World. As early as the eighteenth century someone had figured out a link between the forest and rain, the language of the parliamentary act that created the Kings Hill Reserve, said the land was "reserved and appropriated for the purpose of attracting the rain and the clouds."

11 Silvo- Pastoral Systems

Pastures with either scattered trees or lines of trees growing in them are called silvo-pastoral systems. Grasslands with scattered trees are called savanna or sabanas, to use the Spanish spelling. It turns out that sabana is originally a Taino word meaning "area of either short or tall grass" (American Heritage Dictionary 2003; Rouse 1992). Trees left growing in pastures imitate the (not-so) "natural" savanna ecosystem. Periodic fires or heavy grazing are necessary to prevent trees from recolonizing the grassland and changing it to forest. Before we write about the benefits of silvo-pastoral systems we will discuss their historical context in the Caribbean.

Early Spanish accounts have the Tainos burning the sabanas to catch Hutias, a type of forest rodent that has several genera and each genus has several species. Many species of Hutia are now extinct. The bigger islands of Cuba and Hispaniola (Haiti/Dominican Republic) and the islands that make up the Bahamas also have indigenous species of Pine (*Pinus caribea, Pinus cubanesis, Pinus occidentalis*) that are all adapted to light, frequent fires that would keep the understory clear of brush, promote the growth of grasses and allow regeneration of the pine trees (Rojas and Ortiz 1991). There is evidence (analysis of charcoal and pollen found in lake sediments) from the nearby islands of the Bahamas, that pine forests were more widespread, most likely maintained by human fires in pre-Columbian times, and their distribution declined after European invasion and settlement (Myers et al 2004). An open landscape would also permit the cultivation of crops amongst the trees.

An open landscape of grasses and scattered trees would have appeared familiar to the new Spanish settlers. Most came from the dry regions of Andalucía and Extremadura where livestock raising was the main economic activity (Knight 1990). These two regions of Spain still have large expanses of open oak woodland or parkland grazed by cattle and pigs, called dehesa (Joffre et al 1988).

Soon upon the arrival of settlers very early on in the sixteenth century herds of cattle, pigs and other livestock were introduced to the Greater Antilles. In reference to Hispaniola, Knight (1990) writes "The animals introduced by the Spanish multiplied so rapidly and roamed so freely that they soon became pests, greatly outnumbering the human population and often destroying the conucos or cultivated parcels of the Indians and Spanish. Wild pigs took over the forests, and wild cattle and horses dominated the plains in such profusion that pioneers found abundant meat in the interior, as did later, runaway slaves. Even the Indians became carnivorous." However, this description holds true for the other Caribbean islands particularly Jamaica, Puerto Rico, and Cuba. To quote Knight (1990) again… "Indeed, the ready availability of these animals gave rise to the transient, trans frontier society of buccaneers during the late sixteenth and early seventeenth centuries." Buccaneers started off as (French) frontiersmen who hunted wild cattle on Hispaniola and smoked the meat on *buccan*, another Taino word meaning wooden frame for smoking meat. They would then sell the smoked meat and hides to passing ships. Only later, when the Spanish owners of Hispaniola expelled them, did the buccaneers shift their trade to piracy (Knight 1990).

Silvo-pastoral systems can be either accidental in the sense of trees left over when the original forest is cleared for pasture, or intentionally established. The trees provide direct benefits such as fodder for cattle, firewood and lumber. Livestock are more productive and comfortable when they receive some shade from the sun and protection from the wind. The grass underneath a tree is more nutritious due to nutrient cycling and regular fertilization by the livestock. Trees improve biodiversity just by existing and providing habitat for birds, insects and smaller vertebrates. Fence rows of trees allow animals to travel between patches of wild habitats. (Dagang and Nair 2003).

In the Dominican Republic, trees of Samán (*Saman samanea* F. Muell) are left in pastures. Samán or Rain Tree is a giant wide crowned tree that produces sticky sweet seed pods that are eaten by cattle. The crown provides shade under the tropical sun. The leaves close up at night and during rainstorms allowing dew and rain to reach the ground under the tree. Nodules in the roots capture atmospheric nitrogen. Cattle gather under its crown for shade and fertilize the tree and the grasses growing with them (Geilfus 1994).

Another tropical silvopastoral system is the association of coconut palms and pasture. The tall palms do not cast enough shade to impede the growth of grass. The farmer can receive two products from the same piece of land: cattle and coconuts. The cattle keep the coconut plantation clean and free of bushes and fertilize the plantation with their manure. If a nitrogen fixing ground cover of Kudzu or Clover is added, the coconut palms receive fertilization and the animals receive protein-rich fodder. The number animals the coconut plantation can carry depend on the spacing and the age of the plantation. For example, in the Solomon Islands of the Pacific Ocean, one cow per hectare can be carried in coconut plantations between 6 and 13 years (215 palms per hectare). Two cows per hectare is the carrying capacity of old coconut plantations of 50 or 60 years old (138 palms per hectare) (Geilfus 1994).

Guazuma ulmifolia is a small tree that grows spontaneously in pastures throughout the Caribbean basin. Farmers in the Dominican Republic protect some of the trees that get established in their pastures (Geilfus 1994). The leaves and fruit of woody black balls are eaten by cattle. They enjoy the shade. The farmer can use the wood for charcoal and firewood. The branches make good poles and stakes (CATIE 1991). Another system from Barbados combines mahogany trees for shade, protein banks of closely spaced and frequently pruned *Leucaena leucocephala* trees, a nitrogen-fixing ground cover and black-bellied sheep (Vyvey 2011). Trees and livestock were and still are a productive combination.

Conclusion

The association of trees, crops, and animals is a very old practice and should not be seen as something new. It has been practiced on the Caribbean islands since before the arrival of European and African peoples in 1492. Agroforestry is a form of low-input agriculture in response to local conditions. It produces many benefits both economic and environmental.

Acknowledgements This project was funded by a McIntire-Stennis grant. Thanks to Sarah Workman who forwarded some papers and Dr. Dilip Nandwani who extended the invitation to write the chapter.

References

Aide TM, Zimmerman JK, Pascarella JB, Rivera L, Marcano-Vega H (2000) Forest regeneration in a chronosequence of tropical abandoned pastures: implications for restoration ecology. Restor Ecol 8(4):328–338

American Heritage Dictionary (2003) Definition Savanna. Houghton Mifflin, Boston

Brown N, Olander J, Endreny T, Wray R, Scott C, Mancarella P (1991) Analysis of agroforestry at Plan Sierra, Dominican Republic. In: Buck LE, Lassoie JP (eds) Agroforestry planning and evaluation in marginal areas: exercises in program design. Cornell University, Ithaca, pp 11–53

CATIE (1991) Guácimo (*Guazuma ulmifolia*) especie de árbol de uso múltiple en América Central. CATIE, Costa Rica

Dagang ABK, Nair PKR (2003) Silvopastoral research and adoption in Central America. Agrofor-est Syst 59(2):149–155

Francis JK (1992) *Melicoccus bijugatus* Jacq. Quenepa. US Department of Agriculture, New Orleans

Fraser A (2006) Mountain, bush and garden: the historical context of environmental conflict in St Vincent. In: Anderson RS, Grove R, Hiebert K (eds) Islands, forests and gardens in the Caribbean: conservation and conflict in environmental history. Macmillan Caribbean, Thailand, pp 174–182

Geilfus F (1994) El árbol al servicio del agricultor: manual de agroforestiera al servicio del agricultor. CATIE (Centro AgriculturaTropical de Investigación and Enseñanza), Turialba

Geographic Consulting, (2010) August Tree Hazard Report (of Roadside Trees in St. Croix) http://www.geographicconsulting.com/wp-content/uploads/2011/05/August-Tree-Hazard-2010-Report2.pdf. Accessed 2 Aug 2014

George A, Krauss U (2012) The rehabilitation and expansion of the cocoa industry in St Lucia. In: Proceedings from the 47th meeting of the Caribbean food crops society, PP 26–331

Grove R (2006) The British Empire and the origins of forest conservation in the Eastern Caribbean. In: Anderson RS, Grove R, Hiebert K (eds) Islands, forests and gardens in the Caribbean: conservation and conflict in environmental history. Macmillan Caribbean, Thailand

Honychurch L (2006) "To have a garden is to be a man": Caribs and the transformation of landscape. In: Anderson RS, Grove R, Hiebert K (eds) Islands, forests, and gardens in the Caribbean. Macmillan Caribbean, Thailand

Inga Foundation (2014) Inga Alley Crops. http://www.ingafoundation.org/congo-madagscar-haiti-going-global. Accessed 2 July 2014

Joffre R, Vacher J, De Los Llanos C, Long G (1988) The dehesa: an agrosilvopastoral system of the Mediterranean region with special reference to the Sierra Morena area of Spain. In: Nair PKR (ed) Agroforestry systems in the tropics. Kluwer Academic, Dordrecht, pp 427–456

Knight FW (1990) The Caribbean, the genesis of a fragmented nationalism. Oxford University Press, New York

McDonald MA, Stevens PA, Healey JR (2003) Soil fertility under *Calliandra calothyrus* hedgerows and other land-use treatments following forest clearance in Jamaica. Agroforest Syst 57(2):127–135

Murray GE, Bannister ME (2004) Peasants, agroforesters, and anthropologists: a 20-year venture in income-generating trees and hedgerows in Haiti. Agroforest Syst 61/62(1–3):383–397

Myers R, Wade D, Bergh C (2004) Fire management assessment of the Caribbean pine (Pinus caribea) forest ecosystems on Andros and Abaco Islands, Bahamas. GFI United States of America, The Nature Conservancy, 17 p

Rocheleau D, Weber F, Field-Juma A (1988) Agroforestry in dryland Africa. ICRAF (International Council for Research in Agro-Forestry), Nairobi, 331 p

Rojas F, Ortiz E (1991) Pino caribe (Pinus caribe), especie de árbol de uso múltiple en América Central. CATIE (Centro agronómico tropical de investigación y enseñanza), Costa Rica, 8:59

Rouse I (1992) The Tainos: rise and decline of the people who greeted Columbus. Yale University Press, New Haven

Rubenstein H (2006) Bush, garden and mountain, on the leeward coast of St Vincent and the Grenadines, 1719–1995. In: Anderson RS, Grove R, Hiebert K (eds) Islands, forests and gardens in the Caribbean. Macmillan Caribbean, Thailand, pp 194–213

Thomas T, Devine B (2005) Island peak to coral reef, a field guide to the plant and marine communities of the Virgin Islands. University of the Virgin Islands, USA. 214 p

Vyvey L (2011) Agroforestry in Barbados: integrating mahogany trees and black belly sheep to form a sustainable land use system. Caribbean Food Crops Society Proceeding from 47th Annual Meeting 47:1

Weaver PL (2006) Estate Thomas Experimental Forest, St. Croix, U.S. Virgin Islands: research history and potential. USDA Forest Service International Institute of Tropical Forestry. United States of America, USDA Forest Service, 62 p

Role of Horticulture in Biodiversity Conservation

Ashwani Kumar

1 Introduction

Rampant use of natural resources is leading to the depletion of vegetation, denudated and degraded lands, and lowering of water table. In order to survive and provide enough food for over 8 billion people in coming years, it is important that sustainable development has to take place. World Commission on Environment and Development (Brundtland Commission 1987) defined Sustainable Development as: "Development that meets the needs of the present without compromising the ability of future generations to meet their own needs" (Brundtland 1987). Sustainable development at local, national regional and global level depends on integration of social, economic and environmental development. Sustainable development requires increment and conservation of all natural resources. Sustainable development also involves diminution of waste. Renewable sources of energy, utilization of water at sustainable level is important as we are mining fossil fuels, water and other natural resources that are in limited amount below the earth's surface (Chichilinisky 1997). If biodiversity is not allowed to regenerate itself in a time frame, it becomes extinct. The present rate of extinction is around 100 plant species per annum. UN Agenda 21 proposed in a conference, "United Nations Conference on Environment and Development (UNCED)" at Rio de Janeiro, Brazil (1992) submits that sustainable development is based on the satisfaction of basic needs in developing countries. This can only be achieved when sustainable development of business is measured on the triple bottom line of people (society), planet (ecology) and profit/prosperity (economy).

A. Kumar (✉)
Department of Botany, University of Rajasthan, 302004 Jaipur, India
e-mail: ashwanikumar214@gmail.com

D. Nandwani (ed.), *Sustainable Horticultural Systems,* Sustainable Development and Biodiversity 2, DOI 10.1007/978-3-319-06904-3_7

2 Horticulture

The branch of agricultural sciences that concerns with intensively cultivated plants is termed as horticulture. These plants are widely used by people for food, medicinal purposes, as ornamental or for aesthetic satisfaction. Thus horticultural plants can be divided into (1) edible plants, (2) culinary or medicinal purposes used plants, and (3) ornamental or aesthetic purposes used plants. Apart from this, horticulture also plays an important role in preservation of biodiversity and sustainable development. Horticultural plants can be differentiated from other plants as per their subsequent use and at the management level which is applied for their production.

2.1 Types of Horticulture

In 1893, U.S. Supreme Court legally established the characterization of plants by common usage as complete agreement decision in the case of Nix vs. Hedden. To increase plants production in a profitable manner and to maintain this gain, agriculture, including horticulture becomes dependent on science and technology. Horticulture can be studied scientifically into various disciplines. That branch of horticulture which deals with fruit and tree nut production like apple, orange, peach, raspberry, strawberry and blueberry is termed as Pomology. Olericulture, the other branch of horticulture, deals with vegetables and herbs production. The production of greenhouse-grown or field-grown plants for their showy leaves or flower is termed as Floriculture. Environmental horticulture includes the plant's production for ornamental use for both indoors and outdoors. This industry of horticulture asserts and improves the plant's functional use in populated areas. All the commercial production of nursery and floriculture commodities and the non-production uses of plants can be included under environmental horticulture.

2.2 Role of Horticulture in Sustainable Development

After the Green Revolution in the 1960s and 1970s, people became more concerned in horticulture. Their interest grew with time as it not only provides economic value to local communities but also to the country. Horticulture covers 17.2 million hectares of land which is around 8.5% of the gross cropped area of the country. It becomes one of the industries which provide high-quality food for well-being, thus providing nutritional security to the people. It can also help in improving the productivity of land which improves economic conditions of the farmers and entrepreneurs, enhancing exports and finally generating employment.

Horticulture has excellent driving force of their monetary value, gardeners, farmers and breeders, to increase biological diversity in cultivated plant communities,

Fig. 1 Gardens are rich depository of plant species of different habit and habitats and flora ranging from aquatic plants to desert plant and from minutest plants to the giant plants nearing extinction

Fig. 2 Different varieties of *Chrysanthemum* spp. at display in Palmen Garten, Frankfurt, Germany will help in biodiversity conservation

researchers have domesticated, selected, and bred many new horticultural plants (Pittenger et al. 1991; Zhang et al. 2004; Kulkarni 2012). Some of the significant plants which were multiplied in vitro are *Pinus strobus, Magnolia grandiflora* stem cuttings and *Cephalotaxus* seed dormancy. It is to be noted that plant tissue culture provides mass propagation method of horticultural crops (Neumann, Kumar and Imani 2009; Kumar and Shekhawat 2009; Kumar 2010; Kumar and Sopory 2010; Kumar and Roy 2011; Fernandez et al. 2010; Roy and Kumar 2011).

Gardens and Botanical gardens are the rich depository of varieties and species of a particular plant and have a role in sustainable development. Like *Aloe vera* is a common food and drug for health benefits. Author personally visited the Palmen Garten (in German), Frankfurt, Germany and it is an example how gardens can be a source of conserving biodiversity of different habit and habitat and different flora of the world all at one place (Figs. 1 - 4).

2.2.1 Rare and Endangered Plants

Plant nearing extinction includes many primitive plant species of order Cycadales of Gymnosperms and *Dioon edule* and *Cycas rumphii* are such examples (Figs. 5 and 6).

Fig. 3 Different varieties of *Chrysanthemum* sp. at display in Palmen Garten Frankfurt Germany will help in biodiversity conservation

Fig. 4 *Selaginella involens* (Pteridophya)

Fig. 5 *Cycas rumphii*—A living fossil and plant nearing extinction produces largest ovules on the earth. It is rare to see such a large number of seeds in nature but in Palmen Garten Frankfurt, Germany it is seen with large number of seeds and could be used to make this plant sustainable

2.2.2 Food and Medicine

Asparagus racemosus is commonly used for medicinal purposes and as food but other species of *Asparagus* could also be investigated. *Asparagus retrofracus* is an example of such species in Palmen Garten of Frankfurt, Germany (Figs. 7–10). For example, *Aloe* has a number of species in which *Aloe vera* is commonly used in food and drink globally.

2.2.3 Desert Plants

Acacia drepanolobium (Leguminosae) is a plant of xerophytic habitat and can be used in different environmental conditions to greening of deserts globally. (Figs. 11 and 12)

Fig. 6 *Dioon edule* is one of the rare plant of Cycadales of Gymnosperm, is grown here

Fig. 7 Several species of Aloe are presented here

Fig. 8 *Aloe jacksoni*, another species of *Aloe* which could have potential medical uses

Fig. 9 *Aloe rivae*

Fig. 10 *Asparagus retrofractus* (Liliaceae), species of Asparagus which can have potential benefits but is not cultivated at present. It could be of potential food and medicinal use

Fig. 11 *Acacia drepanolobium* (Leguminosae)

Fig. 12 *Agave macroacantha* (in the middle) and *Echinocactus grusonii* on both the sides.

Fig. 13 *Citrus auranteum*

2.2.4 Fruit Plants (Figs. 13–16)

3 Ecosystem and Biodiversity

Nature has a number of plant and animal varieties and these living organisms differ from each other in their origin, like terrestrial and aquatic ecosystems. These differences in ecological complexes or the variation of life are termed as Biodiversity. Thus, biodiversity refers to ecosystem variation, genetic diversity and species variation between different species of an area, biome, or planet.

Fig. 14 *Citrus limon*

Fig. 15 *Citrus microcarpa*

Fig. 16 *Echinocactus grusonii* (Schwiegermuttersitz i.e. stepmothers stool in German language)

There is no doubt in the fact that for its survival and better health, humans always have to depend upon the other species and the key ecosystems (Osvaldo et al. 2009). But the pervert behaviour of humans make these keystone species disappear which finally results in habitat destruction. The conservation of genes (specific information), their libraries (species), and habitats (support systems) are in urgent concern because of the current changes in rate of extinction caused by human activities and impacts. Increase in population has enhanced a rapid growth in consumption of resources which results in biodiversity exploitation at a faster rate, thus a great loss of biodiversity on the planet. World has 18 'biodiversity hotspots' and India, one of the mega biodiversity centers in the world, has two of them which are located in the Eastern Himalayas and in the Western Ghats (Myers 1997; Roy and Kumar 2011). All of this makes biodiversity as a priority environmental issue for the United Nations which resulted in the declaration of 'International Year of Biodiversity' to save the environment.

3.1 *Role of Biodiversity in Human Health*

Biodiversity plays an important role in the availability of medicinal resources and drug discovery (Mendelsohn and Balick 1995). Directly or indirectly, a substantial proportion of drugs are derived from biological sources. More than 85 % of traditional medicine requires plant extracts and more than 80 % of the world population depends on medicines from nature and thus trusts on traditional medical systems for primary healthcare (Chivian and Bernstein 2008). Plants constitute about 74 % of the total drug resources in traditional as well as modern medical practice. It is estimated that in the US market, more than 50 % of the pharmaceutical compounds are derived from micro-organisms, plants and animals (Chivian and Bernstein 2008). Moreover, human beings also use drugs from bacteria (5 %), fungi (18 %), and snakes (3 %).

3.2 *Biodiversity for Natural Products*

As far as we can remember, natural ecosystems are serving themselves by their goods and service to the human population. Natural ecosystems produce vegetation as crops, fuel wood, fiber, timber, pharmaceuticals and industrial products that are directly used by humans. Over the course of history, human beings have utilized around 70,000 plants having any of the edible parts and around 7,000 plant species for food (Wilson 1989). Fruits, spices, nuts, essential oils and flavorings, gums and exudates, dyes, tannins, resins and oleoresins, all extracted and obtained from hundreds of plants species are used by the humans for their benefits (Leung and Foster 1996). In developing countries, it is estimated that fuel-wood and other plant materials supplies about 15 % of the world's energy consumption and this biomass accounts for nearly 40 % of energy consumption (Hall et al 1993).

Not only plants, but animals are also playing an important role in human welfare. Animals like asses, camel, mules and bullocks are used for labor, while cattle, goats, sheep, and cow are used for meat, milk, leather and wool. Apart from this, waterfowl, deer, bear, moose, rabbits, snakes and monkeys are hunted as game animals or for cultural sports. Moreover, fishes and its products are one of the valuable items in economic markets. The annual world fish catch amount is about 100 million metric tons and its value ranges between $ 50 billion to $ 100 billion. Over 20 % of the population in Asia and Africa depends on fish as it is the leading source of animal protein for them (UNFAO 1993). However due to the ongoing habitat conversion, most of these natural products are in decline phase.

3.3 Loss of Biodiversity

For present and future generations, biological diversity is a global asset of enormous value. During the last century, significant decline has been observed in biodiversity. In 2007, Sigmar Gabriel, German Federal Environment (GFE) Minister summoned that by 2050, up to 30 % of all the species will be extinct (Leakey and Lewin 1996). Out of which, about one-eighth of known plant species are threatened with extinction. Based on Species-Area Theory, it is estimated that species extinction may reach as high as 140,000 species per year (Reid 1995).

Rapid changes in environment cause mass extinctions. Estimate says that < 1-3 % of the species are extinct now which once existed on Earth (Raup 1994). This reduction is mainly caused by human activities and their impacts, particularly habitat devastation is called as Holocene extinction. The decrease in biodiversity is due to the synthetic breeding programmes and human dependency on major cereal crops like wheat, maize and rice. Earlier there were around 40 varieties of rice that were in use for cooking and now just four or five are there. The human population and the industrialization pressure is the main reason for threat to ecosystems due to which a large number of species are on verge of extinction or have already become extinct.

3.4 Importance of Biodiversity Conservation

Every living organism or thing has its own place in balancing of nature and any kind of disturbance or stress can have untold effects which represent a single fact, we are the only losers. Human activities are changing climate at a fast rate which is affecting biodiversity. Ozone depletion is placing a great threat to biosphere to be recovered. Droughts, excessive rains, floods etc. are all resulting in global warming.

Biodiversity plays an important role of cultural heritage in human society and plants and animals are the vital part of this tradition and culture. Biodiversity is a hidden treasure that enriches all our lives. But extinction of these plants and animals cannot be restored. The Protea, India's National Emblem, emblazoned on the jerseys of rugby players. Huge Southern Right Whale in the wild and the tiniest leaf chameleon is at their low density.

Biodiversity is not a luxury but is essential for sustaining life. It acts like our Living Bank Balance which is necessary for our standard of living. We have to remember that it is the only source for humans for obtaining economic benefits, health and safety as well as recreational or aesthetic enjoyment.

3.5 *The Convention on Biological Diversity*

In 1987, United Nations Environment Programme (UNEP) recognized the necessity of protecting biodiversity at International level. So, for the conservation of all the aspects of Earth's biodiversity like ecosystems, species and genetic resources and for the sustainable use of these biological resources, a global agreement was signed by nations at the UNCED Earth Summit at Rio de Janeiro, Brazil (1992), named as The Convention on Biological Diversity (CBD), which came into force in 1993. Presently, there are 193 Parties to this Convention.

The main objectives of the Convention are (1) Conservation of biodiversity, (2) Sustainable use of the biodiversity components and (3) Fair and equal sharing of all the benefits arising from the genetic resources use.

3.6 *Biodiversity and Sustainable Development*

A measurement of industrialization and planning system and thus measuring biodiversity as a result or as nature's balance is regarded as Sustainable Development. It consists of three major elements: (1) natural resource base; (2) technology; and (3) socio-economic elements. Out of these, socio-economic elements determine sustainable development and its dimensions provide the pathways for this development whereas for biodiversity conservation, natural resource database is quite essential.

Conclusion

Biodiversity acts as a foundation or the backbone upon which human civilization has been established. Humans have exploited the environment as common good. Loss of biodiversity in the form of species and ecosystems has caused transformation of the earth. In the last few decades, over-harvesting of plants and animals for food and fodder has resulted in most of the extinctions and this all is mainly due to the commercial hunting. But, if the biological diversity is well-managed, economic sustainable development could be achieved. Biodiversity conservation helps in achieving poverty alleviation. If government of different countries along with concerned organizations make policies and agreements then it will be easy in achieving goal of balance between economic development and sustaining ecosystem services.

Acknowledgement Corresponding author is grateful to Alexander von Humboldt Foundation for grant of fellowship and Professor Dr Sven Schubert, Institut fur Pflanzenernaehrung der Justus Liebig Univesitatat, Giessen for enabling us to visit Palmen Garten Frankfurt and providing his valuable guidance. We also thank Ms Franziska Faust, Mr Stefan Jung and Professor Vijay R. Kumar for cooperation during the visit and photographs.

References

Brundtland GH (1987) The U.N. World Commission on environment and development: our common future. Oxford University Press, Oxford

Chichilinisky G (1997) What is sustainable development? Land Econ. 73: 467–491

Chivian E, Bernstein A (2008) Sustaining life: how human health depends on biodiversity. Center for Health and the Global Environment. Oxford University Press, New York.

Fernandez H, Kumar A, Revilla BMA (2010) Working with ferns: issues and applications. Springer, Germany

Hall DO, Rosillo-Calle F, Williams RH, Woods J (1993) Biomass for energy: supply prospects. In: Johansson TB, Kelly H, Reddy AKN, Williams RH (eds) Renewable energy: sources for fuels and electricity. Island, Washington DC, pp 593–651

Kulkarni A (2012) Biodiversity and sustainable development: a critical analysis. Int J Sci Eng Res 3:1–9

Kumar A, Roy S (2011) Plant tissue culture and applied plant biotechnology. Avishkar, Jaipur

Kumar A, Shekhawat NS (2009) Plant tissue culture and molecular markers: their role in improving crop productivity. IK International, New Delhi

Kumar A, Sopory S (2010) Applications of plant biotechnology: in vitro propagation, plant transformation and secondary metabolite production. IK International, New Delhi

Kumar A (2010) Plant genetic transformation and molecular markers. Pointer, Jaipur

Leakey RE, Lewin R (1996) The sixth extinction: biodiversity and its survival. Phoenix, London, pp 137–142

Leung AY, Foster S (1996) Encyclopedia of common natural ingredients used in food, drugs, and cosmetics. Wiley, New York

Mendelsohn R, Balick, MJ (1995) The value of undiscovered pharmaceuticals in tropical forests. Econ Bot 49:223–228

Myers N (1997) The world's forests and their ecosystem services. In: Daily G, (ed) Nature's services: societal dependence on natural ecosystems. Island, Washington DC

Neumann K, Kumar A, Imani J (2009) Plant cell and tissue culture—a tool in biotechnology basics and application. Springer, Germany

Osvaldo ES, Laura AM, Camille P (2009) Biodiversity change and human health: from ecosystem services to spread of disease. Island, Washington DC 3–5

Pittenger DR, Victor D, Gibeault A, Steve EI, Cockerham T (1991) Environmental horticulture: "growth" industry in California. Calif Agr 45:15–17

Raup DM (1994) The role of extinction in evolution. Proceedings of the National Academy of Sciences 91:6758–6763

Reid W (1995) Reversing the loss of biodiversity: an overview of international measures. The Arid Lands Newsletter, Spring/Summer, 37

Roy A, Kumar A (2011) Ethnobotanical studies on medicinal plants of Himalayas: traditional medicine system of Himalayan region. Lambert Academic, Germany

United Nations Food and Agriculture Organization (UNFAO) (1993) Marine fisheries and the law of the sea: a decade of change. Fisheries Circular No. 853, Rome

Wilson EO (1989) Threats to biodiversity. Sci Am 261(3):108–116

Zhang D, Huang H, Hu D (2004) Significance of horticultural plants on biodiversity conservation. HortScience 39:867

Web resources

http://www.eolss.net/Sample-Chapters/C16/E1-50-14.pdf
http://tuscolaagriculture.weebly.com/uploads/8/3/8/9/8389114/importance_of_hort.pdf
http://www.globalchange.umich.edu/globalchange2/current/lectures/biodiversity/biodiversity.html
http://www.nbaindia.org/introduction.htm
http://www.buzzle.com/articles/importance-ofbiodiversity.html

Avocado History, Biodiversity and Production

Tomas Ayala Silva and Noris Ledesma

1 Introduction

The name avocado originates from the Aztec word "Ahuacatl" *and* "*ahoacaquahuitl*". Ancient Aztec, Olmec and Maya cultures praised avocado as one of the 'gifts of God'. The prehistoric fruit was under cultivation in North, South and Central Americas for thousands of years. From there, avocado dispersed to Europe and the rest of the world. Seed remains found in ancient human settlements in the Tehuacan Valley suggest that the avocado could have been used as early as 8000–7000 BC and possibly domesticated at least 5000 BC by Mesoamerican groups (Smith1966, 1969). The most ancient evidence of the use of the avocado tree (*Persea Americana*; Lauraceae) in Mesoamerica is about 10,000 years ago, in Coaxcatlan, Puebla (Mexico) (Knight 2002; Galindo-Tovar et al. 2007; Landon 2009).

1.1 Avocado History

The avocado (*Persea gratissima* or *P. americana*) gets its name from the Latin American Nahuatl word *ahuacatl* meaning "*testicle*," an obvious reference to the shape of the fruit. It was discovered in Mexico approximately 291 BC. The more easily-pronounced name of *avocado* is attributed to Sir Henry Sloane, who created it in 1669. The word itself first appeared in American print in 1697 (Galindo-Tovar et al. 2007). Early Spanish explorers discovered the Aztecs enjoying avocados, but it was long considered a tasteless food. The Aztecs also used avocados as a sexual stimulant. It was the Spanish explorers who brought the avocado to the English.

T. Ayala Silva (✉)
United States Department of Agriculture, Agricultural Research Service,
Subtropical Horticulture Research Station, National Germplasm
Repository System, 13601 Old Cutler Road, Miami, FL 33158, USA
e-mail: tomas.ayala-silva@ars.usda.gov

N. Ledesma
Fairchild Tropical Botanic Garden, 14885 SW 248 Street, Homestead, FL 33032, USA

D. Nandwani (ed.), *Sustainable Horticultural Systems*, Sustainable Development and Biodiversity 2, DOI 10.1007/978-3-319-06904-3_8

The first Florida crops are credited to horticulturist Henry Perrine who planted groves in 1833. However, avocados did not become a commercial crop until the early 1900s (Knight and Campbell 1999; Knight 2002). Except in California, Florida and Hawaii where they were commonly grown. In early 1950s, the avocado became accepted as a salad item and consumption became more prevalent (Scora et al. 1970).

The diversity of the avocado has also been known since pre-Hispanic times. Benavente, in his "Historia de los Indios de la Nueva España" ([1536] 2003), made a distinction among different avocado types: "the ones common in all this land and all the year, are like early figs. Other avocados are as big as large pears and are as good as best fruit in the New Spain (Mexico). There are others as big as a small pumpkin; ones with a big seed and little flesh and others with more flesh". Sahagún ([1570] 2002) also described three different types of avocado: the *ahuacatl* or *ahuacacahuitl* has dark green leaves and the fruit is black in the outside and white and green in the inside; the *tlacazolahuacatl* is like the former but bigger; and the *quilahuacatl* is green in the outside and very good to eat. Other chroniclers not only described the different avocado types but situated them geographically (Galindo-Tovar et al. 2007). Friar José Acosta, in 1590, in his "Historia Natural y Moral de las Indias" differentiated between the Mexican avocado and the one from Peru. He described the avocados from Peru as big fruits with a hard shell that peels easily and the ones from Mexico as mostly small with a thin shell that peels like an apple (Acosta [1590] 1985). These descriptions agree, respectively, with the West Indian and Mexican avocados described by Bergh and Ellstrand (1986). It is also interesting the way Friar Bernabé de Cobo, in his book "Historia del Nuevo Mundo" ([1653] 1956), described three different avocado types: "The palta in Yucatan is a tree of very attractive appearance, of the size of a large fig tree; its leaf is similar to that of the mulberry and its fruit is one of the finest in the Indies; in some regions it becomes as big as a small squash or large citron. The palta has a thin skin, more tender and flexible than that of a Ceuta lemon, green externally and when the fruit is quite ripe, peeling readily. It has large seed, either in the Indies or Europe, spindle shaped, a reddish white substance, tender like the meat of a chestnut and covered with a grayish parchment. It has the flavor of bitter almonds and when pressed it yields an oil like that of the almond. Between the seed and the outer skin is the meat, slightly thicker than one's finger except at the neck where it is very thick. It is of whitish green color, tender, buttery and very soft. Some people eat it with sugar or salt, others just as it comes from the tree, it being of such good flavor that it requires no seasoning. And although it is very pleasant to taste, it should be eaten in moderation because it is considered to be heavy and indigestible. The best paltas come from hot, dry regions; in Peru they grow in the Valley of Lea. A second kind of paltas is a large, round one that is produced in the province of Guatemala and which does not have as smooth skin as the first. The third is a small palta found in Mexico, which in size, color and form resembles a breva fig (first fruit of the fig tree); some are round and others elongated, with skin as thin and smooth as that of a plum. In some regions they cut the immature Palta in small bits and put it in brine, to take the place of olives. The tree wood is useful in building and for fuel". These avocado descriptions also resemble the ones made by Bergh and Ellstrand in 1986: the palta in Yucatan would be the West Indian avocado; the second kind of palta the Guatemalan type; and the third the Mexican type. Current varieties and rootstocks for avocado cultivation in the world are the products of various breeding programs based on exploration, collections, conservation and evaluation trials throughout their regions of origin and dispersion, (Mijares and López 1998; Knight 2002). Indeed, many modern commercial plantings are new varieties and cultivars, obtained by hybridization of various materials collected in Mexico and Central America. The description of avocado cultivars, methods of vegetative propagation, cultivation and varietal improvement are well documented (Knight 2002; Lahav and Lavi 2002).

1.2 Origin and Distribution

The genus *Persea* belongs to the Lauraceae family, one of the oldest known flowering plants. At present, about 81 species of *Persea* are recognized as valid. Most of the *Persea* species originated in the New World, but *P. indica* apparently originated in the Canary-Madeira-Azores islands. Some species originated in Southeast Asia. *Persea* has been divided further into subgenus *Persea* (includes *P. americana*, the commercial avocado) and subgenus Eriodaphne (a group of species of which most are immune to avocado root rot, but unfortunately are graft incompatible to avocado). Nearly all have tiny fruits lacking palatable flesh (Kopp 1966; Storey et al. 1986). Nearly all species within the genus *Persea* are native to the Americas (Kopp 1966). The species are scattered from northern Mexico through the southeastern United States, east through the West Indies, south through Central America, Colombia, Venezuela, the Guyana's and Brazil as well as Ecuador, Peru, Bolivia and Chile. How one species (*P. indica*) got to the Canary, Madeira and Azores Islands and how several *Persea* got to southeastern Asia, is not well understood (Kopp 1966; Williams 1976; Bergh and Ellstrand 1986; Scora and Bergh 1992; Landon 2009). From its natural habitat it has spread worldwide to almost all tropical and subtropical regions of the world. Plant distribution and taxonomic evidence are compatible with the assumption that the avocado originated in south central Mexico or nearby region (Storey et al. 1986). Speciation has occurred mainly in the highlands, mountain regions of Mexico and Guatemala (North of Guatemala City and Antigua). Distribution of wild species of *Persea americana* expands from the Mexican highlands, through the lowlands of Central America (e.g. Guatemala and Costa Rica) (Popenoe and Williams 1947; Ben-Ya'acov et al. 2003; Knight 2002) (Fig. 1). Systematic botanists and ethnobotanists agree that the avocado was unknown off the continental mainland before discovery of the New World by Columbus. None of the early chroniclers of voyages in the West Indies mentioned it and the various tribes of Amerindians inhabiting the islands had no native name for it.

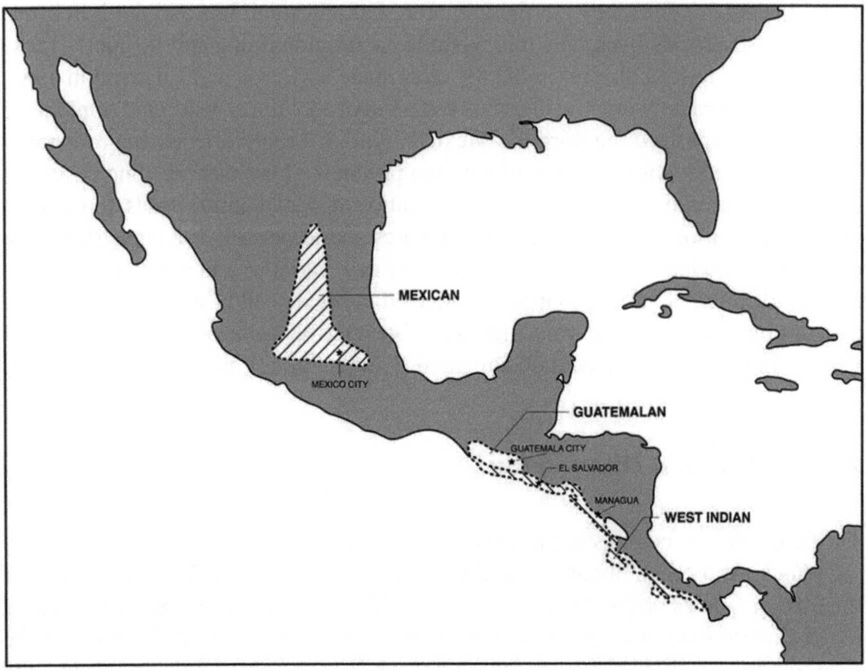

Fig. 1 Centers of origin for the three avocado subspecies (races)

Three main proposed centers of origin (center of biodiversity) for cultivated avocado i.e. Mexican center, Guatemalan center and the hot and humid lowlands of Central or northern South America center (Popenoe 1934; Storey et al. 1986; Galindo-Tovar et al. 2007; Chen et al. 2009).

1.3 Mexican Center

The primary gene center for *Persea americana* comprises the central highland regions of Mexico. The important species originating from this tract are *P. drymifolia, P. flocossa, P. floccosa, tolimanensis* and *P. zentmyerii*. Most of the Mexican avocado cultivars/varieties belong to *P. drymifolia* which were domesticated from wild *P. americana* found in México (Kopp 1966; Scora et al. 1970). However, the progenitors of these avocado varieties probably migrated from west Gondwana, Southern hemisphere (Scora and Bergh 1992; Galindo-Tovar et al. 2007; Landon 2009). The Mexican race, *P. americana var. drymifolia*, is thought to be the primordial race of commercial avocado.

Fig. 2 Avocado trees at the USDA National Germplasm Repository gene bank in Miami, FL

1.3.1 The Mexican Race

There are no pure Mexican varieties grown commercially for fruit in California (Bender 2012). However, genes from the Mexican race are important components in the Mexican-Guatemalan hybrid cultivars such as 'Hass' and 'Fuerte'. Two important traits from the Mexican race were passed on to the hybrids. These are the addition of more cold hardiness to the Guatemalan race and move on the harvest season of the Guatemalans by half a year (Popenoe 1919). This is the hardiest type in cultivation and presently the only one in California that ripens during late fall and early winter. It was introduced to California from Mexico, where it appears to be by far the common and abundant type. Some varieties have unusually rich flavor and excellent quality; in others there is an objectionable amount of fiber in the flesh. The oil content runs as high as 33% and averages considerably higher than other types. The tree (Fig. 2) usually growths vigorously and is, very resilient, withstanding in some instances temperatures as low as 18° or 20° without injury. Because the fruits are usually under one-half pound in weight, the tree is able to carry an immense number of them, 4000 sometimes being produced in a single crop. The fruits are usually oval or pear-shaped, ranging from 3 to 10 ounces in weight and green or dark purple in color. The skin is about as thick as that of an apple. The seed is sometimes loose in the cavity, with loose seed coats. The plant and fruits are characterized by an anise-like odor. If picked at the proper time, fruits of this type can be shipped reasonable distances without difficulty, but they do not hold up so well in market as do the thick-skinned. They are excellent fruits for home use, and because they ripen at a season when no other type is available in the market, they are commercially attractive. The tree comes into bearing earlier than other types, sometimes at 2 or 3 years from seed, and in the case of budded trees, usually within 2 years. Seeds from the Mexican race of avocado have been used as rootstocks in California since the beginning of the industry. Nurserymen like the big seeds and fast-growing qualities and growers have found that Mexican rootstocks usually have better and more consistent production than the Guatemalan and West Indian rootstocks. Of the three races, Mexican seedling rootstocks perform best in the colder soils. Mexican rootstocks are the least tolerant to soil salinity. There are no pure Mexican varieties grown commercially in California or Florida (Bender 2012; Crane et al. 2013). However, genes from the Mexican race are important components in the Mexican-Guatemalan hybrid cultivars such as 'Hass' and 'Fuerte' (Bender 2012). Two important traits from the Mexican race are imparted into the hybrids. These are the addition of more cold hardiness to the Guatemalan race and advancing the harvest season of the Guatemalans by half a year (Bergh and Ellstrand 1986; Bender 2012).

1.4 Guatemalan Center

This center includes Guatemalan mountains tracts and lowlands. *P. nubigena* and *P. nubigena var guatemalensis*, belong to this center (Popenoe 1935; Storey et al.1986). The Guatemalan race apparently developed in the highland region of Guatemala north of Guatemala City and Antigua (Popenoe 1917).The Guatemalan race of avocado is native to the highlands of Central America and is less cold tolerant than the Mexican race. Their leaves have no anise scent, and the young foliage is often reddish. The seed is almost never loose in the cavity. A characteristic of the Guatemalan race is the much longer time to fruit maturity (compared to the other races). Guatemalan cultivars such as 'Nabal' and 'Reed' may take 15 months or more from bloom to maturity (Bergh and Ellstrand 1986). Historically, this characteristic was used in California to advantage to stretch out the harvest season: hybrids with strong Mexican traits are harvested in the winter; Guatemalan cultivars such as Reed, Nabal, Dickinson, Queen and Anaheim are harvested 6–9 months later, in the summer; and Hass (a mostly Guatemalan hybrid) fills in between the two seasons (Bender 2012). A disadvantage to the Guatemalan cultivars is their thicker, woody skins (not all cultivars have this trait). One of the problems with the thick skins is that the consumer cannot tell when the fruit had softened enough for eating (Popenoe 1935; Schieber and Zentmyer 1980); (Bergh and Ellstrand 1986). The hard-shelled 'Dickinson' cultivar is tested for softening by inserting a toothpick into the stem end of the fruit (Bender 2012). The rigidity of the peels does not allow easy peeling; peels are "chipped-off" instead of peeled. The popular 'Reed' cultivar has a thinner skin, but is still very shell-like (Popenoe 1935).

1.4.1 The Guatemalan Race

Commercially, this is doubtless the most valuable type cultivated in California and is the most extensively planted variety, which is peculiar in that it carries its fruits through the winter and into the following summer, thus requiring 12–16 months to ripen (Bender 2012). While the Mexican type blooms in winter and ripens its fruits the following summer and the West Indian type blooms in spring and ripens its fruits in summer, the Guatemalan type blooms in late spring and hold its fruits over the following winter, sometimes as late as September and October of the following year (Bergh and Lahav 1996). The varieties of this type cultivated in California ripen from February to September. For fall and early winter other cultivars must be grown, unless afterward, cultivars of the Guatemalan type which will extend the season are acquired (Bender 2012). The California representatives of this type originated in Guatemala and in Southern Mexico, mainly in the vicinity of Atlixco, state of Puebla, Mexico, at an altitude of nearly 6000 ft above sea level (Popenoe 1919, 1935; Bergh and Lahav 1996). A large proportion of the varieties that originated in California came from seeds imported from Atlixco by John Murrieta of Los Angeles about 1900. Mr. Murrieta's work has had a more profound influence on California avocado culture than any other. Because of the Mexican origin of numerous Guatemalan varieties, some think this name to be inappropriate (Popenoe 1915). Inasmuch as these thick-skinned Mexican varieties belong to the Guatemalan type, however, they should certainly be called by this name, as it serves to show their relationship to other varieties of the same type. The characteristics that distinguish this type from the others are consistent, although it is sometimes difficult to distinguish one of the thinner skinned Guatemalan fruits with an almost smooth surface from a fruit of the West Indian type (Popenoe 1920, 1935). When the surface of the skin cannot reliably identify a fruit, the Guatemalan can usually be distinguished by the color of the fruit and by the character of the seed and its coats (Popenoe 1919, 1935; Bergh 1975). The tree is easily distinguished from that of the Mexican type by the absence of anise like fragrance in the leaves (Popenoe 1935; Kopp 1966; Bergh 1975; Bergh and Lahav 1996). The type seems to be about equidistant in hardiness between the West Indian and the Mexican. There is a slight difference in toughness among the different Guatemalan cultivars. The fruits of this type have a thick skin that is often woody and brittle. The surface is usually rough, sometimes covered with wart-like protrusion around the base, but in a few cases almost smooth. The flesh is usually free from fiber and of good flavor. It is scarcely as rich and oily as the average fruit of the Mexican type, but is very pleasant and satisfactory overall. The seed is usually not large in proportion to the size of the fruit and is almost never loose in the cavity (Popenoe 1919, 1935; Bergh 1975; Bergh and Lahav 1996; Nakasone and Paull 1998; Ospina 2002; Janick 2005).

1.5 West Indian Center

This center includes the hot and humid lowlands from Guatemala through Costa Rica and northern South America (Peru, Ecuador, Colombia and Costa Rica). According to Kopp (1966) *Persea americana var americana* Mill. is widespread in this center. *P americana var americana* is known by many names (*syn. P. gratissima, P. edulis, Persea persea, Persea lyogina* and *Persea americana var. angustifolia.* In spite of its name, this is the most tropical of the three races, and is assumed to have developed somewhere in this area.

1.5.1 The West Indian Race

West Indian avocados are prevalent throughout the lowlands of Tropical America. Naturalized seedlings occur in suitable habitats, particularly in association with human activity, along roadsides, or as remnant trees left after the clearing of forests. Although of local interest and potential genetic value, collecting these highly variable and often horticulturally inferior individuals is not realistic in light of the archeological evidence for the selection for specific horticultural traits across the last 10,000 years (Smith 1966, 1969). The leaves have no anise scent, and fruit size ranges from small to very large. Seeds are relatively large and are sometimes loose in the cavity. The fruits have relatively low oil content and are frequently reported to be "watery" by customers (Popenoe 1935). This race is the most cold-sensitive of the avocado races; and pure cultivars do not grow well in California (Bender 2012). Selections of this race appear to have greater salt tolerance and may be useful as rootstocks if selections can be found that can tolerate the colder soils in California. Trees of the West Indian type are more prone to frost than those of the Guatemalan type; vastly more so than the thin-skinned, small fruited Mexican type (Popenoe 1934, 1935). Its foliage it is often difficult to distinguish from the Guatemalan, but as a rule the leaves are somewhat smaller in California, they are also crowded more closely together on the branches and the wood is of lighter in color (Bender 2012; Boris et al. 2013). The foliage, too, is usually a lighter shade of green. The fruits of this type vary in form, being pear-shaped, oval or round. In color they are usually yellowish green or maroon (Popenoe 1935). The skin is leathery and separates readily from the flesh, but it is not as thick or woody as in the Guatemalan type. The flesh is often rather watery in seedlings and not as richly flavored as in the Guatemalan type, but in some of the best varieties the quality is good (Popenoe 1952). There is one defect that is not usually found in the Guatemalan type, i.e., the seed is large and often loose in the cavity. It seems that this type would not become of commercially important in California (Popenoe 1928). It is the principal one cultivated in Florida, the well-known varieties, 'Trapp' and 'Pollok' being representatives of it (Campbell and Malo 1979; Knigth and Campbell 1999; Crane et al. 2013). It is the one cultivated in Cuba and other West Indian Islands and along the coast of Central and South America. West Indian cultivars and hybrids are well-adapted to southern Florida and provide fruit on the market just ahead of the California winter fruit (Campbell and Malo 1976; Knigth and Campbell 1999; Crane et al. 2013). Selections have been made in Florida to stretch out their harvest season, but a given cultivar will be on the market for just a few weeks. Thus, the Florida industry relies on early, mid and late season West Indian cultivars (Wolfe et al. 1949; Campbell and Malo 1976; Knigth and Campbell 1999; Crane et al. 2013).

2 Taxonomy

The genus is of African Gondwanaland origin and its ancestral species was distributed to Asia and via Europe to North America and via Antarctica to South America probably in Palaeocene times (Scora 1992); Galindo-Tovar et al. 2007). When the Americas joined in the late Neocene, the genus was again united. Mountain building in Central America created new habitats in which, speciation could take place (Kopp 1966; Scora and Bergh 1992). In subgenus *Persea* three species are recognized, *P. schiedeana* Nees, *P. parvifolia* Williams and *P. americana* Mill. The latter species is polymorphic and consists of several separate taxa that may be considered botanical varieties or subspecies, and which are referred to as 'horticultural' races in the popular literature (Scora et al. 2002). Within this group are the varieties that make up the commercial avocados, namely *P. americana* var. *americana* Mill., the West Indian or Lowland (Scora and Bergh 1992) avocado; var. *drymifolia* (Schlect+Cham.) Blake, the Mexican avocado; and var. *guatemalensis* Williams, the Guatemalan avocado; all three are regarded as geographical ecotypes. Additionally, var. *nubigena* (Williams) Kopp, var. *steyermarkii* Allen, var. *zentmyerii* Schieber and Bergh and var. *tolimanensis* Zentmyer and Schieber are recognized as separate varieties of *Persea americana* Mill (Bergh and Lahav 1996). These latter four are believed to have contributed to the ancestry of var. *guatemalensis* (Schieber and Bergh 1987). Another wild botanical variety is *floccosa* Mez. Doubtless more taxa will be described as germplasm exploration continues (Ben-Ya'acov et al. 1992). One new taxon is the endemic form of *P. americana* var. *americana* in Costa Rica, where the typical vars. *drymifolia* and *guatemalensis* are almost absent. This endemic '*aguacate de montana*', which is intermediate between the Guatemalan and West Indian (Lowland) avocados, should be recognized as *P. americana* var. *costaricensis* according to Ben-Ya'acov et al. (1995).

3 Biodiversity and Genetic Resources

Biological diversity is the variety in life on the earth that includes all genes, species and ecosystems. Biodiversity provides recreational, psychological, emotional and spiritual enjoyment to human beings (Tuxill and Nabhan 2001; Borokoni 2013).

Plant genetic resources of cultivated crop plants as well as wild relatives have significant value to mankind, as they provide food, fuel, shelter and industrial products (Gross and Olson 2009; Vaughan et al. 2007). Furthermore, plant breeders require genetic variation (genotypes) for plant improvement. Genetic diversity in wild relatives is very essential because they contain genes resistance to biotic and abiotic stresses. Thus, all unique accessions need to be collected, characterized and preserved (Engelmann 1991; Ben-Ya'acov et al. 1992; Vaughan et al. 2007; Engelman 2013). Genetic diversity is the key component of any agricultural production system. The diversity in plant species consists of traditional varieties, wild relatives and other wild species (Altieri et al. 1987; Merrick LC 1990; Hoisington et al. 1999). Many factors are relevant to the role of landraces in agriculture and biology such as physical environment (latitude, altitude, and climate) and social environment (monoculture, deforestation, population etc.) and species and its utilization (Tuxill and Nabhan 2001; Agora and Rao 1998; Padulosi et al. 2012; Borokini 2013). Further, avocado landraces practically disappeared and cultivation of the traditional varieties increasingly declined first in orchards and gardens and eventually in all areas Mexico and Central America. To meet the demand for more food and to provide a reservoir of genetic variation to the breeders for finding particular characters such as resistance of genes for diseases, pests and for adaptation to wider ecological conditions, it will be necessary to make better use of a broader range of the world's plant genetic diversity (Merrick 1990; Cooper et al. 2001; Vaughan et al. 2007; Miller and Gross 2011).

Germplasm resources of fruit plants are in some cases threatened to the point of extinction (Bennett 1965; Agora and Rao 1998; Normah et al. 2002). Such reductions have serious implication for food security in the long term. Therefore, conservation and sustainable use of genetic resources is important to meet the demand for future food security (Baranski 2013; CGIAR 2013)

3.1 Status of Genetic Uniformity/Variability in Genus Persea

The degree of genetic uniformity of avocado in Mexico and Central America is extremely high, and thus avocado production is extremely vulnerable to extinction (Ben-Ya'acov et al. 1995; Ben-Ya'acov et al. 2003). Most of the major cultivars belong to a single Mexican avocado species (*P. drymifolia*) (Popenoe 1935, 1963; Scora and Bergh 1992; Bender 2012). Approximately 85–90 % of commercial production is accounted for by a single cultivar, 'Hass' (Knight 2002; Bender 2012). Moreover, high genetic similarity exists among several Mexican and Guatemalan avocado cultivars ('Hass', 'Bacon') probably because they have the same progenitors and share the same genetic pool (Lavi et al. 2003; Schnell et al. 2003; Taah et al. 2004; Borrone et al. 2007). On the other hand, *P. drymifolia and guartemalensis* and subspecies that belong to Mexican and Guatemalan avocado have rich diversity. These are naturally occurring interspecific hybrids that are also found near the center of diversity/origin. Molecular studies of genetic diversity and relationships among limited numbers of *P. drymifolia and guatemalensis* varieties indicate a relatively high degree of genetic similarity (Clegg 1992; Sharon et al. 1997); however, the degree of similarity among major *P. americana* varieties is relatively low (Lavi 1991; Mhameed et al. 1997). The techniques of protein electrophoresis and isoelectric focusing of isozymes have been used to characterize genotypes. DNA-based molecular markers (e.g. RAPD, AFLP, RFLP etc.) or unique DNA primers (SSR, ISSR, SCAR, CAPS, SNPs etc.) are being used to assess genotypic diversity, genetic relationships and to identify unique genotypes (Schnell et al. 2003; Borrone et al. 2007; Mhameed et al. 1997; Sharon 1997; Fiedler et al. 1998; Abraham and Takrama 2013). Further recently Dr. David Kuhn at the USDA-ARS in Miami, Florida, has developed an avocado single nucleotide polymorphism (SNP)-array based on polymorphisms in the gene coding regions of Hass, Bacon, Simmonds and Tonnage (Arpaia and Harley 2012). This array contains 6000SNPs that have been positioned on the avocado genetic map. In addition, Dr. Luis Herrera's research group at the Centro De Investigacion y Estudios Avanzados, CIEA, Mexico, is continuing to work on the sequencing, assembling and annotation of the avocado genome. Using the initial genome sequence as a reference, additional avocado genotypes are being targeted for sequencing (Arpaia and Harley 2012). The potential of genus *Persea* remains largely unexplored in Mexico and Central America. Thus, a range of genetic variation exists in these subspecies. Like other subtropical and tropical fruit species, avocado exists in mountainous regions and lowlands of Mexico and Central America in cultivated, wild, naturalized and indigenous forms and a need to be explored (Ben-Ya'acov et al. 1992, 2003).

3.2 Threats to Persea feral germplasm

The genetic resources of avocado are very important for breeding programs of new rootstocks and cultivars, so it is urgent to rescue many types of avocados and related species from the rapid devastation of the forests and jungles. The genetic resources of plants, particularly fruit plants, are vanishing rapidly worldwide, resulting in a large-scale depletion of variability (Merrick 1990; Cruz-Cruz et al 2013; Olsen and Wendel 2013) There are many reasons for such depletion, including deforestation, road laying, urbanization and the introduction of new varieties. Fruit trees in Brazil, Mexico, Central and South America are also threatened by logging and fuel wood gathering (Miller and Gross 2011; Miller and Schall 2006). The Red List of the International Union for the Conservation of Nature (IUCN) currently contains only six *Persea* taxa threatened in different countries including Azores, Madeira and Canary Islands in Macronesia (*P. indica*), Ecuador (wild populations of *P. campii and comferta*,) (Muriel and Pitman 2004) although efforts have been made to maintain these genetic resources by *in situ* and *ex situ* preservations in these countries. Genetically uniform modern varieties are replacing the highly diverse local varieties and landraces in traditional agro-ecosystem, and primitive varieties are being discarded after termination of breeding projects. Bowman (1992) revealed that a low level of variability in cultivated avocado genotypes highlights the need to widen the genetic base of avocado germplasm as to preserve an adequate level of genetic extension in avocado breeding programs. *Persea* germplasm is a rich source of genetic variability that has accumulated through hybridization, mutation and seed-based propagation. Wild populations of *Persea* species are threatened in their natural habitat, with about several avocado cultivars lost in the nineteenth century, continuing even today (Ben-Ya'acov et al. 1992). Moreover, due to various pests and pathogens, a valuable and highly resistant to various abiotic stresses, avocado germplasm has been lost (Ben-Ya'acov et al. 1992). Such germplasm represents basis for tolerance to extreme cold, heat, drought, diseases and pests. Detailed attention to safety and preservation of genetic resources is needed on purely scientific lines for the diverse and rich avocado germplasm present in Mexico, Central America and northen South America (Pliego-Alfaro et al. 1987). Study to understand the diverse germplasm present in the area, its characterization through morphological and advanced molecular genetic studies, is also desired.

4 Characterization and Assessment of Germplasm

Genetic variation is indispensable for effective management and use of genetic resources (Vaughan et al. 2007). Diversity in fruit species is assessed on the basis of phenotypical morphological traits (Arora and Rao 1998) and molecular techniques based on isozyme patterns and protein profile like SDS-PAGE and by using DNA based molecular techniques like RFLPs, AFLPs, RAPDs, ISSRs; SSRs and SNPs, etc. The information regarding genetic diversity in *Persea* germplasm characterized by these methods is useful for efficient management, preservation and also for breeding purpose (Ashworth and Clegg 2003).

4.1 Morphological Characterization

Polygenic morphological and pomological characteristics serve as genetic markers for germplasm characterizations (Ahmed et al. 2009). The evaluation of morphological characters unaided to evaluate genetic diversity may not be very effective. Wide genetic variability in avocado has already been observed in characterization of seedlings for morphological, physiological and phenological traits (Ayala-Silva et al. 2007; Guillen et al. 2011; Nkansah et al. 2013). The diversity based on phenological and morphological characters usually varies with environment. Evaluation of these traits requires ample characterization of genotypes prior to identification (Altieri and Merrick 1987). Morphological and phenological characters aid in the collection of basic information for improvement through breeding program and further evaluation (Ayala-Silva et al. 2004; Postman et al. 2006; Ayala Silva et al. 2007; Sato et al. 2010). Systematic characterization and evaluation of plant genetic resources are fundamentals for effective use of the materials through conventional methods or modern techniques.

Polygenetic and environmental factors need to be evaluated, as has been done in many fruit crops, such as sweet cherry (Hansche and Beres 1966), peach (De Souza et al. 1998) and grapes (Sato et al. 2000). Variability in morphological characteristics is very important for evaluation of avocado genotypes. A majority of the wild avocado genotypes are highly heterogeneous and a number of ecotypes can be distinguished, for instance variability in leaves, flowering, fruits, nutritional composition and resistance against insects and diseases (Williams 1976). Each ecotype needs to be characterized for all existing variation within genotypes (Chen et al. 2009; Borrone et al. 2007; Ahmed et al. 2009). Quantitative characteristics of leaves alone are sufficient to identify avocado species (Bergh 1969; Bergh 1980; Storey et al. 1986). Genotypes of avocado are classified into two groups based on the time of blooming, i.e. early, medium and late flowering types and flowering is based on the type of flower (A or B). Although blooming time is under phylogenetic control (Peterson 1956), variability exists in flowering period (Bergh and Ellstrad 1986; Bergh and Lahav 1996). Flowering time is also affected by environmental factors like mean temperature, differences in chilling prerequisite for breaking bud dormancy, age and vigor of tree and catastrophes, like heavy rainfall and hail storms (Peterson 1956). Diversity in fruit shapes of wild and cultivated varieties exists in both mexican (*P. drymifolia*) and Guatemalan (*P. guatemalensis)* avocados (Popenoe 1920, 1935; Berg and Ellstrand 1986). Fruit morphology is reportedly mostly under polygenetic control (Storey et al. 1986). However, variation in other quantitative and qualitative characteristics of fruit like nutritional components (sugars, soluble contents, vitamins, organic acids) and other quality parameters (aroma, texture, flesh color, flavor and keeping quality) also exist among genotypes (Knight 2002).

Significant differences in morphological characters of several cultivars of avocado (Popenoe 1919, 1920; Popenoe and Williams 1947) have been observed. Storey and Berg (1986) found that several cultivars exhibited remarkable diversity in their tree, foliar, flowers, fruit characteristics, harvesting time and shelf life. The assessment of the genetic diversity in fruit species of seedling origin might assist in the conservation of original habitats and also new sources of genes for resistance and fruit quality, thought to be present within these materials and be available for breeding program (Ben-Ya'acov et al. 2003; Ben-Ya'acov 1998). Gomez-Lopez 1999 and Gomez-Lopez 2002 characterized 190 accessions of wild avocados collected from different agro-ecological zones of Mexico for preliminary evaluation and long conservation in the field gene bank. Ben-Ya'acov et al. (1992) explored various areas in Mexico and Central America for several wild and local varieties of avocado and thought that high-yielding varieties could be selected and a useful gene bank could be developed in Mexico and Israel for breeding purposes. Morphological characterization provides a base and a thought of germplasm of specific plant species about variability and genetic diversity. These characters can be used as a foundation for additional molecular level studies to confirm these morphological characters of the studied germplasm (Arpaia 1997; Cooper et al. 2001; Schnell 2003, 2007).

5 Conservation of Germplasm Introduction

Worldwide awareness of the value of plant genetic resources has increased over the last few decades, along with the acknowledgement that they are under considerable threat (Normah et al. 2002; Miller and Schaal 2006; Drew et al. 2007; Padulosi et al. 2012). Several major international conventions, commencing with the Convention on Biological Diversity (CBD) in 1992 (CBD 1992), have sought to provide a forum for the concerned parties to bring issues of conservation of plant biodiversity into sharper focus at many levels. By 1995, more than 4 million plant accessions had been collected worldwide, consisting mostly of commercial plant species and their wild relatives (FAO 2013). Although this is impressive, several limitations remain. Over 90% of collections worldwide are stored as seed and over 50% consist of grains (Baranski 2013; Walters et al. 2013). Fruit crops and particularly tropical fruit species such as avocado, mango, mangosteen, litchi that are difficult to store as seed, are inadequately represented in national and international collections (Normah et al. 2002). The wild relatives of these species are nearly absent from most collections. Currently, the most appropriate method for conserving these genetic resources is by establishing them on field ex situ in genebanks under living field collections or in situ, where they remain protected (Altieri and Merrick 1987; Normah et al. 2002). Worldwide there are hundreds of tropical fruit species that are important for income generation, nutrition, medicine, timber, fuel and livestock feed (Cooper et al. 2001; Drew et al. 2007; Vaughan et al. 2007). In addition to the commercial fruit industry, there is also a wide range of native tropical and subtropical fruit species that have commercial potential are currently in a status of low economic importance (Arora and Rao 1998; Normah et al. 2002). Conservation of tropical fruit species is often difficult or impossible by traditional methods, such as seed banks or field genebanks. Many species have recalcitrant seeds that are high in moisture causing loss of viability when they are desiccated and chilled for storage purposes (Chen et al. 1989; Walters et al 2013). The seeds of other species have either no natural dormancy mechanism (i.e avocado, mango and rambutan), or have a short life of only a few days (mangosteen). Some species have no seeds (banana and breadfruit). *Nephelium lappaceum* and *N. ramboutan-ake* are seasonal fruit species that produce recalcitrant seeds. Thus, their seeds do not tolerate freezing temperatures and are desiccation sensitive (Normah et al. 2002). Currently, there is no single method available for the long-term storage of many of these seeds. Recently developed biotechnological techniques offer alternative strategies, including slow growth in vitro (in tissue culture) for medium-term conservation and cryopreservation (in liquid nitrogen, − 196 °C) for long-term conservation (Solarzano 1989; Shibli et al. 2006; Reed 2008; Volk 2012) on the other hand, banana is the only tropical fruit species for which large numbers of accessions are currently held in vitro genebanks. For the most part tropical fruit species are held in field collections, which are vulnerable to disease, insect attack and natural disasters, such as typhoons (hurricanes), floods, cold weather and volcanic activity (Ayala-Silva and Schnell 2010; Ayala-Silva et al. 2013). Procedures to preserve them by in vitro culture and cryopreservation are mostly emerging.

Thus, the lack of appropriate methods for long-term and sustainable conservation is the major hurdle in effective conservation and use of tropical fruit species genetic resources (Arora and Rao 1998; Normah et al. 2002). Plant genetic resources in horticultural crops and their wild relatives are of immense value to mankind as they provide food, fuel, shelter and industrial products (Drew et al. 2007; CGIAR 2013). The plant breeders require reservoir of genetic variation (gene pools) for crop improvement. The larger the reservoir, the better are the chances of finding particular characters such as resistance genes for diseases, viruses, fungus and insects or for adaptation to wider ecological amplitudes and stress conditions (Prance and Nesbit 2005). However, in the wake of spread of high yielding varieties, this genetic variability encompassing landraces is increasingly getting eroded resulting in the large scale weakening of variability. This situation thus demands priority action to conserve such germplasm (Prance and Nesbit 2005; Vaughan et al. 2007; Gross and Olsen 2009). Worldwide apprehension about loss of precious genetic resources has encouraged international action to save prized food sources and breeding materials (Drew et al. 2007; CGIAR 2013; Hummer et al. 2011). Various countries have realized and paid attention to collect and preserve genetic resources. Consequently, repositories have been established in China, USA and France and the germplasm has been evaluated for agricultural and biological characters (CGIAR 2013). The process of genetic erosion in *Persea* germplasm has been recognized, and attempts have been made to conserve such a precious germplasm with minimal cost and resources and appropriate methods (Campbell and Malo 1976; Ben-Ya'acov et al. 1999; Ben-Ya'acov et al. 2003; Crane et al. 2013).

In vitro culture method has high degree of genetic steadiness and symbolizes an aseptic resource of plant multiplication (Reed 1990; 1995; Sahijram and Rajasekharan 1996; Reed et al. 2004; Reed 2008). A number of fruit plants are normally propagated vegetatively, and the clonal material carries variable gene combinations. These clones are maintained in field gene banks of which the traditional methods are expensive due to: (i) high labor costs, (ii) vulnerability to environmental hazards and (iii) requirement for large amount of space (Towill and Ross 1989). The most serious problem is the vulnerability of such clones to pests and pathogens, or natural disasters like earth quake and landslides, to which they are almost continuously exposed (Altieri and Merrick 1987). This can lead to unexpected loss of priceless germplasm. Such field gene banks do not represent the entire range of genetic variability within the respective genus (Ben-Ya'acov et al. 1995; Hoisington et al. 1999; Gross and Olsen 1999; Janick 2005). This has led to the consideration of *in vitro* techniques and cryopreservation for germplasm conservation (Janick 2005; Cruz-Cruz et al. 2013).

5.1 *In Vivo Conservation*

(Ex *situ)* conservation includes germplasm banks, botanical gardens, seed banks, DNA banks and techniques involving tissue culture, cryopreservation; incorporation of disease, pest and stress tolerance traits through genetic transformation and ecological restoration of rare plant species and their populations (Merrick 1990; Lavi et al. 2003; Miller and Schaal 2006; Miller and Gross 2011). *Ex situ* conservation has gained international recognition with its inclusion in Article 9 of the Convention on Biological Diversity (CBD 1992). Much of the diversity present in domesticated avocados is currently maintained in germplasm repositories and amateur collections, including a broad spectrum of the selection and propagation of superior genotypes derived from open-pollination progenies (i.e., 'chance seedlings') (Ben-Ya'acov et al. 1992; Postman et al. 2006; Ayala-Silva and Schnell 2010; Ayala-Silva et al. 2013). Avocado genetic resources are maintained ex situ in field repositories at great cost and are always under the threat of bad weather, pests and disease (Darda and Litz 2003; Ayala-Silva et al. 2004, 2010; Drew et al. 2007; Ayala-Silva and Schnell 2010; Ayala Silva et al. 2013). Although vast diversity of cultivars exists, avocado production worldwide is now largely based on the cultivation of a small number of commercial and edible cultivars, which are grafted onto less than a dozen different clonal rootstocks (Ben-Ya'acov et al. 1992; Ben-Ya'acov et al. 1995, 1999; Knight 2002; Bender 2012). Indeed, the major cultivars being grown in Florida, Hawaii and some other areas with similar climates are first or later generation hybrids of Guatemalan and West Indian races (Knight 2002; Crane et al. 2013). The main cultivars grown in California, Israel, South Africa and Australia and similar less tropical climates are hybrids of the Guatemalan and Mexican races (Bergh 1976; Knight 2002). Conservation of old varieties and native wild avocado species serves as a stock and source for gene banks for fruit propagation (Ben-Ya'acov et al. 1992; Ben-Ya'acov et al. 2003; Guillen et al. 2011). Identification, collection and preservation have particular importance in their maintenance (Ayala-Silva and Schnell 2010; Ayala-Silva et al. 2013). *In vivo* management of fruit trees in the form of orchard/ field trees or by establishing their nurseries allow confirmation of their morphological identity. This germplasm stock can be evaluated for resistance to pest, diseases and adverse environments (Ben-Ya'acov et al. 2003; Ayala Silva et al. 2004; Ayala Silva et al. 2007; Ayala Silva et al. 2010). Establishment of fruit nurseries of local germplasm and archaic cultivars is a way of fruit germplasm maintenance, in addition to botanical gardens, park, reserves and orchards. It is an effective method of germplasm collection and preservation to develop a fruit repository (Arora and Rao 1998; Normah et al. 2002; Ayala-Silva et al. 2004; Ayala Silva et al. 2007, 2010). *In vivo* conservation of genetic diversity of the existing collections in Mexico, Guatemala and other Central American countries (i.e. Nicaragua, Costa Rica, Honduras, El Salvador) needs to be reviewed for germplasm conservation and collection (Ben-Ya'acov et al. 1992; Ben-Ya'acov et al. 2003; Ayala-Silva et al. 2004; Ayala Silva et al. 2007; 2010; Guillén et al. 2011). The National Clonal Germplasm Repository (NCGR) in Miami, Florida holds three major taxa of *Persea* and their hybrids (Ayala-Silva et al. 2004; Ayala Silva et al. 2007, 2013; Ayala-Silva and Schnell 2010; Ayala Silva et al. 2013). Wild species around the world are represented by clonal collections and by *in vitro* preservation. Clonal genotypes including varieties cultivated for fruit and used as rootstocks are preserved as living trees, with nearly 5 % of the clonal material duplicated as *in vitro* cultures (Sandoval and Villalobos 2002; Witjatsono et al. 2004). At the Miami germplasm repository, over 330 accessions of avocado cultivars are maintained; another 300 are maintained by the Fairchild Tropical Gardens Farm in Homestead Florida and by the University of California at the South Coast station in Irvine, California. In addition, Mexico and Israel maintain their own collections (Ben-Ya'acov et al. 2003; Ayala-Silva et al. 2004, 2007, 2010, 2013; Guillén et al. 2011).Their seedlings and offspring have been selected and were evaluated for different useful characteristics (Schnell et al. 2003). Different agro-ecological habitats in Mexico and Central America are very genetically diverse in tropical fruits, especially avocado. Collection and preservation *in vivo* are essential to preserve the genetic diversity for further use (Ben-Ya'acov et al. 2003; Guillén et al. 2011).

5.2 *In vitro Conservation*

The term tissue culture is commonly used in a very wide sense to include in vitro culture of plant cells, tissues as well as organs. Plant tissue culture is a method or technique to isolate parts of plants (protoplasm, cells, tissues, and organs) and grow them on artificial media in aseptic conditions in a controlled space so that parts of these plants can grow and develop into complete plants. Among other applications, tissue culture has been successfully linked to conservation and exchange of germplasm in horticulture. Avocado genetic resources are maintained ex situ in field repositories at great cost and always under threat of inclement weather, pests and diseases (Ben-Ya'acov et al. 2003; Ayala-Silva et al. 2004, 2010, 2013; Guillén et al. 2011). Cryopreservation is an important alternative method for long-term conservation of plant genetic resources (Reed 1990, 2002, 2004, 2008; Witjaksono et al. 2004). Moreover, it is an important storage method for biotechnology research, in which experimental materials (i.e., embryogenic cultures) lose morphogenic competence relatively quickly and cannot be stored reliably in vitro (Sahijram and Rajasekharan 1996; Reed 2001; Efende and Litz 2003; Witjaksono et al. 2004; Reed et al. 2005). The conventional methods of maintaining avocado germplasm under field, nursery and orchard conditions require extensive space and labor (Sandoval and Villalobos 2002; Ayala-Silva et al. 2010; Padulosi et al. 2012).

It is very difficult to conserve avocado germplasm by typical methods. *In vitro* tissue culture and cryopreservation techniques have been developed in the past decades; in vitro techniques have been widely developed for over thousands of different species (Reed 1990, 2004; Hummer et al. 2012). Tissue culture systems allow for axenic propagation of plant material with high multiplication rates. Virus-free plants can be obtained through meristem culture, in combination with thermotherapy (Reed 2008; Kapai et al. 2010; Hung and Trueman 2011; Volk 2012). The miniaturization of explants reduces space requirements and consequently labour costs for maintenance of germplasm collections (Mohamed-Yasseen 1993; Zulfiqar et al. 2009). Organized culture systems have a high degree of genetic stability and are more likely to be important for germplasm storage, especially of shoot tips or meristem cultures (Towill and Ross 1989).

In vitro techniques are employed to eliminate diseases. While most fungal and bacterial diseases are eliminated during surface sterilization and culture, viruses and viroids survive through successive multiplication if the mother plant is infected (Reed 2002; Kaviani 2011; Volk 2012). Most can be eliminated by meristem or shoot tip cultures, perhaps in combination with heat or cold therapy (Reed 2004; Volk 2012). Generally, two approaches are used to maintain germplasm collections *in vitro:* (i) minimal growth and (ii) cryopreservation (Reed 2002, 2004, 2008). *In vitro* culture has yielded positive results when juvenile material is used (Cooper 1987; Reed 1990, 2002, 2005; Barceló-Muñoz et al. 1990; Sahijram and Rajasekharan 1996; Reed 2002; Reed 2005); however, low results are obtained with adult explants (Cooper 1987; Pliego-Alfaro et al. 1987). Breeding of avocado rootstocks has been executed almost exclusively by seed. However, this process cannot guarantee the genetic homogeneity of the material because avocado breeds by cross-pollination, which produces highly heterozygotic seedlings (Peterson 1956; Sauls 1994; Lavi and Lahav 2003; Chen et al. 2009). For this reason, clonal propagation is essential when standardized plant rootstocks are required. Through this technique, rootstocks can be propagated that are resistant to adverse conditions such as drought, floods, saline, alkaline or acidic soils, or to diseases such as *Phytophthora* root rot (Pliego-Alfaro et al. 1987; Guillén et al. 2011).

Tissue culture propagation of avocado using axillary buds of adult avocado plants has been described (Barceló-Muñoz et al. 1999; Mohamed-Yasseen 1993; Efendi 2003); however the morphogenetic ability is better in immature tissues than in adult material (Pliego-Alfaro 1988; Mohamed-Yasseen 1993; Efendi 2003). Shoot tips from avocado seedlings and juvenile tissues have been used for in vitro establishment of different avocado materials (Cooper 1987; Pliego-Alfaro 1988; Barceló-Muñoz et al. 1990; Barceló-Muñoz and Pliego-Alfaro 2003). However, the use of juvenile tissues has the disadvantage of genetic variability due to seed propagation of stock plants (Pliego-Alfaro 1988; Zulfiqar et al. 2009), a problem that can be avoided by the effective regeneration from mature avocado explants. Minimal growth conditions for short- to medium-term storage can be performed by reducing temperature from normal incubating temperature to below up to 0 °C depending on the species, addition of growth 22 retardants or by osmotic stress with sucrose or mannitol and maintenance of cultures at a reduced nutritional status (Efendi 2003; Efendi and Litz 2003). The advantage of this approach is that cultures can be readily brought back to normal culture conditions to produce plants on demand. Because methods of micropropagation are now well recognized (Towil 1989; Stushnoff 1991; Witjaksono et al. 1998; Kapai et al. 2010), it is possible to preserve genetic resources of woody species as shoot tip cultures *in vitro* under conditions of minimal growth storage (also called "slow growth" storage or growth suppression) or cryopreservation, which represents a reliable source for maintenance of orchards as a practice of fruit tree conservation (Forsline et al. 1998; AI and Luo 2005; Ellis et al. 2006; Hung and Trueman 2011; Kapai et al. 2010; Kaviani 2011; Kaczmarczyc et al. 2012).

Cryopreservation at ultralow temperatures, e.g. in liquid nitrogen (LN) (−196 °C) offers the possibility for long-term storage with maximal phenotypic and genotypic stability (Towill 1989; Ellis et al. 2006; Benson 2008; Reed 2008). This method is relatively convenient and economical for maintaining a large number of genotypes. To secure avocado germplasm collections, short to medium and long term, *in vitro* storage methods like minimal growth and cryopreservation may be required (Efendi 2003; Effendi and Litz 2003; Drew et al. 2007; Guzmán-Garcia et al. 2012). The first report of successful *in vitro* propagation of avocado was of disks from mature avocado fruits grown on agar nutrient media (Scroeder 1955). *In vitro* conservation techniques, using slow growth storage, have been developed for a wide range of species, including temperate woody plants, fruit trees, horticultural and numerous tropical species (Forsline et al. 1998; Normah et al. 2002; Barceló-Munoz and Pliego-Alfaro 2003; Benson 2008; Guzmán-Garcia et al. 2012). Low temperature (5–12 °C) is suitable for *in vitro* storage of meristem culture of many tropical and subtropical species (Mohamed-Yasseen 1993; Normah et al. 2002; Reed 2004; Guzmán-Garcia et al. 2012). These preservation techniques (slow growth) are less costly and more safe to conserve germplasm (Reed 2008). *In vitro* storage based on slow growth techniques represent alternative strategies to conserve plant genetic resources (Shibli et al. 2006; Rai et al. 2008; Reed 2008; Hung and Trueman 2011). This approach is particularly useful when seed banking is not possible, such as with vegetatively propagated plants, recalcitrant seed species and plants with unavailable or non-viable seeds due to damage of grazing or diseases and large and fleshy seeds (Kaviani 2011; Walters et al. 2013). Slow growth techniques for germplasm preservation have been studied in many countries with respect to various fruit genotypes that are propagated when desired (Ahmed et al. 2010). A large number of tropical and subtropical fruit crop species have highly recalcitrant seeds that often exhibit viability times of just a few weeks (Baranski 2013; Walters et al. 2013). Several vegetatively propagated tropical fruits, including woody perennials, need to be conserved with their genetic fidelity intact. Maintenance and preservation of germplasm of fruits in field genebanks is difficult, expensive and labor-intensive (Ben-Ya'acov et al. 2003; Ayala-Silva et al. 2004, 2010, 2013; Postman et al. 2006; Ayala Silva et al. 2010; Guillén et al. 2011). In addition, the connection may also be vulnerable to natural perturbations. Although conservation of seeds of economically important crops has been practiced since ancient times, it suffers from severe limitations, like low seed viability and heterozygosity.

6 The Avocado Tree

The avocado is a fruit native to the tropical America and subtropical regions of North and South America, where it has been used for thousands of years as a high-priced article of food (Fig. 2). From its natural habitat it has spread worldwide to almost all tropical and subtropical regions of the world (Popenoe 1963; Williams 1976; Nakasone and Paull 1998; Knight 2002; Ospina 2002; Janick and Paul 2008;

Nkansah et al. 2013). *Persea americana* Mill. (Avocado) is a tree crop which originated from the tropics of the western hemisphere and has three general ecological races: Mexican, Guatemalan and West Indian adaptable to a wide range of climatic conditions (Bergh 1969; Morton 1987; Knight 2002; Ospina 2002; Janick and Paul 2008). Avocado trees grow well in areas with over 150 mm of annual rainfall and between 55 and 550 m above sea level. The tree reaches a height of 9–18 m and trunk diameter between 30 and 60 cm. Avocado is a plant of periodic growth where growth rates are affected by local conditions. In areas of constant humidity, avocado grows all year. In drier or cold regions, the tree can go through four annual growth stages and during certain periods can lose a lot of foliage; the main stage usually coincides with flowering (Berg and Ellstrand 1986; Sandoval and Villalobos 2002; Knight 2002; Crane et al. 2013). The new growth of shoots or sprigs occurs only on certain parts of the tree. In years with greater growth, the fruit harvest will be reduced; and in several cultivars the yield is markedly biennial (Condit 1916; Popenoe 1920; Kopp 1966; Williams 1976; Knight 2002; Guillén et al. 2011). The shoots or sprigs are cylindrical or prismatic and have alternative leaves which have axillary buds. The shape of the leaves varies considerably depending on the position. Wild species can reach 20 m in height. It grows well in soils that are loose, well-drained, slightly acid, and rich in organic matter. The tree grows at elevations from sea level to 2400 m, with average temperatures of 16–24 °C and annual precipitation of 800–1700 mm. Wild trees have a spherical crown, the flowers emerge in panicles that sprout from the new growth on the apex of the sprigs or from the axil of the leaves (Popenoe 1920, 1935; Wolfe et al. 1949; Williams 1976; Nakasone and Paull 1998; Ospina 2002; Janick and Paul 2008; Ashworth 2011; Crane et al. 2013). The axis of the panicle is strong and pubescent and carries several deciduous bracts. The tree yields many thousands of flowers per plant. The panicles open up for long periods of weeks or months. However, the number of flowers that yield fruit is 5 % or more. *Persea americana* may be propagated by seedling or grafting. Grafting is recommended for commercial plantings because the fruits of grafted trees have uniform characteristics in size and shape. The terminal bud graft is the easiest and most successful. To produce healthy and vigorous trees, seeds should be selected from good-sized fruits. These seeds should have a higher coefficient of germination and the subsequent seedlings should grow faster. To prevent dehydration, seeds should be planted immediately after extraction from the fruits. Seeds may be preserved in wooden trays with humid sand between 5 and 7 °C. To prevent disease, seeds should be disinfected in hot water (49 °C for 15 min); the ground should be treated with water vapor (90 °C for 4 h or 60 °C for 6 h); and all tools should be treated, possibly with sodium hypochlorite (Wolfe et al. 1949). Avocado is frequently cultivated on hillsides because minimum temperatures are higher than those on flat ground (Bender 2012) while cultivated trees, originating from graft are more tolerant. The avocado tree may be erect, usually to 9 m but sometimes 18 m or more, with a trunk 30–60 cm in diameter, (bigger in very old trees) or it may be short and scattering with branches beginning close to the ground. Almost evergreen, being shed briefly in dry seasons at blooming time. Mexican race are strongly anise-scented. Seedling trees of Guatemalan (var. *guatemalensis*) and West Indian (var.

americana) ecotypes, especially in their native rainforest environments, can reach heights exceeding 30 m, while Mexican (var. *drymifolia*) seedling trees are shorter at 15 m.

6.1 Leaves

Leaves are approximately 7.6–41.0 cm in length and variable in shape (elliptic, oval, lanceolate). They are often hairy (pubescent) and reddish when young, then become smooth, leathery, and dark green when mature (Kopp 1966). The leaves are alternate, dark-green and glossy on the upper surface, whitish on the underside; variable in shape (lanceolate, elliptic, oval, ovate or obovate), 7.5–40 cm long (Morton 1987; Janick and Paul 2008).

6.2 Inflorescence

The avocado has perfect flowers, each capable of producing pollen and developing fruit. But a pollination peculiarity makes it undesirable to plant solid blocks of single cultivars (Wolfe et al. 1949). The many-flowered lateral inflorescences (structures that hold the flowers) are borne in a pseudoterminal position. The central axis of the inflorescence terminates in a shoot. Flowers are perfect, yellowish-green, and 3/8 to 1/2 inch (1–1.3 cm) in diameter. The flowers are hermaphroditic, actinomorphic, greenish-whitish, with short and pubescent pedicels. The perianth is made up of one involucrum, which has interpreted itself as a calyx consisting of six parts that are acute, yellow, pubescent on both surfaces, and arranged in two groups of three. The exterior parts are the largest. It is in fact three sepals and three petals of very similar appearance. There are 12 stamens in four cycles; the first two are external and simple filaments whose anthers open up through four pores located toward the center of the flower. The third cycle consists of three stamens with the pores opened outward; its filaments have, at the base, an orange gland or nectary. The fourth cycle, the innermost one, is made up of staminodia. The pistil is made up of an ovoid, monocarpic, superior, monospermic, unilocular, white, and pubescent ovary, which ends in a short style with a globose stigma. Low fruit yield occurs because the stigmatia receive a few grains of fecundating pollen when the stamens and the pistils in each flower do not mature uniformly (Kopp 1966; Schroder 1952; Peterson 1956; Williams 1977; Bergh and Ellstrand 1986; Scora and Bergh 1990, 1992; Scora et al. 2002).

The flower of avocado is protogynous, e.g. its pistil is receptive before pollen shedding. The flower opens twice for several hours each time and each opening is separated by at least one overnight period. The flower is female during the first opening and male during the second opening. Avocado cultivars fall into one of two flowering groups: A or B (Wolfe et al. 1949; Peterson 1956; Lahav and Gazit 1994). Group A—the first (female opening) starts in the morning and ends before noon. Second (male) opening occurs in the afternoon of the next day; and Group B is the reverse pattern; the female opening happens in the afternoon and male opening next morning (Lahav and Lavi 2002). Length of flowering varies with cultivar and climate. The cooler the temperature, the longer the flowering period. Guatemalan cultivars mostly bloom later than the Mexicans or West Indians (Wolfe et al. 1949; Lahav and Lavi 2002).

6.3 *Fruit*

The fruit is a berry, consisting of a single large seed, surrounded by a buttery pulp. Indian and Mexican cultivars mature their fruit 150–240 days after flowering, whereas the Guatemalan type take more than 260 days. The skin is variable in thickness and texture. Fruit color at maturity may be green, yellow-green, black, purple or reddish, depending on cultivar. Generally the pulp is entirely pale to rich-yellow, buttery and bland or nutlike in flavor. In shape, the fruit is usually pyriform to oval and round, and the fruit weigh from a few ounces to 5 lbs (2.3 kg). Oil content diverge from 7.8 to 40 % on a fresh weight basis. The seed is oblate, round, conical or ovoid, 2–2 ½ inch (5–6.4 cm) long hard and heavy, ivory in color but enclosed in two brown, thin, papery seed coats often adhering to the flesh cavity. Depending on race type (i.e. Mexican, Guatemalan) the seed may be tight or loose (Popenoe 1935; Wolfe et al. 1949; Morton 1987; Ospina 2002; Janick and Paul 2008).

The time between flowering and harvesting of fruits depends on the race; the Guatemalan cultivars between 8 and 10 months, Mexican cultivars 7–10 months, whereas the West Indian cultivars are between 5–6 months. The fruit does not generally ripen until it falls or is picked from the tree (Campbell and Malo 1979; Morton 1987; Ospina 2002; Janick 2005; Crane et al. 2013). In Florida, the fruit is considered adequately mature for harvest when it reaches a specified calendar date and weight or size. The specific dates, weights, and sizes used to establish maturity vary by cultivar (Wolfe et al. 1949; Campbell and Malo 1976; Crane et al. 2013). Some cultivars may be stored on the trees and harvesting may be manipulated according to marketing schedule. However, West Indian cultivars have little or no storage time, whereas the Guatemalans and Mexican cultivars have a longer tree-storage capability (Janick 2005). The characteristics of the fruit are very variable, depending on the race and the variety. Pear shaped fruits prevail, but spherical and ovoid fruits also exist. They are usually irregular, and the side with more fibers or vascular bundles is thicker. The pericarp is made up of a cortex whose thickness and color vary from yellowish green to purple or almost black; the surface varies from smooth and shiny to corrugated and opaque. The mesocarp is a pulpy, soft mass, yellowish-greenish-white in color, with green pigmentation close to the cortex. The ovoid seed occupies a large part of the fruit; it is made up of two pulpy cotyledons and a small embryo; it contains no endosperm. The time between flowering and harvesting of fruits depends on the race: for the West Indian race it is between 5 and 6 months, for the Guatemalan race between 8 and 10 months, and for the Mexican race between 7 and 10 months (Popenoe 1935; Morton 1987; Ospina 2002; Janick and Paul 2008). The fruits are collected manually using ladders and scissors or knives. Pulling the fruit off can injure and damage it. Because the fruits are delicate, they should not be put on the ground without protection. The stem should be cut close to the fruit to prevent damage to other fruits when packed. Fruits harvested early should be placed in the dark and refrigerated (Crane et al. 2013). The humidity content in seeds with harvest ripeness is approximately 65 % (Ospina 2002; Janick and Paul 2008). The seeds are recalcitrant and lose their viability 2–3 weeks after removal from the fruit. However, the fruits can be stored for periods of more than 8 months in a dry room at 5 °C (Popenoe 1920; Wolfe et al. 1949). Viability can also be maintained for several months by covering the seeds with a powdered fungicide and storing them in wet sawdust or peat in polyethylene bags at 4–5 °C (Popenoe 1920). The critical humidity content (the point to which one can lower the humidity of the seed without losing its viability) is roughly 57.6 % for slow drying and 57.4 % for fast drying (Ospina 2002). The seeds should be pretreated by immersion in water at room temperature for 24 h. About 70 % of the seeds germinate underground in an average of 21 days (Popenoe 1920; Wolf et al. 1949).

6.4 Pollination

Avocado flowers are bisexual, they have a unique flowering behavior and all cultivars and wild species irrespective of race, fall into of two categories. The female and male flower parts function at different times of the day. Varieties are classified into A and B types according to the time of day when the female and male flower parts become reproductively functional (Peterson 1956). Self-pollination occurs during the second flower opening when pollen from the anthers is transferred to the stigma of the female flower parts. Cross-pollination may occur when female and male flowers from A and B type varieties open simultaneously. Self-pollination appears to be primarily caused by wind, whereas cross-pollination is caused by large flying insects such as bees and wasps.

Varieties vary in the degree of self- or cross-pollination necessary for fruit set. Some varieties, such as 'Waldin', 'Lula', and 'Taylor' fruit well when planted alone. Others, such as 'Pollock' and 'Booth 8' (both B types) do not and it is probably advantageous to plant them with other varieties (A types) which bloom simultaneously to facilitate adequate pollination and fruit set (Campbell and Malo 1976; Morton 1987; Knight 1999; Ospina 2002; Janick and Paul 2008; Crane et al. 2013).

7 Cultivars

A set of morphological and ecological characteristics distinguishes each of the three botanical races of Avocado. The centers of origin for these three races are distinct (Popenoe 1934; Kopp 1966; Bergh and Ellstrand 1986; Knight 2002), and prior to the arrival of the Europeans to the Americas, it is thought that the genetic purity of each race was mostly conserved. This is attributable to the minimal social interchange between the Amerindian tribes that inhabited Mexico and Central America and by significant geographical barriers that exist between the centers of varieties origin (Ashworth et al. 2011; Galindo-Tovar et al. 2007; Landon 2009; Galindo-Tovar et al. 2008). Cultivars grown in tropical and subtropical environments are from the Mexican, Guatemalan and West Indian races and are more adapted to temperatures that exist in these areas. Exploration of Mexico, Guatemala, Honduras, El Salvador, Nicaragua and other countries of Central and Northern South America during early nineteenth century by horticulturists from the USA, provided superior seedlings to the states in California (CA) and Florida (FL) for commercial exploitation, laid the foundations for genetic improvement of avocado for production in these states (Bender 2012). This was somewhat supported by collecting material from already established orchards in Mexico and Nicaragua (Popenoe 1919, 1947, 1951, 1952, 1957; Schroder 1947; Bergh 1957). Avocado cultivars are selected and named because of particularly desirable fruit and tree characteristics. Every avocado tree grown from seed is a potential candidate for selection as a cultivar, because cross-pollination in avocado's reproduction process ensures genetic variability. Once selected, cultivars are propagated by vegetative methods, such as grafting, to produce genetically identical trees (Wolfe et al. 1949; Campbell and Malo 1976; Knight 2002).

7.1 Cultivar Selection Criteria

Important criteria in selecting avocado cultivars can be grouped into (1) production and (2) fruit quality considerations. For home gardens, production characteristics may not be as important as good fruit quality (Bender 2012; Crane et al. 2013). Whereas for commercial growers, production characteristics could be extremely important, but consumer preferences for fruit appearance and quality are also important in determining whether or not a selection is appropriate (Bergh 1969; Campbell and Malo 1976; Bender 2012).

7.2 Cultivars

'Bacon' Mexican X Guatemalan (MXG) hybrid; originated 1928 in Bueno Park, California; trees tall with pointed crowns, leaves have aniseed smell when crushed, red flecking on wood of new shoots; fruit ovate, medium to large size weighing 170–510 g; skin thin, green and glossy with leathery texture; seed size is large (Fig. 3); early maturing with very pale yellow-green flesh; flower Group B; precocious with consistent production and higher yields than 'Fuerte' in colder areas. Cold tolerance is widely reported from California, Australia, Italy and Corsica; frost tolerance is down to 4.4 °C, thus production is suited to colder regions. Susceptible

Fig. 3 'Bacon' (MXG) originated as a seedling tree on the ranch of James E. Bacon in Buena Park, California. The fruit is *dark green*, oval in shape of medium quality and 7–14 ounces in weight. Flower Group B, thin skin, early maturing with very *pale yellow* skin. Considered the most cold-hardy of the commercial varieties

to insect attack, extremely susceptible to anthracnose, unsuitable for humid subtropical areas; skin is susceptible to wind scarring, in severe cases fruit splits exposing seed. 'Bacon' has been used as a pollinator for 'Hass'.

'Beta' Guatemalan X West Indian hybrid (GXW); a 'Waldin' seedling selected at the home orchard of W.H. Krome, Homestead, Florida; fruit elliptical, medium to large size weighing 453–680 g, 84–89 mm diameter; skin is green, smooth; seed is medium sized, tight in cavity; harvested mid-season with fair quality flesh; oil content unknown; moderate cold tolerance; flower Group B; recommended post-harvest storage temperature 4–10 °C. Defects include overbearing and limb breakage (Wolfe et al. 1949; Campbell and Malo 1979; Crane et al. 2013).

'Choquette' (G X WI) Guatemalan X West Indian hybrid; seedling of unknown origin selected at the property of R.D. Choquette, Miami, Florida, and first fruited in 1934; flower Group A; fruit elliptical, flattened obliquely toward apex on one side, central pedicel insertion, medium to large size weighing 510–1133 g, 95–111 mm diameter; skin is light to dark green, nearly smooth with some undulations, glossy, somewhat leathery; seed is medium large, fairly tight in cavity (Fig. 4); harvested

Fig. 4 'Choquette' (G X WI) Guatemalan X West Indian hybrid; seedling of unknown origin; first fruited in 1934; flower Group A; fruit elliptical, medium to large size weighing 510–1133 g, 95–111 mm diameter; skin is *light* to *dark green*, glossy, somewhat leathery; seed is medium large, good to excellent quality flesh; oil content is 8–13 %; moderately high cold tolerance

mid-to-late season with good to excellent quality flesh; oil content is 8–13 %; moderately high cold tolerance; recommended postharvest storage temperature 4–10 °C. Defects include some susceptibility to Cercospora spot (Wolfe et al. 1949; Campbell and Malo 1976; Crane et al. 2013).

'Dickinson' (G) Guatemalan hybrid; a California selection, first propagated in 1912; fruit oval to obovate; small to medium; skin dark-purple with large maroon dots, rough, very thick, granular, brittle; flesh of good quality; seed small to medium, tight (Fig. 5); flower Group A; %Ratio Seed/Skin/Flesh: 13:21:66. Season: June-October in CA; February and March in FL; January and February in PR. Tree is a moderate but regular bearer; in Israel 'Dickinson' is described as round, small to

Fig. 5 'Dickinson' Guatemalan hybrid; oval to obovate; small to medium; skin *dark-purple* with large *maroon dots*, rough, very thick, granular, brittle; flesh of good quality; seed small to medium, tight; flower Group A; %Ratio Seed/Skin/Flesh:13:21:66

large, very thick-skinned with very large seed; of poor quality, not worth growing. It is no longer grown as a commercial cultivar in Florida or California.

'Duke' Mexican (M) Tree originated in 1912 from seed planted at Sunny-slope Nursery, Bangor, California. Color green, weight 227–340 g; shape pyriform; skin nearly smooth; quality excellent; oil content 21 %. Seed medium, sometimes loose. Flower group A. Season: September to November in CA; late July or mid-August to mid-September in Israel. Tree is large, symmetrical and wind and cold-resistant, and also highly resistant to root rot, especially when grown from cuttings. Vigorous, very hardy to cold, wind resistant, productive. Fine for home planting in cold

Fig. 6 Fuchsia (WI)—seed of unknown origin planted in Homestead, FL, around 1910; propagated commercially in 1926; pear shaped to oblong, sometimes with a neck; of medium size; skin smooth; flesh *pale greenish-yellow*; 4–6 % oil. Tree not very prolific in Florida; no longer popular in commercial groves

interior districts. It is a poor bearer in some areas of CA; however, has borne 168 lbs (78 kg) annually from the 6th to the 15th year in Israel.

'Fuchsia' (Fuchs) West Indian (WI). Originated on the place of C.T. Fuchs Sr., near Homestead, FL, around 1910; first fruited in 1916. Propagated commercially in 1926; pear shaped to oblong, sometimes with a neck; of medium size; skin smooth; flesh pale greenish-yellow (Fig. 6); oil content 4–6 %; good flavor; seed loose in the large cavity. Flower group A; this cultivar is thrifty, prolific and precocious. Season: early June-August; sensitive to damage during shipping. Tree not very prolific in FL; no longer popular in commercial groves.

'Gottfried' Seed of a seedling on Key Largo planted at USDA, Miami, in 1906; distributed in (1918); pear shaped; medium size; skin smooth, purple; flesh of excellent quality, 9–3 % oil; seed medium. Flower group A; Season: August to October.

Tree prolific in CA; a poor bearer in southern FL and subject to anthracnose, but hardy and desirable for home gardens on west coast of FL.

'Hass' (G) Cultivar, predominantly Guatemalan but with some Mexican genes. A chance seedling of unidentified parenting, selected by Mr. Ruddolph Hass in 1935 in California. Hass makes up more than 85 % of all avocados grown and sold worldwide. Hass avocado is more oval than other varieties and has distinctive pebbly skin which turns a rich purple when ripe. On average, this type has a small seed, weighs about 140–340 g and contains a good amount of edible flesh. %Ratio Seed/Skin/Flesh:16:12:72. Flower Group A. Hass avocados grow almost all year round in different regions of California. Hass is also recognized to have several shortcomings, including poor fruit set in some locations, sensitivity to saline irrigation water, intolerance to cold temperature below 30°F (Bergh 1984), and susceptibility to Persea mites and avocado thrips (Wolfe et al. 1949; Campbell and Malo 1979; Crane et al. 2013).

'Lula' Guatemalan X West Indian (GXWI) hybrid; believe to be a seedling of 'Taft'. Originated at the property of G.B. Cellon (Miami, Florida) and named for his wife, first fruited in 1919; flower Group A; fruit pyriform, fruit apex rounded, level pedicel insertion, medium to large size weighing 397–680 g, 81–105 mm diameter; skin is green to dark green, nearly smooth, slightly rough; seed is large, tight in cavity (Fig. 7); harvested late-season with good to excellent quality (slightly sweet) flesh, 65 % edible pulp, oil content is 6.0–15 %; high cold tolerance. Flaws include fruit highly vulnerable to scab and difficult to control tree size (Wolfe et al. 1949; Campbell and Malo 1979; Crane et al. 2013).

'Mexicola' Mexican (M) Originated brought about 1910 at Pasadena, California and propagated about 1912; very small; skin black; flesh of excellent flavor; seed large. Flower Group A. %Ratio Seed/Skin/Flesh:27:12:61, Season: August to October; grown only in home gardens in California and maintained in germplasm collections for breeding purposes. Producing early and regularly; very heat- and cold-resistant; much used as a parent in CA breeding programs.

'Monroe' Guatemalan (G)—West Indian hybrid; seedling of unknown parentage originating at the orchard of J.J.L. Phillips, Homestead, Florida, first fruited in 1935; flower Group B; fruit elliptical, flattened obliquely toward apex on one side, asymmetrical pedicel insertion, medium to large size weighing 453–1133 g; 92–111 mm diameter; skin is dark green, glossy, slightly rough; seed is medium large, tight in cavity; harvested late-season with good to excellent quality flesh (Fig. 8), oil content is 6.1–14 %; Flower group A; moderate cold tolerance; safest low storage temperature 4–10 °C. Defects include fruit moderately susceptible to scab, Cercospora spot and anthracnose, fruit drop, and limb breakage (Wolfe et al. 1949; Campbell and Malo 1979; Crane et al. 2013).

'Nabal' (G) Introduced from Guatemala by Wilson Popenoe in 1917. Propagated commercially in CA since 1927 but in FL only since 1937. Fruit almost round, of medium size, 400–680 g. Skin nearly smooth, dark green. Flesh yellow, of good flavor. Seed medium small, tight in cavity (Fig. 9). Oil content 10–15 %. Flower Group B. %Ratio Seed/Skin/Flesh:10:10:80. Season January and February in CA,

Fig. 7 'Lula' (GXWI) Guatemalan West Indian hybrid; flower Group A; fruit pyriform, medium to large size weighing 397–680 g, 81–105 mm diameter; skin is *green* to *dark green*, nearly smooth, slightly rough; seed is large, tight in cavity; harvested (slightly sweet) flesh, 65 % edible pulp, oil content is 6.0–15 %; high

November–December FL. In CA has fruited heavily but in FL restricted to some areas in Dade County.

'Semil 34' (GXW) A Guatemalan X West Indian hybrid from a seedling selected at Finca El Semil, Villalba, Puerto Rico in 1947. Fruit (Fig. 10) is obovate/pyriform and flattened, weight 396–700 g and typically 84–92 mm in diameter, but could be larger; the peel is green, slightly bumpy to smooth and glossy (Fig. 10). In flower Group A; the seed (Fig. 10) is medium size and tight with good to excellent quality edible pulp (78 %). Fruit are harvested mid-season, fruit oil content is 7.5–15 %. Considered one of the major cultivars grown in Dominican Republic (Wolfe et al. 1949; Campbell and Malo 1979).

Fig. 8 'Monroe' West Indian hybrid; seedling of unknown; flower Group B; fruit elliptical, medium to large size weighing 453–1133 g; 92–111 mm diameter; skin is *dark green*, glossy, slightly rough; seed is medium large, tight in cavity

'Sharwil' (GXM) Predominantly Guatemalan with some Mexican genes; selected in 1951 by Sir Frank Sharpe at Redland Bay, Queensland, parentage unknown, the name 'Sharwil' being an mixture of Sharpe and Wilson (J.C. Wilson being the first propagator); trees are large and rounded in shape, broad crown, extremely vigorous, forming a large dome, more upright than 'Fuerte'; new leaves (Fig. 11) distinctively red but changing to green when fully grown, red flecking on wood of new shoots; fruit are pyriform to ovate, medium size weighing 245–475 g; skin is medium to thick, green, medium gloss, corky with wrinkled surface; seed size is small, conical; mid-season maturity with buttery to golden-yellow flesh (Fig. 11), excellent quality with rich nutty flavor, and high palatability over the full maturity

Fig. 9 'Nabal' Budwood brought from Guatemala in 1917; propagated in CA since 1927, in FL from 1937 an in Israel since 1934, nearly round; medium to large; skin nearly smooth, thick, granular; flesh of high quality, *green* near skin; 10–15 % oil in Florida, 18–22 % in Queensland

range (21–30 % dry matter); flower Group B; fruit stores well both on tree and postharvest. Has potential to set heavy crops but flowering and fruit set are susceptible to cool temperatures, thus reliable production only occurs in warm subtropical climates; has poor tolerance to sub-zero temperatures. Good tolerance to anthracnose and insect pests due to medium to thick skin. One of the most vulnerable cultivars to boron deficiency. Growth is vigorous but large limbs are brittle with protection from strong winds required (Wolfe et al. 1949; Campbell and Malo 1979).

'Simmonds' West Indian (WI); a seedling of 'Pollock' selected at the United States Department of Agriculture (USDA) Subtropical Horticulture Research Station in Miami, Florida; first fruited in 1913. Flower Group A; fruit obovate, flattened obliquely toward apex on one side, asymmetrical pedicel insertion, medium to large size weighing 453–963 g, 78–98 mm diameter; skin is light green to green, smooth, glossy; seed medium sized, tight in cavity; harvested early season with good to excellent quality flesh, 76 % edible pulp, oil content is 3.3–5.0 %; tree has low cold tolerance; recommended postharvest storage temperature 13 °C. Defects

Fig. 10 'Semil 34' (GXWI)- A Guatemalan X West Indian hybrid from a seedling selected at Finca El Semil, Villalba, Puerto Rico in 1947. In flower Group A. Fruit is obovate/pyriform, weight 396–700 g with good to excellent quality edible pulp

include low cold tolerance, lack of vigor and excessive fruit drop (Wolfe et al. 1949; Campbell and Malo 1979; Crane et al. 2013).

'Taylor' Guatemalan (G); originated of the California Guatemalan cultivar 'Royal' donated to the USDA Subtropical Horticulture Research Station (formerly the United States Plant Introduction Garden), Miami, Florida, named after Dr. Taylor of the USDA, first fruited in 1913; propagated commercially in 1914; flower Group A; fruit obovate to pyriform with rounded base (Fig. 12), pedicel insertion is central, small to medium size weighing 340–510 g, 79–86 mm diameter; skin is dark green, rough-pebbled; seed medium sized, tight in cavity; with excellent quality flesh (Fig. 12), 60–69 % edible pulp, oil content is 12–17 %; flower Group A; harvested mid-to-late-season high cold tolerance; recommended postharvest storage

Fig. 11 'Sharwil' Predominantly Guatemalan with some Mexican genes; selected in 1951 by Sir Frank Sharpe at Redland Bay, Queensland; trees are large and rounded in shape, broad crown, extremely vigorous, forming a large dome, *buttery* to *golden-yellow* flesh, excellent quality with rich nutty flavor

temperature 4 °C. One of the first Guatemalan varieties to be planted commercially, still popular in some places (Chile).

'Tonnage' Guatemalan (G) Seedling of 'Taylor' selected at the orchard of D.M. Roberts, Homestead, Florida, and first fruited in 1921; fruit (Fig. 13) pyriform with rounded base, pedicel insertion very asymmetrical, small to medium size weighing 340–567 g, 76–90 mm diameter; skin (Fig. 13) is pebbled, dark green, glossy, thick; seed (Fig. 13) medium sized, fairly tight in cavity; harvested mid-season with fair quality flesh, oil content is 8–10 %; flower Group B; high cold tolerance; recommended postharvest storage temperature 4–10 °C. Defects include only fair fruit quality, upright growth habit and fruit is moderately vulnerable to scab (Wolfe et al. 1949; Campbell and Malo 1979; Crane et al. 2013).

Fig. 12 'Taylor' (G) Guatemalan; seedling of the California Guatemalan cultivar 'Challenge' or 'Royal'; named after Dr. Taylor of the USDA, flower Group A; fruit obovate to pyriform, excellent quality flesh, 60–69 % edible, oil content is 12–17 %; high cold tolerance

'Waldin' West Indian (WI); seedling of unknown parentage selected at the property of B.A. Waldin, Homestead, Florida, and first fruited in 1913; flower Group A; fruit ellipsoid-spheroid (Fig. 14), flattened obliquely toward apex on one side, central pedicel insertion, medium to large size weighing 397–680 g, 87–102 mm diameter; skin is pale green to greenish-yellow, smooth, leathery (Fig. 14); seed is medium to large, fairly tight in cavity (Fig. 14), seed used for rootstock; harvested mid-season with good to excellent quality flesh, 64 % edible pulp, oil content is

Fig. 13 'Tonnage' (G) Guatemalan seedling of 'Taylor'; flower Group B; fruit pyriform with rounded base, fruit small to medium size weighing 340–567 g, skin is pebbled, *dark green*, glossy, thick; seed medium sized, fairly tight in cavity; with fair quality flesh, oil content is 8–10 %; high cold tolerance

6–12%; tree has low cold tolerance; recommended postharvest storage temperature 13 °C (Table 1).

8 Production

Some of the main avocado producing countries (Mexico, Chile, USA, Australia, South Africa, Israel,) production is discussed in detail. Mexico is the world's largest producer of avocados, representing over one-third of global production (FAO 2013).

Fig. 14 'Waldin' (WI) West Indian; seedling of unknown parent; flower Group A; fruit ellipsoid-spheroid, fruit medium to large size weighing 397–680 g, 87–102 mm diameter; skin is *pale green* to *greenish-yellow*, smooth, leathery; seed is medium to large, fairly tight in cavity

Chile was the second largest producer in 2011 (330,000 t) accounting for nearly 6.5 % followed by Dominican Republic and Indonesia with about 6.0 % each (FAO 2013) Table 2. World production of avocados is expected to rise in the 2013/2014 period owing to improved production by the major avocado exporting countries, improvement in phytosanitary conditions, border restrictions lifted, reductions in trade barriers and advances in transportation and post-harvest technologies. As a result, Mexico and Chile will continue to dominate the export trade, while the United States could remain the number one importer of avocados. The main avocado producing countries (Mexico, Chile and Dominican Republic) as well as the other main producing countries is discussed in detail (Table 2).

Table 1 Comparison of selected characteristics of the three species (races) of *Persea americana* Mill. (Source: Berg and Lahav 1996; Bergh and Ellstrand 1986; Ayala-Silva et al. 2002)

		Race		
Traits		Mexican	Guatemalan	West Indian
Tree				
General	Origin	Subtropical	Subtropical	Tropical
	Climate	Most	Intermediate	Least
	Cold tolerance	Least	Intermediate	Most
	Salt tolerance	Intermediate	Least	Most
	Alternate bearing	Less	More	Less
Form	Internodes	Longest	Long	Shortest
	Twig lenticels	Pronounced	Absent	Absent
	Bark roughness	Less	Less	More
	Stem pubescence	More	Less	Less
Leaf	Size	Smallest	Large	Largest
	Colour	Green	Green	Pale green
	Flush colour	Greenest	Reddest	Yellowish-green
	Anise	Present (usually)	Absent	Absent
	Underside waxiness	More	Less	Less
Fruit				
Flower	Season	Early	Late	Early-Intermediate
	Bloom to maturity	5–7 months	10–18 months	6–8 months
	Perianth persistence	Greater	Less	Less
Stem	Length	Short	Long	Short
	Thickness	Medium	Thick	Thin
	Shape	Cylindrical	Conical	Nailhead
Fruit	Size	Tiny–Medium	Small–Large	Medium–Very large
	Shape	Mostly elongate	Mostly round	Variable
Pulp	Flavour	Anise-like, spicy	Often rich	Sweet, mild
	Oil content	Highest	High	Low
	Distinct fibers	Common	Less common	Intermediate
Skin	Color	Usually purple	Black or green	Pale green/ maroon
	Surface	Waxy coating	Variably rough	Shiny
	Thickness	Very thin	Thick	Medium
	Stone cells	Absent	Present	Slight
	Pliability	Membranous	Stiff	Leathery
	Peeling	No	Variable	Yes
Seed	Seed/Fruit ratio	Large	Often small	Large
	Coats	Thin	Usually thin	Thick
	Tightness in cavity	Often loose	Tight	Often loose
	Surface	Smooth	Smooth	Rough
Oil content		Highest	High	Lowest

Table 2 World avocado production by country from 2008 to 2011. (FAOSTAT Database 2012)

Country	2008	2009	2010	2011	2008	2009	2010	2011	2008	2009	2010	2011
	Area harvested (Ha)				Yield (Hg/Ha)				Production (t)			
Australia	9827	10,249	9800	9500	45,792	37,543	38,061	38,237	45,000	38,478	37,300	36,325
Bolivia	497	880	871	870	66,519	63,273	63,054	63,253	3306	5568	5492	5503
Brazil	9453	8411	11,111	10,753	155,733	165,366	137,871	149,145	147,214	139,089	153,189	160,376
Cameroon	14,000	13,800	14,200	13,999	39,286	39,130	39,437	49,669	55,000	54,000	56,000	69,532
Chile	33,800	33,500	34,057	36,388	97,929	97,910	96,896	101,288	331,000	328,000	330,000	368,568
China	14,000	15,000	15,000	16,000	67,857	66,667	68,000	67,813	95,000	100,000	102,000	108,500
Colombia	18,470	19,255	21,592	24,514	99,604	98,171	95,148	87,744	183,968	189,029	205,443	215,095
Costa Rica	5117	5431	5810	5615	40,934	40,447	43,453	48,012	20,946	21,967	25,246	26,959
Côte d'Ivoire	4700	4600	5300	5200	58,936	58,913	59,811	59,615	27,700	27,100	31,700	31,000
Democratic Republic congo	8766	8886	9008	8881	74,401	74,400	74,396	93,694	65,220	66,112	67,016	83,210
Dominican republic	5832	7183	10,558	10,649	322,598	256,644	273,430	277,109	188,139	184,357	288,684	295,081
Ecuador	2900	2950	2957	2776	95,472	94,915	98,025	98,667	27,687	28,000	28,986	27,390
El Salvador	1450	1481	440	639	14,724	14,733	42,759	45,180	2135	2182	1880	2887
Ethiopia	6473	5067	5694	7212	40,331	64,046	66,124	79,450	26,106	32,452	37,651	57,299
Ghana	1685	1754	1770	1817	47,478	48,683	47,458	46,780	8000	8539	8400	8500
Guatemala	9293	9363	9435	9246	103,304	105,350	99,932	98,915	96,000	98,639	94,286	91,457
Haiti	8568	9692	10,595	9731	52,521	45,448	46,282	53,105	45,000	44,048	49,036	51,676
Honduras	380	390	413	414	43,421	43,590	43,535	48,502	1650	1700	1798	2008
Indonesia	19,802	19,979	20,507	21,653	123,328	128,956	109,367	127,443	244,215	257,642	224,278	275,953
Israel	6270	6480	6565	6780	84,737	131,124	105,933	111,043	53,130	84,968	69,545	75,287
Jamaica	307	338	354	364	60,293	68,195	64,633	67,335	1851	2305	2288	2451
Kenya	7900	10,053	10,320	11,246	131,042	144,438	196,021	179,155	103,523	145,204	202,294	201,478
Madagascar	3026	2639	2938	3202	79,313	97,620	88,768	80,050	24,000	25,762	26,080	25,632
Mexico	112,479	121,491	123,403	126,598	103,346	101,322	89,717	99,855	1,162,429	1,230,973	1,107,135	1,264,141
Morocco	1972	1863	1868	2539	97,632	139,560	180,112	132,017	19,253	26,000	33,645	33,519

Table 2 (continued)

Country	2008	2009	2010	2011	2008	2009	2010	2011	2008	2009	2010	2011
	Area harvested (Ha)				Yield (Hg/Ha)				Production (t)			
New Zealand	4000	4117	4000	3976	42,500	49,794	49,740	48,250	17,000	20,500	19,896	19,184
Panama	960	1001	1094	1098	41,771	41,039	38,464	93,980	4010	4108	4208	10,319
Paraguay	2411	2400	2691	2446	55,006	58,333	55,622	58,418	13,262	14,000	14,968	14,289
Peru	14,370	16,292	17,750	19,314	94,852	96,621	103,870	110,209	136,303	157,415	184,370	212,857
Portugal	11,600	11,067	11,000	10,981	14,009	14,457	15,311	14,659	16,250	16,000	16,842	16,097
Puerto Rico	811	918	1004	922	24,661	21,329	21,713	24,913	2000	1958	2180	2297
Réunion	69	72	80	87	79,710	80,139	75,000	68,736	550	577	600	598
South Africa	16,000	14,500	15,000	16,346	52,209	52,914	55,469	46,028	83,534	76,726	83,204	75,237
Spain	10,023	10,016	10,434	10,558	73,416	71,816	72,508	79,017	73,585	71,931	75,655	83,426
United States of America	29,473	26,819	24,253	24,261	35,704	100,978	65,208	98,324	105,230	270,813	158,150	238,544
Venezuela	6396	7000	7647	7673	112,212	107,143	95,530	106,334	71,771	75,000	73,052	81,590

Not in ranking order

8.1 Mexico

The avocado production forecast for MY 2012/13 (July/June) is 1.3 million metric t (MMT). The forecasted increase over MY 2011/2012 is based on expected good weather and a mild winter. Implementation of phytosanitary pest control programs has also helped boost production. Production estimates for MY 2011/2012 are 1.26 MMT, higher than previously expected due to a mild winter and good weather conditions (FAS 2013). Total area planted for MY 2012/2013 is forecast at 150,000 ha, an increase of more than 5 % over MY 2011/2012, as growers in different states in Mexico are interested in increasing area due to good domestic and international demand for Mexican Hass avocados. The area planted and harvested for MY 2011/2012 increased compared to MY 2010/2011 (FAS 2013). Growers forecast avocado exports higher for MY 2012/2013 compared to MY 2011/2012 or about 500,000 MT. Exports have been increasing due to a good international demand and year-round market access to all 50 U.S. states. According to the Global Trade Atlas, exports for MY 2011/2012 are estimated at 409,761 MT, however, the industry estimates exports at about 460,000 MT. U.S. imports are estimated at about 360,923 MT according to the Census Bureau. According to the Global Trade Atlas, avocado exports for MY 2010/2011 were estimated at 318,462 MT, however the industry estimates exports at about 360,000 MT. U.S. imports were estimated at 281,671 MT according to the Census Bureau (FAS 2012). The Unites States is the top export market for Mexico, consuming 75 % of total exports. Japan and Canada are strategic market niches where Japan has about 10 % of the market and Canada about 7 %. Currently, 34 packers in Michoacan are eligible to export Mexican avocados to the United States (FAS 2012).

8.2 Chile

Chilean avocado production is expected to increase during the current 2013/2014 production season compared to last season. Normal weather conditions are responsible for the increase in production and subsequently exports for this season. For the 2014/2015 season, output is forecasted to increase, as an abundant blooming in most production areas forecasts strong production if weather conditions remain normal. A little over 98 % of all Chilean commercial avocado trees are planted in the central area of the country from Region IV through Region VI. Almost all the planting expansion has been of the Hass variety in the last decade, and there are over 20 other varieties planted in Chile. Of the total planted area in Chile, around 30,000 ha are planted as e Hass avocado variety, which represents almost 100 % of total exports (99.8 %) (FAS 2013). Although declining in importance, the United States is still largest export market for Chilean avocados, followed by the Europe. The US market received 45 % of Chile's total avocado exports in 2012, down from over 66 % in the previous year. Exports to the EU have increased over the last few years as a result of a big industry effort made to diversify their markets (FAS 2013).

8.3 Dominican Republic

The Government's support to avocado producers of the Dominican Republic has generated a substantial increase in exports of fruit to international markets, with a volume of 20,000 t per year and a foreign exchange contribution that exceeds 30 million dollars. A report from the Department of Fruit of the Ministry of Agriculture indicates that the Dominican Republic has a stable production of avocado with a tendency to increase since, in 2012, 1000 containers of 35,000 units each were exported while, this year, more than 800 containers were exported. The average production per year is 250,000 t of avocado, close to 20 % of which is exported, especially the Semil-34, Jass, Carla and Pollock varieties. The support of the Ministry of Agriculture to avocado growers consists of the delivery of supplies, training, certification of good agricultural practices, market management and the repair and maintenance of roads. Besides, the Special Fund for Agricultural Development (FEDA), through the Agricultural Bank, has funded avocado growers from Cambita, Garabito, in San Cristóbal, with 90 million pesos. The largest fruit production areas are located in San José de Ocoa, Cambita, in San Cristobal; Villa Trina, in Espaillat; Altamira and Mamey, of Puerto Plata; Amina, in Valverde; The Calimetes, in Elías Piña; Padre Las Casas, in Azua; Los Arroyos, in Pedernales; in Peravia and in Neyba.

8.4 United States of America

The value of U.S. avocado production increased from $ 479.1 million in 2010 to $ 492.1 million in 2011. The total volume amounted to 226,450 t, an increase of more than 52,120 t from 2010. According to NASS (2012), the California avocado crop jumped to 195,000 t, while the Florida crop rose to 31,100 t. The number of acres under production stabilized at 59,950 and the yield per acre dramatically increased to 3.8 t. The value of avocado imports into the United States has increased substantially over the past two decades, reaching nearly $ 913 million in 2011. Of that amount, more than $ 17.2 million was for imported organic avocados. Once again, Mexico supplied most of the avocados imported into the United States in 2011, followed by Chile (FAS 2011).

Conclusion

Avocado has been the favored fruit of many people in Mesoamerica dating back to 10,000 BC. It is unique amongst the fruits and it is neither sweet nor bitter. Some superior cultivars have anise or nutty flavor. Today avocado brings pleasure to hundreds of millions of people worldwide. At the beginning of the twentieth century, the Avocado made its first steps on the United States (CA and FL) market, and total consumption on last year was close to 600,000 t. Now a day's quality enhancement of the fruit allows for long distance shipping in shorter periods of time. Avocado is consume now worldwide and the forecast is for consumption to increase. The increase in productivity has increased in the last decades and work including new biotechnology such as molecular markers, preservation in tissue culture and cryopreservation will assist with the breeding and improvement of the fruit. The nutritional importance of the avocado and its worth on promoting good health and the increasing use of its oils and secondary products for cosmetics will benefit the prospects for a significant increase in avocado consumption.

The outlook of the avocado breeding as a whole is increasing with new development of techniques and its full prospective have not yet been exploited given all genetic resources and new technology available.

Acknowledgements The authors express their gratitude to the National Germplasm Repository, Miami, Florida for their support. We also would like to thank Dr. Ed Boza and Dr. Seth Finley for their useful comments and suggestions and Tomas Ayala Garcia for the graphics and pictures.

References

Abraham JD, Takrama JF (2013) Morphological and genetic diversity of *Persea americana* mill. (Avocado) in two regions of Ghana. Tropentag 2013, Stuttgart, Germany, 17–19 Sept 2013

Acosta Jose [1590] (1985) Historia Natural y Moral de las Indias. 2nd ed. Fondo de Cultura Económica. México

Ai P, Luo Z (2005) Cryopreservation of dormant vegetative buds and genetic stability of regenerated plantlets in persimmon. Acta Horticult 685:85–92

Altieri MA, Merrick LC (1987) In situ conservation of crop genetic resources through maintenance of traditional farming systems. Econ Bot 41:86–96

Arora RK, Rao VR (eds) (1998) Tropical fruits in Asia: diversity, maintenance, conservation and use. Proceedings of the IPGRI-ICAR-UTFANET regional training course on the conservation and use of germplasm of tropical fruits in Asia held at Indian Institute of Horticultural Research, Bangalore, India, 18–31 May 1997

Arpaia ML (1997) Avocado germplasm preservation. California Avocado Society, avocado research project plan and grant requirement. University of California, Riverside

Arpaia ML, Harley S (2012) 2011–12 Production research report California avocado plant improvement planning meeting: taking advantage of what we learned from avocado brainstorming 2011

Ashworth VETM, Clegg MT (2003) Microsatellite markers in avocado (*Persea americana* Mill.): genealogical relationships among cultivated avocado genotypes. J Hered 94:407–415

Ayala-Silva T, Schnell II RJ, Meerow AW, Goenaga RJ, Zee FT (2004) Current status of the subtropical and tropical germplasm repositories of the national plant germplasm system. Proceedings of Florida State Horticultural Society 117:182–187

Ayala-Silva T, Schnell II RJ, Goenaga RJ, Zee FT, Irish BM (2013) The national plant germplasm system: the subtropical and tropical fruit genebanks. Acta Horticult 975:43–54

Ayala-Silva T, Schnell II RJ, Meerow AW, Winterstein MC (2007) Phenotypic characterization of fifteen commercial and rootstock avocado varieties. Proceedings VI World Avocado Congress, Viña del Mar, Chile, 12–16 Nov 2007

Baranski M (2013) Seed collection and plant genetic diversity, 1900–1979. *Embryo Project Encyclopedia* (2013-11-01). ISSN: 1940-5030. http://embryo.asu.edu/handle/10776/6468

Barceló-Muñoz A, Pliego-Alfaro F (2003) Micropropagation of avocado. In: Jain SM, Ishii K (eds) Micropropagation of woody trees and fruits. Kluwer Academic, Dordrecht, pp 519–542
Benavente FT Mde [1536] (2003) Historia de los Indios de la Nueva España Special edDastin. S.L. España
Bender GS (2012) Avocado production in California. A cultural handbook for growers, 2nd edn. Univ Calif Exp Program, Ch. 2 pp 1–32
Benson E (2008) Cryopreservation of phytodiversity: a critical appraisal of theory & practice. Crit Rev Plant Sci 27(3):141–219
Ben-Ya'acov A, Solis-Molina A, Bufler G (2003) The mountain avocado of Costa-Rica. Persea americana var. costaricensis, a new sub-species. In: Proceedings V world avocado congress 27–33, Consejeria de Agricultura y Pesca. Malaga, España, 19–24 Oct 2003
Ben-Ya'acov A (1998) The taxonomy of the avocado: A proposed new classification of the Persea, subgenus Persea, In: Proceedings of the Third World Avocado Congress, Tel Aviv, Israel, p 75
Ben-Ya'acov A, Bufler G, Barrientos A, De La Cruz Torres E, Lopez L, Degani H, Solis Molina A (1990) A study of the avocado genetic resources. GIARA Project 1–15
Ben-Ya'acov A, Bufler G, Barrientos-Priego A (1992) A study of the avocado germplasm resources (1988–1990) I, General description of the international project and its findings. In: Lovatt C, Holthe PA, Arpaia ML (eds) Proceedings of the Second World Avocado Congress, vol 2. University of California, Riverside, pp 535–541
Ben-Ya'acov, Solis A, Peri E (1995) Progress of the study on the avocado genetic resources, II, The avocado genetic resources in Costa Rica, Abstracts 3, ABY–10
Bergh B (1969) Avocado. In: Ferwerda FP, Wit F (eds) Outlines of perennial crop breeding in the tropics. Institute of plant breeding, Landbouwhogeschool, Wageningen, pp 23–51
Bergh B (1975) Avocados. In: Janick J, Moore JN (eds) Advances in fruit breeding. Purdue University Press, West Lafayette, pp 541–567
Bergh B (1992) The origin, nature, and genetic improvement of the avocado. Calif Avocado Soc Yearb 76:61–75
Bergh B (1993) Avocado Persea americana (Lauraceae) In: Smart J (ed) Evolution of crop plants. Longman, New York, p 531
Bergh B, Ellstrand N (1986) Taxonomy of the avocado. Calif Avocado Soc Yearb 70:135–145
Bergh B, Lahav E (1996) Avocados. In: Janick J, Moore JN (eds) Fruit breeding, vol I: tree and tropical fruits. Wiley, West Lafayette, pp 113–166
Blake SF (1920) A preliminary revision of the North American and West Indian avocados (Persea spp.). J Wash Acad Sci 10:9–21
Borokoni TI (2013) The state of ex-situ conservation in Nigeria. Conservation Science, vol 4. Issue 2, p 197
Brooks CI (1929) Avocados. Proc Flo State Horticult Soc 42:123–129
Campbell CW, Malo SE (1976) A survey of avocado cultivars. In: Sauls JW, Phillips RL, Jackson LK (eds) The avocado. Proc. 1st Sub-Tropical Fruit Short Course, Univ Florida Coop Ext Serv, pp 20–24
CGIAR (2013) 1971–1996 database: 25 years of food and agriculture improvement in developing countries. http://www.worldbank.org/html/cgiar/25years/25cover.html. Accessed: Sept 2013
Chen H, Morrell PL, de la Cruz M, Clegg MT (2009) Tracing the geographic origins of major avocado cultivars. J Hered 100:56–65
Cobo Bde [1653] (1956) Historia del Nuevo Mundo. Biblioteca de Autores Españoles. España
Condit IJ (1916) History of the avocado in California. Report of the Calif Avocado Soc Yearb 2:105–106
Convention on Biological Diversity (1992) http://biodiv.org/convention/articles.asp
Cooper HD, Spillane C, Hodgkin T (2001) Broadening the genetic base of crops: an overview. In: Cooper HD, Spillane C, Hodgkin T (eds) Broadening the genetic base of crop production. CABI Publishing, CAB International, Wallingford, pp 1–23
Cooper PA (1987) Advances in the micropropagation of avocado (*Persea americana* Mill.). Acta Horticult 212:571–575

Crane JH, Balerdi CF, Maguire I (2013) Avocado growing in the Florida home landscape. http://edis.ifas.ufl.edu/MG213. Accessed 15 Oct 2013

Crisosto CH, Ferguson L, Thompson M, Byers P, Zee F (2012) Emerging fruit crops. In: Badenes, ML, Byrne D (eds) Fruit breeding. Springer, New York, pp 97–147

Cruz-Cruz CA, González-Arnao MT, Engelmann F (2013) Biotechnology and conservation of plant biodiversity. Resources 2(2):73–95

Diario Libre (2013) Dominican republic: avocado exports generate more than US$ 30 million. http://www.freshplaza.com/article/107278/Dominican-Republic-Avocado-exports-generate-more-than-US-30-million#SlideFrame_1

Drew R, Ashmore S, Somsri S, Noor N, Thi Hoa T, Damasco O, Rao R (2007) Advanced technologies for germplasm conservation of tropical fruit species. Acta Horticult 760:91–98. http://www.actahort.org/books/760/760_10.htm

Efendi D (2003) Transformation and cryopreservation of embryogenic avocado (Persea americana Mill.) cultures. Dissertation, University of Florida

Efendi D, Litz R (2003) Cryopreservation of avocado. Proceedings V World Avocado Congress (Actas V Congreso Mundial del Aguacate), pp 111–114

Ellis D, Skogerboe D, Andre C, Hellier B, Volk G (2006) Implementation of garlic cryopreservation techniques in the National Plant Germplasm System. CryoLetters 27:99–106

FAO (2013) Chile Production Forecast for Marketing Year 2013/114 Avocado Annual Report. http://gain.fas.usda.gov/Recent%20GAIN%20Publications/Avocado%20AnnualSantiago-Chile11-27-2013.pdf

FAS (2012) Mexico Production Forecast for Marketing Year 2012/13 Avocado annual report. http://gain.fas.usda.gov/Recent%20GAIN%20Publications/Avocado%20Annual_Mexico%20City_Mexico_11-26-2012.pdf

Fiedler J, Bufler G, Bangerth F (1998) Genetic relationships of avocado (*Persea americana* mill) using RAPD markers. Euphytica 101:249–255

Forsline PL, Towill LE, Waddell JW, Stushnoff C, Lamboy WF, McFerson JR (1998) Recovery and longevity of cryopreserved dormant apple buds. J Am Soc Horticult Sci 123:365–370

Galindo-Tovar ME, Arzate-fernandez AM, Ogata-Aguilar N, Landero-Torres I (2007) The avocado (Persea americana, Lauraceae) crop in Mesoamerica: 10,000 years of history. Harv Papers Bot 12(2):325–334

Galindo-Tovar ME, Ogata-Aguilar N, Arzate-Fernandez AM (2008) Some aspects of avocado (Persea americana Mill.) diversity and domestication in Mesoamerica. Genet Resour Crop Evol 55:441–450

Gómez-López VM (1999) Characterization of avocado (*Persea americana* Mill.) varieties of low oil content. J Agric Food Chem 47:2707–2710

Gomez-Lopez VM (2002) Fruit characterization of high oil content avocado varieties. Sci Agric (Piracicaba, Braz.) 59(2):403–406. ISSN 0103-9016. http://dx.doi.org/10.1590/S0103–90162002000200030

Gross BL, Olsen KM (2009) Genetic perspectives on crop domestication. Trends Plant Sci 15:529–537

Guillén H, Rodríguez A, Pedraza M, Vidales I, Morales JL, Tapia M (2011) Morphological characterization of avocados (Persea americana MILL) native to Tacambaro, Michoacan, Mexico. VII World Avocado Congress, Cairns, Queensland Australia, 5–9 Sept 2011

Guzmán-García E, Bradaï F, Sánchez-Romero C (2012) Cryopreservation of avocado embryogenic cultures using the droplet-vitrification method. Acta Physiol Plant. doi:10.1007/s11738-012-1062-3

Hoisington D, Khairallah M, Reeves T, Ribaut JM, Skovmand B, Taba S, Warburton M (1999) Plant genetic resources: what can they contribute toward increased crop productivity? PNAS 96(11):5937–5943. doi:10.1073/pnas.96.11.5937

Hummer KE, Pomper KW, Postman J, Graham CJ, Stover E, Mercure EW, Aradhya M, Hung CD, Trueman SJ (2011) Encapsulation technology for short term preservation and germplasm distribution of the African mahogany *Khaya senegalensis*. Plant Cell Tissue Organ Cult 107:397–405. doi:10.1007/s11240-011-9990

Hung CD, Trueman SJ (2011) In vitro propagation of the African mahogany Khaya senegalensis. New Forestry 42:117–130
Janick J (2005) The origins of fruits, fruit growing and fruit breeding. Plant Breed Rev 25:255–321
Janick J, Paul RE (2008) Persea americana—Avocado. The encyclopedia of fruits & nuts. CABI CAB International, Cambridge, pp 439–449
Kaczmarczyk A, Funnekotter, B, Menon A, Phang PY, Al-Hanbali A, Mancera RL, Bunn E (2012) Curr Issues Plant Cryo, Curr Front Cryobiol, Prof. Igor Katkov (ed.), ISBN: 978-953-51-0191–8
Kapai VY, Kapoor P, Rao IU (2010) In vitro propagation for conservation of rare and threatened plants of India—a review. Int J Biol Tech 1(2):1–14
Kaviani B (2011) Conservation of plant genetic resources by cryopreservation. Aust J Crop Sci 6(5):778–800
Knight Jr RJ (1999) Genetic diversity in avocado. In: Arpaia ML, Hofshi R (eds) Proc avocado brainstorming 99, University of Florida, Homestead, 27–29 Oct 1999
Knight Jr RJ (2002) History, distribution and uses. In: Whiley AW, Schaffer B, Wolstenholme BN (eds) The avocado: botany, production and uses. CAB International, Wallingford, pp 1–14
Kopp LE (1966) A taxonomic revision of the genus Persea in the Western hemisphere (Persea-Lauraceae). Mem NY Bot Gard 14:1–117
Labat JB (1964) The avocado—in 1694. Calif Avocado Soc 1964 Yearbk 64:48–50
Lahav E (1994) Israeli avocado industry. Calif Avocado Soc Yearb 78:51–52
Lahav E, Gazit S (1994) World listing of avocado cultivars according to flowering type. Fruits 49:299–313
Lahav E, Lavi U (2002) Genetics and classical breeding. In A. Wiley, B. Schaffer., and B. Wolstenholome (eds.). The avocado: Botany, Production, and uses. CABI, New York, pp 39–69
Landon AJ (2009) Domestication and Significance of Persea americana, the avocado, in Mesoamerica. Nebraska Anthropologist 47. http://digitalcommons.unl.edu/nebanthro/47
Lavi U, Sa'ada D, Ragev I, Lahav E (2003) Avocado genetics and breeding-present and future. World Avocado Congr VA 42:134–135
López LL, Barrientos-Priego AF, Ben-Ya'acov A (1999) Variabilidad genética de los bancos de germoplasma de aguacate preservados en el Estado de México. Rev Chapingo Sci Horticult 5:19–23
Merrick LC (1990) Crop genetic diversity and its conservation in traditional agroecosystems. In: Altieri MA, Hecht SB (eds) Agroecology and small farm development. CRC, Boca Raton
Mhameed S, Sharon D, Kaufman D, Lahav E, Hillel J, Degani C, Lavi U (2011) Genetic relationships within avocado (*Persea americana* Mill) cultivars and between *Persea* species. Theor Appl Genet 94:279–286
Mijares P, López L (1998) Variedades de aguacate y su producción en México Memoria Fundación Salvador Sánchez Colín CICTAMEX S.C. México
Miller A, Gross BL (2011) From forest to field: perennial fruit crop domestication. Am J Bot 98:1389–1414
Miller AJ, Schaal B (2006) Domestication and the distribution of genetic variation in wild and cultivated populations of the Mesoamerican fruit tree *Spondias purpurea* L. (Anacardiaceae) Mol Ecol 15(6):1467–1480
Mohamed-Yasseen Y (1993) In vitro propagation of Avocado (*Persea americana* Mill.). Calif Avocado Soc Yearb 77:107–111
Morton JF (1987) Fruits of warm climates. Florida Fair Books, Miami, p 505
Muriel P, Pitman N (2004) Persea conferta. 2006 IUCN Red List of Threatened Species
Nakasone HY, Paull RE (1998) Tropical fruits. CAB International, New York, pp 76–102
NASS (2011) Noncitrus Fruits and Nuts 2011 Summary, National Ag Statistics Service (NASS), USDA
Normah MN, Clyde MM, Cho EG, Rao VR, Drew R (2002) Ex situ conservation of tropical rare fruit species. Proc. IS on Trop. & Subtrop. Fruits Acta Horticult 575:221–230
Oduro G, Kwabena N, Ofosu-Budu G, Williams A (2013) Genetic diversity among local and introduced avocado germplasm based on morpho-agronomic traits. Int J Plant Breed Genet 7:76–91

Olsen KM, Wendel JF (2013) A bountiful harvest: genomic insights into crop domestication phenotypes. Annu Rev Plant Biol 64:47–70
Ospina JA (2002) Persea americana Mill. Lauraceae (Laurel Family). In: Vozzo JA (ed) Tropical tree seed manual. USDA, Forest Service, South For Exp Stat Bot Gard 14:1–120
Padulosi S, Bergamini N, Lawrence T (2012) On-farm conservation of neglected and underutilized species: status, trends and novel approaches to cope with climate change: Proc Int Conf. Friedrichsdorf, Frankfurt, 14–16 June 2011. Bioversity International, Rome, Italy
Peterson PA (1955) Dual cycle of avocado flowers. Calif Agric 9:6–7
Peterson PA (1956) Flowering types in the avocado with relation to fruit production. Calif Avocado Soc Yearb 40:174–179
Pliego-Alfaro F (1988) Development of an *in vitro* rooting bioassay using juvenile-phase steam cutting of *Persea americana*. J Horticult Sci 63:295–301
Pliego-Alfaro F, Encina CL, Barceló-Muñoz A (1987) Propagation of avocado rootstocks by tissue culture. South African Avocado Growers' Assoc Yearb 10:36–39
Popenoe W (1919) The avocado in Guatemala USDA Bulletin 743
Popenoe W (1920) Manual of tropical and subtropical fruits, excluding the banana, coconut, pineapple, citrus fruits, olive and fig. Macmillan, New York
Popenoe W (1924) Manual of tropical and subtropical fruits. MacMillan, New York
Popenoe W (1926) The parent 'Fuerte' tree. Annual report. Calif Avocado Assoc 1925–26:24–33
Popenoe W (1927) Wild avocado. Calif Avocado Soc Yearb 1927:51–54
Popenoe W (1934) Early history of the avocado. Calif Avocado Soc Yearb 1934:106–110
Popenoe W (1935) Origin of the cultivated races of avocados. Calif Avocado Soc Yearb 1935:184–193
Popenoe W (1941) The avocado: a horticultural problem. Trop Agric 18:3–7
Popenoe W (1951) 'Toltec', another new horticultural variety of avocado. Ceiba 1:225–227
Popenoe W (1952) The avocado. Ceiba 1:269–367
Popenoe W (1963) Early history of avocado. Calif Avocado Soc 1963 Yearb 47:19–24
Popenoe W, Williams LO (1947) The expedition to Mexico of October 1947. Calif Avocado Soc Yearb 1947:22–28
Popenoe W, Zentmyer GA, Schieber E (1997) The avocado has many names. Calif Avocado Soc 1997 Yearb 81:155–162
Postman JD, Hummer KE, Stover EW, Krueger R, Forsline PL, Grauke LJ, Zee FT, Ayala-Silva T, Irish BM (2006) Fruit and nut genebanks in the us national plant germplasm system. HortSci 41(5):1188–1194
Prance G, Nesbit M (2005) The cultural hstory of plants. Routledge, London
Rai MK, Jaiswal VS, Jaiswal U (2008) Encapsulation of shoot tips of guava (*Psidium guajava* L.) for short-term storage and germplasm exchange. Sci Horticult 118:33–38. doi:10.1016/j.scienta.2008.05.017
Reed BM (1990) Survival of *in vitro*-grown apical meristems of *Pyrus* following cryopreservation. HortScience 25:111–113
Reed BM (2001) Implementing cryogenic storage of clonally propagated plants. Cryo Letters 22:97–104
Reed BM (2002) Implementing cryopreservation for long-term germplasm preservation in vegetatively propagated species. In: Towill LE, Bajaj YPS (eds) Biotechnology in agriculture and forestry: cryopreservation of plant germplasm II, vol 50. Springer, Berlin, pp 22–36
Reed BM (2008) Plant cryopreservation: a practical guide. Springer, New York, ISBN 978-0-387–72275–7
Reed BM, Engelmann F, Dulloo ME, Engels JMM (2004) Technical guidelines for the management of field and *in vitro* germplasm collections. Handbook for genebanks, no 7. IPGRI/SGRP, Rome
Reed BM, Schumacher L, Dumet D, Benson EE (2005) Evaluation of a modified encapsulation-dehydration procedure in corporating sucrose pretreatments for the cryopreservation of *Ribes* germplasm. *In Vitro*. Cell Dev Biol Plant 41:431–436

Reed BM, Tanprasert P (1995) Detection and control of bacterial contaminants of plant tissue cultures. A review of recent literature. Plant Tissue Cult Biotech 1:137–142

Sahagún FB [1570] (2002) Historia general de las Cosas de la Nueva España Cien de México. México

Sahijram L, Rajasekharan PE (1996) Tissue culture strategies applicable to *in vitro* conservation of tropical fruit crops. In: Tropical Fruits in Asia: Conservation and Use; IBPGR Progress Report. International Board for Plant Genetic Resources, Rome, pp 113–111

Sandoval JA, Villalobos AVM (2002) Avocado. IPGRI Tech Bull 7:61–64. Background and methods for in vitro collection of avocado germplasm

Schalls S (1987) Propagation of avocado (*Persea americana* Mill. cv. Fuerte) by microcutting in vitro. Fruits 42(3):171–176

Schieber E, Bergh B (1987) *Persea zentmyerii*: a new species from Guatemala. Calif Avocado Soc Yearb 71:199–203

Schieber E, Zentmyer GA (1973) Collecting Persea in Central America and Mexico. Calif Avocado Soc Yearb 56:94–101

Schieber E, Zentmyer GA (1975) Exploring for Persea in Matagalpa, Nicaragua. Calif Avocado Soc Yearb 59:118–120

Schieber E, Zentmyer GA (1978) Hunting for *Persea steyermarkii* in the mountains of Guatemala. Calif Avocado Soc Yearb 62:67–71

Schieber E, Zentmyer GA, Johnson ELV (1974) Collecting Perseas in the highlands of Guatemala. Calif Avocado Soc Yearb 57:128–134

Schnell RJ, Brown JS, Olano CT, Power EJ, Krol CA (2003) Evaluation of avocado germplasm using microsatellites markers. J Am Soc Horticult Sci 128:881–889

Schroeder CA (1955) Proliferation of mature fruit pericarp tissue slices in vitro. Science 122:601

Schroeder CA (1968) The longevity of avocado tissue in vitro. Calif Avocado Soc Yearb 52:128–130

Schroeder CA (1980) Avocado tissue culture *in vitro*. Calif Avocado Soc Yearb 64:139–141

Scora RW, Bergh B (1990) The origins and taxonomy of avocado (*Persea americana*) Mill. Lauraceae. Acta Horticult 275:387–394

Scora RW, Bergh B (1992) Origin and taxonomic relationships within the genus Persea. In: Lovatt C, Holthe PA, Arpaia ML (eds) Proceedings of the Second World Avocado Congress, vol 2. University of California, Riverside, pp 505–574

Scora RW, Bergh BO, Storey WB, Kumamoto J (1970) Three-factor analysis of essential leaf oils in selected Persea species. Phytochemistry 9:2503–2507

Shibli RA, Mohamad A, Shatnawi MA, Subaih S, Mohammed M, Ajlouni MM (2006) In vitro conservation and cryopreservation of plant genetic resources: a review. World J Agric Sci 2(4):372–382

Smith Jr CE (1966) Archaeological evidence for selection in avocado. Econ Bot 20:169–175

Smith Jr CE (1966) Archeological evidence for selection in avocado. Econ Bot 20:169–175

Smith Jr CE (1969) Additional notes on pre-conquest avocados in Mexico. Econ Bot 23:135–140

Solorzano D (1989) Propagation *in vitro* of rootstocks of avocado. Calif Avocado Soc Yearb 73:149–151

Storey WB, Bergh BO, Zentmyer GA (1986) The origin, indigenous range and dissemination of the avocado. Calif Avocado Soc Yearb 70:127–133

Storey WB, Bergh B, Zentmyer GA (1987) The origin, indigenous range and dissemination of the avocado. Calif Avocado Soc Yearb 70:127–133

Stushnoff C (1991) Cryopreservation of fruit crop genetic resources. Implications for maintenance and diversity during conservation. HortScience 26:518–522

Towill LE, Roos EE (1989) Techniques for preservation of plant germplasm. In: Knutson L, Stoner AK (eds) Biotic diversity and germplasm preservation. Global imperatives. Kluwer Academic, Dordrecht, pp 379–403

Tuxill J, Nabhan GP (2001) People, plants and protected area. Earthscan. London

Vaughan DA, Balazs E, Heslop-Harrison JS (2007) From crop domestication to super-domestication. Ann Bot 100(5):893–901. doi:10.1093/aob/mcm224

Volk GM (2012) Use of cryotherapy to eradicate pathogens from horticultural crops. Meeting Abstract. http://ashs.org/abstracts/m/abstracts12/abstract_id_12207.html
Volk GM et al (2005) Ex situ conservation of vegetatively propagated species: development of a seed-based core collection for Malus sieversii. J Am Soc HorticultSci 130(2):203–210
Walters CT, Berjak P, Pammenter N, Kennedy K, Raven P (2013) Preservation of recalcitrant seeds. Science 339:915–916
Williams LO (1950) Two new Perseas from Central America. Ceiba 1:55
Williams LO (1976) The botany of the avocado and its relatives. In: Sauls JW (ed) Proceedings of the First International Tropical Fruit Short Course, Miami Beach, FL, pp 9–15
Williams LO (1977) The avocado, a synopsis of the genus *Persea*, subg. *Persea*. Econ Bot 31:315–320
Witjaksono, Litz RE, Pliego Alfaro F, Efendi D (2004) *Persea americana* Avocado. In: Litz RE (ed) Biotechnology of fruit and nut crops. CAB International, Wallingford, pp 326–349
Wolfe HS, Toy LR, Stahl AL, Ruehle GD (1949) Avocado production in Florida. Bulletin 141. Agricultural Ext Service, Gainesville, p 124
Zentmyer GA, Schieber E (1990) Persea tolimanensis: a new species for Central America. Acta Horticult 275:383–386
Zentmyer GA, Schieber E, Popenoe W (1987) Early history of the avocado during the time of the Conquistadores. South African Avocado Growers' Assoc Yearb 10:11–12
Zulfiqar B, Akhtar N, Abbasi NA, Touqeer A, Ahmad T, Hafiz IA (2009) Effect of explant sources and different concentrations of plant growth regulators on in vitro shoot proliferation and rooting of avocado (Persea americana Mill.) CV. "FUERTE". Pak J Bot 41(5):2333–2346

Orchid Fruit Diversity at Puebla Mexico: A New Insight into the Biodiversity of a Fragmented Ecosystem with Need for Conservation and Potential for Horticultural Exploitations in Future

William Cetzal-Ix, Raúl Alvarez-Mora, Saikat Kumar Basu, Jaqueline Cosme-Pérez and Eliana Noguera-Savelli

1 Introduction

Biodiversity is an important aspect of measuring the health of any ecosystem. It provides a gross estimate of the general health of any fragile ecosystem by indicating the diversity of species successfully existing within it. The orchid members could be looked upon as flagship species while assessing and estimating the quality of the environmental health of a specific ecosystem since most orchids are epiphytic in nature and they are dependent on the survival of other host plants (Newman 2009; Franck et al. 2010). If the diversity of orchid species are high that will also indicate an overall positive health and diversity parameter for the existing ecosystem

W. Cetzal-Ix (✉)
Herbarium CICY, Centro de Investigación Científica de Yucatán,
A. C. (CICY), Calle 43. No. 130. Col. Chuburná de Hidalgo,
Mérida, Yucatán 97200, México
e-mail: rolito22@hotmail.com

R. Alvarez-Mora · J. Cosme-Pérez
Instituto Tecnológico Superior de Zacapoaxtla,
Zacapoaxtla, Puebla, México
e-mail: biol.ram21@gmail.com

R. Alvarez-Mora
Jardín Botánico Xoxoctic, Cuetzalan del Progreso,
Puebla, México

J. Cosme-Pérez
e-mail: jaky_coper_9104@hotmail.com

S. K. Basu
Department of Biological Sciences, University of Lethbridge,
Lethbridge, AB T1K 3M4, Canada
e-mail: saikat.basu@uleth.ca

E. Noguera-Savelli
Francisco de Montejo, Mérida, Yucatán 97203, México
e-mail: eliananoguera@gmail.com

D. Nandwani (ed.), *Sustainable Horticultural Systems,* Sustainable Development and Biodiversity 2, DOI 10.1007/978-3-319-06904-3_9

too; hence this new approach. In addition quality plant biodiversity can be easily correlated to the positive biodiversity of different zoo taxa within that ecosystem. If the plant biodiversity is high, the animal species directly and indirectly dependent on them are also bound to be abundant, diverse and healthy. Also to the best of our knowledge and efforts, there are hardly any existing reports that have extensively studied and evaluated fruit biodiversity in an exhaustive manner on a specific plant family as is reported in the current study from a rich and biodiverse ecosystem representing the Mesoamerican region (Ramos-Ordoñez et al. 2012).

We looked upon a new innovative approach of estimating the ecological health of a specific ecosystem by looking into the reproductive biodiversity and abundance of a specific plant family within that target ecological zone. We have predominantly focused in estimating the fruit biodiversity of Orchidaceae family at Cuetzalan del Progreso, Puebla, Mexico to get an understanding of the diversity of this specific plant family and also to collect morphometric and ecological biodiversity of the members of the family.

Hence, we humbly report this to be the first technical research report on the fruit biodiversity of Orchidaceae from Cuetzalan del Progreso, Puebla, Mexico. The predominant fruit type in Orchidaceae is capsular in nature. However, wide diversity in shapes and sizes are observed among different species even within the same genus. Our study records the diverse orchid species from this unique ecosystem along with photographic, dimensional and morphometric details for the first time in technical terms. Several of this wild orchid species have also been identified to have great horticultural potential and hence in need of appropriate conservation for judicial exploitation in the future.

2 Material and Methods

Study Area Cuetzalan del Progreso is one of the municipalities of the State of Puebla in Mexico that still has conserved fragments of tropical montane cloud forests "bosque mesófilo" and tropical rain forest "selva alta perennifolia". This municipality is located in the northeastern Puebla between the limits of the Eastern Sierra Madre and the Gulf of Mexico Provinces (Veracruz border) (Morrone 2001, Fig. 1). These ecosystems is located at an altitude ranging from 180 to 1600 m, with semi-warm to warm humid climate and rainfall throughout the year (INEGI 1987; PIGEUM 2009). The ecosystem also constitutes the pine–oak and tropical deciduous forests. However, these forests have been severely fragmented, unfortunately reducing them to minor fragments of vegetation due to land use change for coffee production and for establishing grazing areas for the local livestock and dairy herds (INEGI 1985). The diversity of the vegetation types of the Gulf of Mexico lowlands and the mountainous parts of the Sierra Madre Oriental make this municipality ecologically rich with high orchid biodiversity.

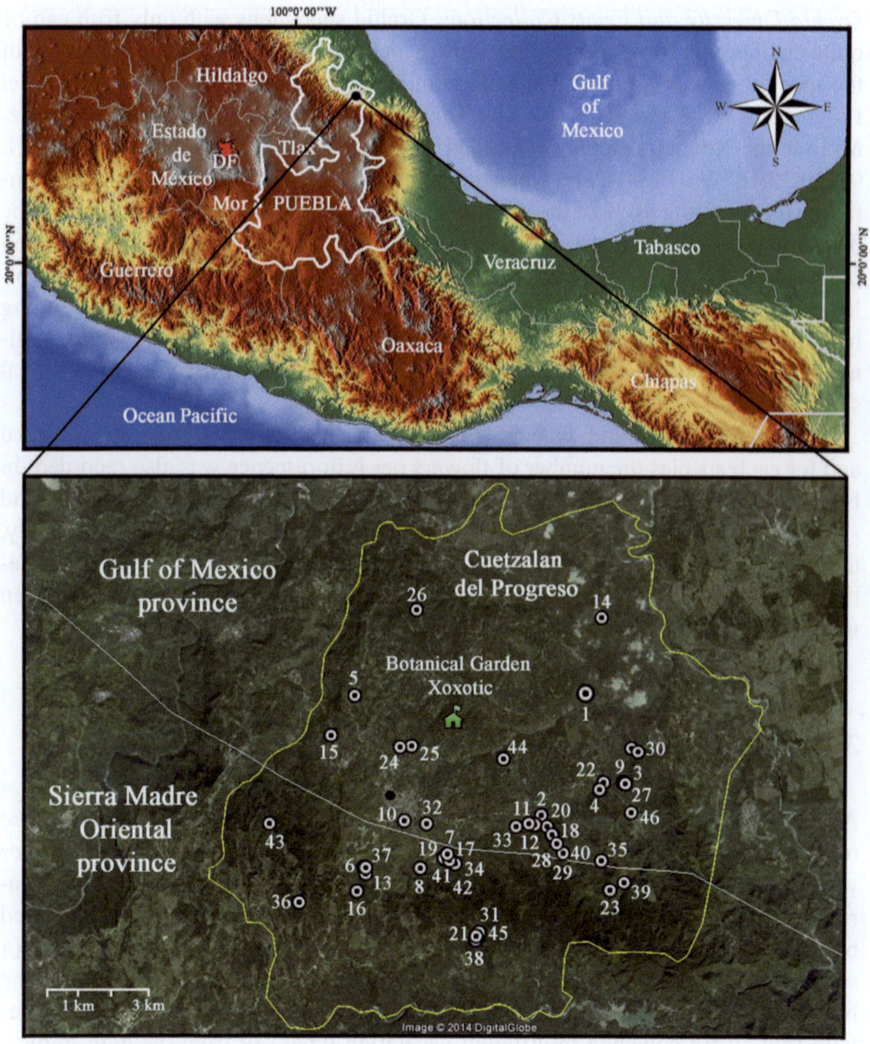

Fig. 1 Study area and distribution of orchid species in Cuetzalan del Progreso, Puebla, Mexico. *1. Campylocentrum micranthum, 2. Brassia verrucosa, 3. Camaridium cucullatum, 4. Dichaea glauca, 5. Dichaea muricatoides, 6. Elleanthus cynarocephalus, 7. Encyclia gravida, 8. Epidendrum atroscriptum, 9. E. cardiophorum, 10. E. eustirum, 11. E. melistagum, 12. E. ramosum, 13. E. veroscriptum, 14. Eulophia alta, 15. Gongora galeata, 16. Heterotaxis maleolens, 17. Isochilus major, 18. Jacquiniella equitantifolia, 19. J. teretifolia, 20. Lycaste aromatica, 21. L. deppei, 22. Masdevallia floribunda, 23. Maxillariella variabilis, 24. Nidema boothii, 25. Notylia barkeri, 26. Oncidium sphacelatum, 27. Ornithocephalus inflexus,* 28. *Platystele stenostachya, 29. Pleurothallis cardiothallis, 30. Polystachya lineata, 31. Ponera juncifolia, 32. Prosthechea cochleata, 33. P. ochracea, 34. P. pseudopygmaea, 35. P. pygmaea, 36. P. vitellina, 37. Sobralia macrantha, 38. Stanhopea oculata, 39. S. ruckeri, 40.* S. *tigrina, 41. Stelis pachyglossa, 42. S. platystylis, 43. S. rubens, 44. S. veracrucensis, 45. Stenorrhynchos speciosum, 46. Vanilla planifolia*

Orchid Diversity and Fruits Collections Orchid specimens with only fruit can be quite challenging for identification at the genera or species level. The collection of the specimens was conducted in and around the surroundings of the Cuetzalan del Progreso in 2012 and 2013 (Fig. 1). These plants were cultivated by R. Alvarez-Mora in the Xoxoctic Botanical Garden in Cuetzalan del Progreso (20°2′22.63″N, 97°30′32.16″W) (Fig. 1); and when the plants later flowered, 92 species were identified (representing 46 % of the orchids of Puebla, Alvarez et al. unpublished). The determination of flowering specimens was performed based on the digital catalog of Orchids of Mexico (Soto-Arenas et al. 2007). The habitat where they were collected have been quite close (e.g., ca. 500–1000 m of distance) to the surrounding human settlements and have been found to be severely deforested for indiscriminate agricultural activities and/or illegal timber collection for firewood by the local settlers.

During May 2013 and January 2014, 92 species with flowers were recorded and for each species the number of flowers per inflorescence was also recorded. In February 2014, the species with mature fruits was recorded, measured (long and short axes) and simultaneously photographed (Figs. 2, 3, 4, Table 1). Additionally, the number of fruits and slits, habit (epiphytic, terrestrial and litophytic), vegetation type where it was collected, altitude and the month of bloom have also been included (Table 1).

3 Results and Discussions

A total of 47 orchid species with fruits belonging to 13 subtribes and 30 genera were recorded from the habitat (Table 1). The subtribes with higher number of species were represented by Laeliinae (14 species), Pleurothallidinae (7 species), Maxillarinae and Oncidiinae (5 species). The genera with highest species were represented by *Epidendrum* L. (6), *Prosthechea* Knowles & Westc. (5 species) and *Stelis* Sw. (4 species). These three genera are represented by numerous species in Mexico and the Neotropics, 1435, 118, and 887 species, respectively (The Plant List 2013). Species with fruits were collected in different vegetation types, 29 from the high tropical rain forest, 16 from the tropical montane cloud forests, and two from the pine–oak forest.

This first report regarding the presence of orchid fruit biodiversity is a good sign that the species are being regularly visited by their pollinators in its natural habitat in this highly sensitive section of fragmented vegetation. It is important therefore to take measures to conserve these forest fragments that host a high biodiversity of species. The conservation of this fragile ecosystem is also important considering the fact that it is located within a priority area for conservation by the cloud forest and the evergreen tropical forests by the Comisión Nacional para el Conocimiento y el Uso de la Biodiversidad (Arriaga et al. 2000).

Orchids for being one of the most diverse families in species and vegetative and floral morphology it is considered one of the most important horticultural groups in

Fig. 2 Fruits biodiversity in Cuetzalan del Progreso, Puebla. **a** *Campylocentrum micranthum*. **b** *Eulophia alta*. **c** *Encyclia gravida*. **d** *Epidendrum atroscriptum*. **e** *Epidendrum cardiophorum*. **f** *E. eustirum*. **g** *E. melistagum*. **h** *E. ramosum*. **i** *E. veroscriptum*. **j** *Jacquiniella equitantifolia*. **k** *J. teretifolia*. **l** *Prosthechea cochleata*. **m** *P. ochracea*. **n** *P. pseudopygmaea*. **o** *P. pygmaea*. **p** *P. vitellina*

the world. However, due to the destruction and fragmentation of their habitat and illegal extraction of wild plants for ornamental purposes (Hágsater and Soto-Arenas 1998), their populations are severely threatened and many of its species in Mexico

Fig. 3 Fruits biodiversity in Cuetzalan del Progreso, Puebla. **a** *Lycaste aromatica*. **b** *L. deppei*. **c** *Camaridium cucullatum*. **d** *Dichaea glauca*. **e** *D. muricatoides*. **f** *Heterotaxis maleolens*. **g** *Maxillariella variabilis*. **h** *Brassia verrucosa*. **i** *Nidema boothii*. **j** *Notylia barkeri*. **k** *Oncidium sphacelatum*. **l** *Ornithocephalus inflexus*. **m** *Trichocentrum candidum*. **n** *Masdevallia floribunda*. **o** *Platystele stenostachya*. **p** *Pleurothallis cardiothallis*

and in the world are considered at some level of conservation risk by the NOM-059-SEMARNAT-2010 (SEMARNAT 2010) and IUCN (2010), respectively. In this regard, 29 species in Cuetzalan del Progreso are used as ornamentals based on their

Fig. 4 Fruits biodiversity in Cuetzalan del Progreso, Puebla. **a** *Stelis pachyglossa*. **b** *S. platystylis*. **c** *S. rubens*. **d** *S. veracrucensis*. **e** *Polystachya lineata*. **f** *Isochilus major*. **g** *Ponera juncifolia*. **h** *Elleanthus cynarocephalus*. **i** *Sobralia macrantha*. **j** *Stenorrhynchos speciosum*. **k** *Gongora galeata*. **l** *Stanhopea oculata*. **m** *S. ruckeri*. **n** *S. tigrina*. **o** *Vanilla planifolia*

spectacular vegetative or floral morphology (e.g., *Brassia verrucosa*, *Camaridium cucullatum*, *Elleanthus cynarocephalus*, *Epidendrum melistagum*, *Lycaste deppei*, *Sobralia macrantha*, *Stanhopea ruckeri*, *S. tigrina*, and *S.oculata*, Table 1, Fig. 5).

Table 1 Summary of fruit diversity morphometrics of different orchid species at Puebla, Mexico along with their respective habitat specifications and flowering period

Subtribe	Taxa	Size fruit (cm)	#Fruits/plant	Slits	Habit	Vegetation type	Altitude (m)	Month of bloom	SOP
Angraecinae	*Campylocentrum micranthum* (Lindl.) Rolfe	0.82 × 0.28	8	6	E	TRF	389	June	–
Eulophinae	*Eulophia alta* (L.) Fawc. & Rendl	4.25 × 1.28	1	6	T	TRF	330	June	+
Laeliinae	*Encyclia gravida* (Lindl.) Schltr.	3.31 × 1.04	6	3	E	TMCF	948	July	–
Laeliinae	*Epidendrum atroscriptum* Hágsater	3.69 × 1.08	1	3	L	TMCF	1028	October	+
Laeliinae	*Epidendrum cardiophorum* Schltr.	2.54 × 0.98	4	3	E	TRF	549	June	+
Laeliinae	*Epidendrum eustirum* Ames, F.T. Hubb. & C. Schweinf.	2.42 × 0.91	3	3	E	TRF	1034	October	+
Laeliinae	*Epidendrum melistagum* Hágster	3.40 × 1.94	2	3	E	TRF	732	June	+
Laeliinae	*Epidendrum ramosum* Jacq.	1.05 × 0.50	27	3	E	TRF	735	September	+
Laeliinae	*Epidendrum veroscriptum* Hágsater	3.01 × 1.22	2	3	E	TRF	1092	November	+
Laeliinae	*Jacquiniella equitantifolia* (Ames) Dressler	2.55 × 0.76	13	3	E	TRF	814	November	–
Laeliinae	*Jacquiniella teretifolia* (Sw.) Britton & P. Wilson	2.22 × 0.55	6	3	E	TMCF	949	August	–
Laeliinae	*Prosthechea cochleata* (L.) W.E. Higgins	6.13 × 2.14	4	3	E, L	TRF	963	May	+
Laeliinae	*Prosthechea ochracea* (Lindl.) W.E. Higgins	1.27 × 0.76	9	3	E	TRF	706	May	–
Laeliinae	*Prosthechea pseudopygmaea* (Finet) W.E. Higgins	2.29 × 1.04	3	3	E	TRF	1050	August	+
Laeliinae	*Prosthechea pygmaea* (Hook.) W.E. Higgins	1.49 × 0.90	5	3	E	TMCF	816	August	–
Laeliinae	*Prosthechea vitellina* (Lindl.) W.E. Higgins	3.3 × 1.59	2	3	E	TMCF	1340	October	+
Lycastinae	*Lycaste aromatica* (Graham) Lindl.	4.26 × 1.54	1	6	E	TRF	846	May	+
Lycastinae	*Lycaste deppei* (Lodd.) Lindl.	6.46 × 2.44	1	6	E	TMCF	1435	September	+
Maxillarinae	*Camaridium cucullatum* (Lindl.) M.A. Blanco	5.12 × 1.53	1	3	E	TRF	557	October	+

Table 1 (continued)

Subtribe	Taxa	Size fruit (cm)	#Fruits/plant	Slits	Habit	Vegetation type	Altitude (m)	Month of bloom	SOP
Maxillarinae	*Dichaea glauca* (Sw.) Lindl.	2.26 × 0.63	1	3	E	TRF	594	May	+
Maxillarinae	*Dichaea muricatoides* Hamer & Garay.	2.18 × 0.72	1	3	E	TRF	688	May	+
Maxillarinae	*Heterotaxis maleolens* (Schltr.) Ojeda & Carnevali	6.35 × 1.68	4	6	E	TMCF	1199	June	+
Maxillarinae	*Maxillariella variabilis* (Bateman ex Lindl.) M.A. Blanco & Carnevali	2.58 × 0.47	8	6	E	TMCF	1080	November	–
Oncidiinae	*Brassia verrucosa* Bateman ex Lindl.	6.20 × 1.88	2	3	E	TRF	690	May	+
Oncidiinae	*Nidema boothii* (Lindl.) Schltr.	3.27 × 0.57	1	6	E	TRF	864	January	+
Oncidiinae	*Notylia barkeri* Lindl.	2.50 × 0.70	1	6	E	TRF	839	January	–
Oncidiinae	*Oncidium sphacelatum* Lindl.	4.01 × 1.56	1	3	E	TRF	489	May	+
Oncidiinae	*Ornithocephalus inflexus* Lindl.	1.00 × 0.53	1	6	E	TRF	545	August	–
Oncidiinae	*Trichocentrum candidum* Lindl.	2.51 × 0.77	1	3	E	TRF	582	September	+
Pleurothal-lidinae	*Masdevallia floribunda* Lindl.	2.15 × 0.58	4	6	E	TRF	592	September	–
Pleurothal-lidinae	*Platystele stenostachya* (Rchb. f.) Garay	0.57 × 0.21	6	3	E	TRF	735	October	–
Pleurothal-lidinae	*Pleurothallis cardiothallis* Rchb. f.	5.52 × 0.69	6	6	E	TRF	763	November	–
Pleurothal-lidinae	*Stelis pachyglossa* (Lindl.) Pridgeon & M.W. Chase	2.72 × 0.63	4	6	E	TMCF	961	November	–
Pleurothal-lidinae	*Stelis platystylis* (Schltr.) Solano & Soto Arenas	1.05 × 0.45	1	3	E	TMCF	961	November	–
Pleurothal-lidinae	*Stelis rubens* Schltr.	0.57 × 0.21	2	3	E	TMCF	1002	July	–
Pleurothal-lidinae	*Stelis veracrucensis* Solano	0.73 × 0.48	3	3	E	TRF	667	October	–
Polys-tachyinae	*Polystachya lineata* Rchb. f.	1.23 × 0.42	9	3	E	TRF	364	October	–

Table 1 (continued)

Subtribe	Taxa	Size fruit (cm)	#Fruits/plant	Slits	Habit	Vegetation type	Altitude (m)	Month of bloom	SOP
Poneriinae	*Isochilus major* Schltdl. & Cham.	1.05 × 0.36	17	3	E	TMCF	991	June	+
Poneriinae	*Ponera juncifolia* Lindl.	1.23 × 0.42	1	6	E	POF	1434	November	+
Sobraliinae	*Elleanthus cynarocephalus* (Rchb. f.) Rchb. f.	1.68 × 0.52	46	3	E, L	TMCF	1079	June	+
Sobraliinae	*Sobralia macrantha* Lindl.	8.82 × 1.59	3	6	E, L	TMCF	1077	June	+
Spiranthinae	*Stenorrhynchos speciosum* (Jacq.) Rich. ex Spreng.	1.72 × 0.51	4	3	E	POF	1431	December	−
Stanhopeinae	*Gongora galeata* (Lindl.) Rchb. f.	6.74 × 1.09	7	3	E	TRF	738	September	+
Stanhopeinae	*Stanhopea oculata* (G. Lodd.) Lindl.	10.1 × 2.76	1	6	E	TMCF	1419	June	+
Stanhopeinae	*Stanhopea ruckeri* Lindl.	8.84 × 3.2	1	6	E	TMCF	997	June	+
Stanhopeinae	*Stanhopea tigrina* Bateman ex Lindl.	13.9 × 3.95	1	6	E	TMCF	909	June	+
Vanillinae	*Vanilla planifolia* Andrews	20.5 × 1.41	2	3	H	TRF	277	May	+

Habit: *E* epiphytic, *T* terrestrial, *L* litophytic. Vegetation type: *TMCF* tropical montane cloud forests, *TRF* tropical rain forest, *POF* pine–oak forest. *SOP* species with ornamental potential (based on their spectacular vegetative or floral morphology (+))

Fig. 5 Example of orchids used as ornamentals in Cuetzalan del Progreso, Puebla. **a** *Brassia verrucosa*. **b** *Camaridium cucullatum*. **c** *Elleanthus cynarocephalus*. **d** *Epidendrum melistagum*. **e** *Lycaste deppei*. **f** *Sobralia macrantha*. **g** *Stanhopea ruckeri*. **h** *S. tigrina*. **i.** *S. oculata*

Additionally, we have identified 11 species with ornamental potential for their showy flowers or that are used or sold as ornamental in neighboring states (Table 2).

The ornamental demand in Cuetzaln del Progreso is mainly covered by extracting wild specimens that are sold in the streets and markets of the municipality or are carried neighboring states such as Veracruz or Estado de Mexico. The potential of orchids as ornamental plants depends on their sustainable management for the local communities. As well as knowledge of the biology of the species and their cultivation methods, essential points for study and conservation.

Table 2 Other species with ornamental potential based on their showy flowers in Cuetzalan del Progreso Puebla, Mexico

Subtribe	Taxa	Habit	Vegetation type
Catasetinae	*Mormodes maculata* var. *unicolor* (Hook.) L.O. Williams	E	TMCF
Chysiinae	*Chysis laevis* Lindl.	E	TMCF
Chysiinae	*Chysis bractences* Lindl.	E	TRF
Coeliopsidinae	*Coelia macrosthachya Lindl.*	L	POF
Coeliopsidinae	*Coelia triptera* (Sm.) G. Don ex Steud.	L	TMCF
Laeliinae	*Encyclia candollei* (Lindl.) Schltr.	E	TMCF
Laeliinae	*Epidendrum tuxtlense* Hágsater, García-Cruz & L. Sánchez	L	TMCF
Laeliinae	*Prosthechea radiata* (Lindl.) W.E. Higgins	E	TRF
Oncidiiane	*Lophiaris lurida* (Lindl.) Braem	E	TRF
Oncidiinae	*Oncidium incurvum* Barker ex Lindl.	E	TMCF
Stanhopeinae	*Gongora truncata* Lindl.	E	TMCF

Habit: *E* epiphytic, *T* terrestrial, *L* litophytic. Vegetation type: *TMCF* tropical montane cloud forests, *TRF* tropical rain forest, *POF* pine–oak forest

3.1 Our Study Has the Following Significances:

1. It highlights the significant biodiversity of this region/ecosystem for the first time.
2. It reports for the first time the biodiversity of the members of the Orchidaceae family in this ecosystem.
3. It reports the fruit morphometric diversity across different genera and species of Orchidaceae family for the first time in this ecosystem.
4. It accounts for the dimensional and morphometric variation/variability among the fruits produced under different genera and species within orchid family members in this ecosystem for the first time.
5. It overall indicates the general positive health and status of the current ecosystem by showcasing the biodiversity of the reproductive structure (seed bearing structures) of the flagship species from this ecosystem for the first time.
6. The diversity indicated in our study also suggests that the current fragmented ecosystem needs protection for the purpose of conservation of the wide biodiversity observed in this eco-region.
7. Our study also highlights the vulnerability of this ecosystem rich in biodiversity through our study of the flagship species for the first time form this eco-region.
8. This study has also highlighted for the first time several anthropogenic factors contributing towards the vulnerability of this ecosystem region for the first time.
9. Several Orchidaceae members identified in this study for the first time have great horticultural possibilities for the near future for the first time.

10. Currently we are looking into the different pollinators of the orchid members and the information when presented will also turn into a first report for the pollinator diversity of Orchidaceae family in this ecosystem.

Factors that are vulnerable for the current ecosystem and the successful natural breeding and multiplication of the flagship species (Orchid members) along with some specific recommendations for conservation purposes are presented as follows:

1. Uncontrolled/non-monitored/random visits by poor local inhabitants and house hold members in the restricted ecosystem region for their daily sustenance due to poverty, lack of employment opportunities and women empowerment, poor health facilities, grazing and agricultural productivity and poor socio-economic conditions.
2. Unrestricted grazing of domestic animals in the vulnerable ecosystem region.
3. Sporadic use of fires within the forest by local inhabitants for cooking and other domestic purposes and unmanaged forest fires.
4. Rapid collection of rare species including orchid members and other wildlife species for the illegal wildlife and exotic pet and nursery trades growing in this region.
5. Rapid infiltration of local people into restricted forest areas to avoid confrontation with local lawmakers and peace officers after any unwanted activities or petty crimes.
6. Heavy dependence of the poor local inhabitants on forest resources has been extremely detrimental to the successful maintenance and management of the pristine ecosystem. With increase anthropogenic activities within the forest areas people are coming in close contact with the vulnerable species while moving unrestricted within the forest including the core and buffer areas. As a result, if such activities are not properly monitored in future several species including the diverse genera and species of orchids thriving within this ecosystem has the high potential of becoming critically endangered.
7. People moving up the trees for procuring honey, leaves, fruits and flowers are not only endangering the survival of threatened species of orchids, bromeliads, epiphytic ferns, mosses, lichens and fungal species thriving at or near the canopy but also highly evolved and specially adapted animals and birds; species representing diverse taxa that are directly or indirectly dependent on these plant species for breeding, nesting and food resources are being negatively impacted.
8. Hence it is important to develop a long term, eco-friendly, sustainable environment management plan for the conservation of this vulnerable ecological region housing a wide diversity of species of Mesoamerican origin and many of which are endemic to the region. It will be important to include the local inhabitants as local stakeholder in this management plan. It will be extremely important to improve the socio-economic conditions of the local people, without which no sustainable and successful environment management or species oriented conservation planning would be successful.
9. Several of the wild orchid species identified have great horticultural potential and hence in need of appropriate conservation for judicial exploitation in the future.

10. A local nursery and greenhouse should be developed for year round monitoring and surveillance of the potential species for understanding their horticultural specifications, pollinators, exact pollination mechanisms, breeding specificities, multiplication strategies, study their complete life cycles process, identify specific growing conditions and any naturally occurring biotic and abiotic diseases before looking into possibilities of commercial exploitations.

Acknowledgements The senior author is grateful to Consejo Nacional de Ciencia y Tecnología (CONACyT) by the scholarship for post-doctoral studies. The second author would thank Cristina García Juárez owner of Jardín Botánico Xoxoctic for the facilities provided for growing of orchids.

References

Arriaga L, Espinoza JM, Aguilar C, Martínez E, Gómez L, Loa E (2000) Regiones terrestres prioritarias de México. Comisión Nacional para el Conocimiento y uso de la Biodiversidad. CONABIO, México

Franck R, Michaël K, Piazza C, Laetitia H, Bertrand S (2010) Transfer of one population of a rare orchid in Corsica. Proceedings 7th European Conference on Ecological Restoration, Avignon

Hágsater E, Soto-Arenas MA (1998) Orchid conservation in Mexico. Selbyana 19:15–19

INEGI (1985) Cuetzalan del Progreso, estado de Puebla: cuaderno estadístico municipal. Instituto Nacional de Estadística y Geografía, México

INEGI (1987) Síntesis geográfica, nomenclátor y anexo cartográfico. Instituto Nacional de Estadística y Geografía, México

IUCN (2010) Guidelines for using the IUCN Red List categories and criteria. Version 8.1. Prepared by the Standards and Petitions Subcommittee in March 2010. http://intranet.iucn.org/webfiles/doc/SSC/RedList/RedListGuidelines.pdf. Accessed 21 April 2014

Morrone JJ (2001) Biogeografía de América Latina y el Caribe, vol 3. M&T-Manuales & Tesis SEA, Zaragoza

Newman BJ (2009) Orchids as indicators of ecosystem health in urban bushland fragments. Dissertation, Murdoch University

PIGEUM (2009) Prontuario de información geográfica de los Estados Unidos Mexicanos. Cuetzalan del Progreso, Puebla. Clave geoestadística 21043. Instituto Nacional de Estadística y Geografía, México

Ramos-Ordoñez MF, Arizmendi M del C, Márquez-Guzmán J (2012) The fruit of *Bursera*: structure, maturation and parthenocarpy. AoB PLANTS. doi:10.1093/aobpla/pls027

SEMARNAT (2010) Norma Oficial Mexicana NOM-059-SEMARNAT-2010, Protección ambiental—Especies nativas de México de flora y fauna silvestres—Categorías de riesgo y especificaciones para su inclusión, exclusión o cambio. Lista de especies en riesgo. Diario Oficial de la Federación 2ª Sección, 30 de diciembre de 2010. Secretaría del Medio Ambiente y Recursos Naturales, México

Soto-Arenas MA, Hágsater E, Jiménez R, Salazar GA, Solano R, Flores R, Contreras I (2007) Las Orquídeas de México: Catálogo Digital. Instituto Chinoín, A.C., México

Part III
Section C: Breeding and Improvement in Sustainable Horticultural Systems

Advances in Microbial Insect Control in Horticultural Ecosystem

Shaohui Wu, Gadi V.P. Reddy and Stefan T. Jaronski

1 Introduction

Currently, the management of insect pests in horticulture still heavily relies on the use of chemical insecticides. However, due to the environmental concerns arising from heavy and repeated use of insecticides, i.e. insecticide resistance, impact on pollinators and natural enemies, the desirability of more environmentally friendly practices in pest management is increasing for the sustainable development of cropping ecosystem. Under these circumstances, the development of biorational insecticides like entomopathogens has been given more attention.

The first attempt of microbial insect control started in about 130 years ago, when Metschnikoff used *Metarhizium anisopliae* (Metchnikoff) Sorokin to control the wheat chafer, *Anisoplia austriaca* (Herbst) (Coleoptera: Rutelidae) in Russia (Lord 2005). In the 1890s the Kansas State Department of Agriculture unsuccessfully attempted a large scale introduction of *Beauveria* to control the chinch bug (Snow 1893). While epizootics were widely observed, the insect populations were not controlled. The practical and commercial use of microbial products did not really begin until the development of *Paenibacillus popilliae* (Dutky) (formerly *Bacillus popilliae*) and *Bacillus thuringiensis* Berliner (*Bt*) in the early 1900s. In particular, since its first commercialization in 1938, *Bt* has been relatively broadly used in different cropping systems, and now takes up to 75 % of the biopesticide market in the USA, Canada and Mexico (Bailey et al. 2010). The history of using entomopathogenic nematodes (EPNs) dates back to the 1930s when Rugolf William Glaser (1888–1947) pioneered research on the culture and field application of EPNs, including using *Steinernema glaseri* (Steiner) against the Japanese beetle *Popillia japonica* Newman (Coleoptera: Scarabaeidae) (Gaugler and Kaya 1990). Compared

G. V.P. Reddy (✉) · S. Wu
Western Triangle Ag Research Center, Montana State University, Conrad, MT 59425, USA
e-mail: gadi.reddy@montana.edu

S. T. Jaronski
USDA ARS Northern Plains Agricultural Research Laboratory, Sidney, MT 59270, USA

D. Nandwani (ed.), *Sustainable Horticultural Systems,* Sustainable Development and Biodiversity 2, DOI 10.1007/978-3-319-06904-3_10

with other microbial biocontrol agents, the use of viruses in insecticidal pest control is more recent. Although the role of viruses in insect control was first recognized in the 1940s, the commercialization of viral products did not start until 1975 when *Helicoverpa zea* (Boddie) (Lepidoptera: Noctuidae) nuclear polyhedrosis virus (NPV) was first registered in the USA (Lord 2005). The world commercialization status of these microbial pesticides was recently summarized by Kabaluk et al. (2010).

These biopesticides are good alternatives to conventional insecticides for sustainable management of insect pests. However, high costs, short shelf-life, instability and inconsistency in field efficacies make them much less competitive than chemical insecticides in the market share. Numerous studies have been carried out to explore the potential for increasing their compatibility, e.g. lowering costs in produce, improving field performance and stability, etc. Following is a general review of the biology, current commercialization status, improvements in mass production, formulation, and application techniques of microbial biopesticides used in horticultural cropping systems.

2 Entomopathogenic Bacteria

2.1 Biology

Various spore-forming and nonspore-forming bacteria are pathogenic to insects. The most notable spore-forming bacteria used as biopesticides are *Bt* and *P. popilliae*. *Bt*, in particular, has served as a corner stone for microbial control of insects.

The insecticidal activity of *Bt* relies on a number of protein toxins, consisting of Cry (crystal), Cyt (noncrystalline, cytoplasmic) or VIP (vegetatively expressed insecticidal protein) toxins (Crickmore et al. 1998, 2013). There are currently 72 major Cry protein groups, the specificity of which is not known for many. Cyt toxins consist of three main groups, while VIP toxins fall into four major classes, each with many subclasses (Crickmore et al. 2013). The mode of action for *Bt* is to kill the insect by disruption of the midgut epithelium followed by disruption of the haemocoel equilibrium, and finally septicemia (Raymond et al. 2010). There are several subspecies of *Bt* currently being commercialized for the management of insect pests in horticultural cropping systems. These include, *B. thuringiensis kurstaki* & *aizawai* for the control of lepidopteran pests, and *B. thuringiensis tenebrionis* for beetle pests (Bravo et al. 2011). Another subspecies of *Bt*, *B. thuringiensis japonensis* strain Buibui has good potential for suppressing white grubs, especially Japanese beetle *P. japonica* (Bixby et al. 2007), but it is not yet registered for commercialization. Although the host range of these bacteria were initially known to be restricted to specific insect orders, there have been increasing reports on *Bt* proteins with pesticidal activities across orders and phyla, as reviewed by van Frankenhuyzen (2013). Host range specificity of *Bt* pesticidal proteins need to be more strictly screened to ensure the ecological safety in practical pest control applications.

Paenibacillus popilliae, also called milky spore disease, is another type of bacteria pathogenic to *P. japonica*, and has been used for controlling this insect in the USA for more than 60 years (Lord 2005). Other than these entomopathogenic bacteria, another species *Brevibacillus laterosporus* (Laubach) Shida et al. was recently found to have biopesticidal potential for control of insect pests in different orders including Coleoptera, Lepidoptera, and Diptera (Ruiu 2013).

Compared with spore-forming bacteria, the non-spore-forming types have not received much attention, and only a few of these types of bacteria have been used for insect control. Among those, the most notable example is *Chromobacterium subtsugae* Martin et al., which is the first significant non-*Bacillus* species to be commercialized. Currently, this bacterium is being commercially produced by Marrone Bio Innovations, Davis, CA (USA) for managing a variety of insect and mite pests, e.g. European corn borer *Ostrinia nubilalis* (Hübner) (Lepidoptera: Pyralidae), citrus rust mite *Phyllocoptruta oleivora* (Ashmead) (Acari: Eriophyidae), and stink bugs (Hemiptera: Pentatomidae), on agricultural and greenhouse crops, including vegetables, fruit, flowers, bedding plants, ornamentals, and turf (US EPA 2012). Also, bacterium *Serratia entomophila* Grimont et al. is effective against the grass grub *Costelytra zealandica* (White) (Coleoptera: Scarabaeidae), and has been commercialized since 1990 in several formulations (Johnson et al. 2001; Townsend et al. 2004). Recently, *Photorhabdus* and *Xenorhabdus*, which are symbiotic bacteria associated with EPNs, have also received research attention for their insecticidal activities against many insect pests. Due to the high toxicity against insects and the increasing pressure arising from insecticide and *Bt* resistance, the potential for insecticidal toxins isolated from these bacteria as novel insecticidal proteins for insect control is currently under investigation (ffrench-Constant et al. 2010).

2.2 *Bt* Resistance and Strain Improvement

Insects have evolved resistance to *Bt* Cry toxins, due to high selectivity and toxicity, and the broad-scale planting of *Bt*-transgenic crops. This was first reported in laboratory selection of *Bt kurstaki* (*Btk*) in 1985 by McGaughey (1985) in the Indian meal moth, *Plodia interpunctella* (Hübner) (Lepidoptera: Pyralidae), and in field conditions by Kirsch and Schmutterer (1988) in 1988 in the diamondback moth, *Plutella xylostella* (L) (Lepidoptera: Plutellidae) from the Philippines, followed with subsequent cases reported worldwide for *P. xylostella* in *Bt* products (Tabashnik 1994) and *Bt* transgenic crops (Tabashnik et al. 2008). Other than *P. xylostella* and *P. interpunctella*, the cabbage looper, *Tricoplusia ni* (Hübner) (Lepidoptera: Noctuidae) is another insect species that has evolved resistance to *Bt* (Janmaat and Myers 2003).

To alleviate the resistance problem, new strains or isolates of *Bt* are required for the development of novel biopesticides. Strain selection and improvement can be achieved via genetic tools to increase toxicity against target insects, broaden the host spectra, improve field persistence, and optimize fermentation production, as reviewed by Sansinenea et al. (2010). Although there is little new information on the genetic manipulation, the desirability for discovering new strains and biopesticides to mitigate insecticide and *Bt* resistance cannot be underscored for the sustainable development of agriculture.

2.3 Mass Production and Formulation

Entomopathogenic bacteria, predominantly *Bt*, are mass-produced via liquid fermentation, with selective media combined of carbon, nitrogen and mineral resources (Table 12.2 in Couch and Jurat-Fuentes (2013)). The prevention of contamination, and the loss or exchange of Cry-gene containing plasmids of *Bt* during sub-culturing are the major challenges in fermentation process (González et al. 1982), which could be alleviated by limiting the number of transfers from the parent culture.

Bacterial biopesticides (mainly *Bt*) are often formulated as sprayable powders, which can be stored for years, maintaining viability of the spores and insecticidal efficacy. The bacteria produced from fermentation are concentrated via centrifuging, mixed with aqueous dispersing agents, and then dried to form powders (Behle and Birthisel 2013). Spray drying is the commonly method for drying *Bt* during the formulation process. The mechanism of this technique is spraying an aqueous suspension of propagules into a chamber of heated air that rapidly vaporizes the water, leaving the solids as a powder to be collected. Various ingredients are included in spray dried formulation of *Bt*, for example, tapioca starch, sucrose, milk powder, silica fume, polyvinyl alcohol, Tween 20, rice bran, oil, and antifoam solution, as reported by Teera-Arunsiri et al. (2003). Commercial products may also include other components.

The aqueous liquid formulation is another processing form for bacterial-based products in the industry. Based on the concentrated fermentation slurry (FC), the formulation combines ingredients, including a dispersant, a suspending agent, an antievaporation agent, a stabilizer, and sometimes a thickener when the FC has a very low viscosity, for optimal storage and field delivery. During this process, to prevent secondary fermentation, bacteriostatic and fungistatic agents are added for FC stabilization. Compared with powder formulations, liquid formulations are cheaper to produce and formulate, but they require more strict conditions for storage, as they are more susceptible to temperature extremes (preferably 10–15 °C) and exogenous contamination (Couch and Jurat-Fuentes 2013).

Although commercialization of *P. luminescens*, the symbiotic bacterium of *Heterorhabditis* spp., is still under development, researchers found that the bacterium encapsulated in sodium alginate beads successfully infected *Spodoptera litura* (Fabricius) (Lepidoptera: Noctuidae) larvae exposed in a sterile soil environment (Rajagopal et al. 2006).

3 Entomopathogenic Viruses

3.1 *Biology*

The entomopathogenic viruses important in insect control have been described from four families, the *Baculoviridae* (nucleopolyhedrosis viruses, NPVs; and granuloviruses, GVs), *Reoviridae,* specifically the genus *Cypovirus* (cytoplasmic polyhedrosis viruses, CPVs), *Iridoviridae* (iridescent viruses), and *Poxviridae* (entomopoxviruses, EPVs) (Jaronski 2012). Among them, baculoviruses have received the most attention for commercialization and practical use in pest management (Huber 1986; Miller 1997). The viruses enter insect hosts via ingestion as occlusion bodies, which are dissolved in the insect's alkaline midgut, releasing the virions to infect the columnar cells of the midgut epithelium, and eventually kill the host insect. Although they are reported to infect insects from a number of orders, like Lepidoptera, Diptera, Coleoptera, and Hymenoptera (mainly sawflies), their practical use is limited to lepidopteran insects infesting forest, vegetable and orchard crops.

Among baculovirus described, one of the most studied species is *Autographa californica* multicapsid nucleopolyhedrovirus (AcMNPV), which was originally isolated from the alfalfa looper *A. californica.* The use of AcMNPV protected over 2 million ha of soybean against velvet bean caterpillar, *Anticarsia gemmatalis* Hübner (Lepidoptera: Noctuidae) in Brazil, which has been considered one of the most successful cases of insect control with microbial viruses (Moscardi et al. 2002, 2011). Other than AcMNPV, a number of viruses have been studied extensively, and have been commercialized successfully, e.g. codling moth, *Cydia pomonella* (L.) (Lepidoptera: Tortricidae), granulovirus (CpGV) (Cory and Franklin 2012). These viruses generally have a narrow host range, which makes them safe to use (Gröner 1990), and is also a major drawback limiting their market share. Nevertheless, there are a number of successful products.

3.2 *Strain Improvement*

Slow speed of kill and narrow host range are the major barriers that hinder the massive-scale use of entomopathogenic viruses. This may be achieved by strain improvement via genetic approaches, reviewed by Haase et al. (2013). Speed of kill may be improved by manipulating virus-encoded genes to interfere host hormones. For example, when a diuretic hormone gene was introduced into *bombyx mori* L. (Lepidoptera: Bombycidae) baculovirus (BmNPV) genome, the time of insect killing was increased by 20 % faster (Maeda 1989). Other than introducing hormone genes, another approach is to inhibit the juvenile hormone genes that promote insect growth. After deleting the virus-encoded ecdysteroid glucosyltransferase (egt) gene, infection with the recombinant AcMNPV resulted in a 30 % faster killing of larvae and significant reduction in food consumption (O'Reilly and Miller 1991). Another strategy to improve the speed of kill is the insertion of insect-specific toxins, which is considered one of the most promising tactics that have been explored for strain improvement (Carbonell et al. 1988). The development of baculovirus recombinants with expanded host range may be addressed by the selection of appropriate promoters for the expression of heterologous genes (Haase et al. 2013).

3.3 Mass Production and Formulation

Different from other entomopathogens, insecticidal viruses must be produced in living cells, either *in vivo* by inoculating live insects, or *in vitro* with cultured insect cell lines, propagated using bioreactor technologies, as reviewed by Reid et al. (2013). In particular, *in vivo* methods have been predominant for commercial production of baculovirus, as documented by Harrison and Hoover (2012), Moscardi et al. (2011), and Buerger et al. (2007).

Kabaluk et al. (2010) recently reviewed the use and production of microbial biopesticides, and listed a number of *in vivo*-produced baculoviruses in several countries over the world. Those include baculoviruses for control of heliothine larvae (Lepidoptera: Noctuidae) (*Heliothis virescens* (Fabricius) and *Helicoverpa zea* Boddie in North America; *Helicoverpa armigera* Hübner in China, India, and Australia), *A. gemmatalis* in Brazil and Argentina, *C. pomonella* in Europe, Argentina, Canada, the United States, and New Zealand, and various *Spodoptera* pests (*S. litura* in China, India, and Europe; *Spodoptera exigua* Hübner (Lepidoptera: Noctuidae) in China, Europe, and the United States). Among them, NPV production in Australia using an automatic insect-handling system (Buerger et al. 2007), has been one of the most successful examples of viral production *in vivo*.

As of today, production of baculovirus *in vitro* has not been as successful as *in vivo* methods. The predominant barrier for *in vitro* production is quality control, which depends strongly on the cell line used for amplification and for testing (Reid et al. (2013). The cell culture quality may vary due to the variability in the source material for medium and manufacturing process of hydrolysates. In addition, varying infection levels in various cell lines may be also resulted even from the same batch of viruses. Also, the occlusion derived viruses (ODVs) produced in cell culture have inferior speed of kill compared to ODVs produced in larvae, as shown for AcMNPV (Bonning et al. 1995). Future directions for *in vitro* production include improving chemically defined media (e.g. low-cost serum-free medium) for insect cell culture, increasing virus production yields via genetic modifications of the host cell lines, etc.

Produced baculovirus occlusion bodies are commonly formulated in aqueous liquid, with the addition of glycerol, to limit the growth of contaminating microbes. Compared to dry formulations, aqueous formulation are superior in retaining the insecticidal activity of the virus. However, viral infectivity and viability are highly susceptible to high temperatures (Lasa et al. 2008), and UV degradation (Salamouny et al. 2009), which thus need to be considered during storage and application process. Several plant extracts, e.g. tea, coffee and lignin, hold good potential for protecting baculovirus from UV degradation (El-Salamouny et al. 2009; Salamouny et al. 2009). Also, CpGV encapsulated in particles from gas-saturated solutions (PGSS) displayed no loss of biological activity, when these particles included UV protectants and phago-stimulants (Pemsel et al. 2010). One notable synergist of virus activity against the Lepidoptera has been a class of optical brighteners incorporated into the formulation (Shapiro and Argauer 1997).

4 Entomopathogenic Fungi (EPF)

To date, at least 90 genera and over 700 species of EPF have been described (Goettel et al. 2010). However, commercial development of EPF has been confined to species in the Hypocreales, most notably *Lecanicillium* spp. (previously *Verticillium* spp., i.e. *V. lecanii*), *Beauveria bassiana* (Bals.) Vuill., *B. brongniartii* (Sacc.) Petch, *Metarhizium* spp., and *Isaria fumosorosea* (Wize) (formerly *Paecilomyces fumosoroseus*) (Table 1). In 2006, at least 171 commercial products containing these fungi were identified (De Faria and Wraight 2007).

4.1 Biology

For most EPF, the route of infection is direct penetration of the cuticle wall, not via ingestion. Spores or conidia adhere to the cuticle of susceptible hosts, and germinate hyphae tubes to penetrate the body wall directly, probably with the aid of both enzymatic degradation (e.g. chitinases and proteases) and mechanical pressure. EPF kill the host by a variety of means, such as starvation, nutrient depletion, or body obstruction by the hyphae. Pre-lethal effects, such as reduced feeding or oviposition, diminished response to or production of sex pheromones, etc., may be observed in EPF-infected larvae, as reported in the diamondback moth, *Plutella xylostella* L. (Lepidoptera: Plutellidae), infected with *Zoophthora radicans* Brefeld (Furlong et al. 1997; Reddy et al. 1998).

Some fungi may produce insecticidal toxins, such as destruxins, which are cyclic peptide toxins secreted by *Metarhizium* spp. (Anke and Steiner 2002), *Aschersonia* sp. (Krasnoff et al. 1996), and *Beauveria felina* (DC.) J.W. Carmich (Kim et al. 2002), Beauvericin produced by *B. bassiana*, etc., to assist in pathogenesis (Zimmermann 2007a, 2007b). Currently, more than 35 different destruxins have been described (Liu et al. 2004). In addition to toxicity at high doses, some destruxins may reduce growth and reproduction of the host (Brousseau et al. 1996), act as repellent or antifeedant (Robert and Riba 1989; Amiri et al. 1999; Thomsen and Eilenberg 2000), or be linked to increased host range (Amiri-Besheli et al. 2000). Synthetic analogs of destruxins have been proven to be toxic to lepidopteran insects, and their potential as novel pesticides has been considered (Thomsen and Eilenberg 2000).

The role of EPF in insect control has been extensively reviewed by many authors (Ferron 1985; Carruthers and Soper 1987; McCoy et al. 1988; Samson et al.

Table 1 Entomopathogenic fungi produced commercially, their major market and main target insect pests in horticulture (Bradley et al. 1992; Feng et al. 1994; Shah and Goettel 1999; Copping 2001; Lomer et al. 2001; Wraight et al. 2001; Alves et al. 2003; De Faria and Wraight 2007; Kabaluk et al. 2010; Khan et al. 2012)

Fungus name	Market	Main target pests
Beauveria bassiana (Bals.) Vuill	China	Forest and crop pests
	USA, Mexico, Denmark, Italy, Spain, Sweden, Japan Greece, Switzerland	Aphids, whiteflies, thrips, weevils, white grubs, leaf beetle, mealybug, lepidopteran, grasshoppers
	Czech Republic	Weevils, leaf beetle
	Costa Rica & Panama	Weevils
	Colombia	Coleopteran, dipteran, hemipteran & lepidopteran
	France	Lepidoptera (Crambidae)
	South Africa	Weevils, aphids
	India	White grubs, weevils, coffee berry borer, diamondback moth, thrips, grasshoppers, whiteflies, aphids, codling moth
	Russia	Thrips, whiteflies
	Brazil	Weevils, whiteflies, termites, long-horn beetle
	Venezuela	Lepidoptera (Noctuidae)
Beauveria brongniartii (Sacc.) Petch	Columbia	Coleopteran, homopteran, lepidopteran & dipteran
	Japan	Longhorn beetles
	Europe	European cockchafer & other white grubs
	Peru	Andean potato weevil
Lecanicillium longisporum (Petch) R.Zare & W.Gams (formerly *Verticillium lecanii*)	Finland, Switzerland, UK, Japan	Aphids
	Brazil	Whiteflies, ensign scale insects
Lecanicillium muscarium (Petch) R.Zare & W.Gams (formerly *V. lecanii*)	Netherlands, Denmark, Finland, France, Italy, UK, Switzerland, Turkey, Japan	Whiteflies, thrips
	Russia	Whiteflies, aphids
Lecanicillium sp. (formerly *V. lecanii*)	Switzerland	Whiteflies, aphids, thrips
	India	Whiteflies, aphids, mealybug, scale insects
	Mexico	Aphids, thrips
	Brazil	Aphids, whiteflies, ensign scale insects
	Colombia, Peru, Costa Rica, Honduras	Whiteflies, other hemipteran, dipteran
Metarhizium anisopliae (Metschn.) Sorokin[a]	Brazil	Sugarcane spittle bugs

Table 1 (continued)

Fungus name	Market	Main target pests
	Australia	Red-headed cockchafer
	China	Grasshoppers, locusts
	India	Termites, leafhoppers, beetles, grubs other coleopteran, lepidopteran
	Austria & Italy	White grubs, weevils, sap beetles
	Switzerland	White grubs, weevils
	Germany	Weevils, other coleopteran
	Mexico	White grubs, weevils, grasshoppers, Spittlebugs, termites
	USA	Weevils, white grubs, whiteflies, thrips, dipteran, termites
	Costa Rica & Panama	Spittle bugs
	Colombia	Coleopteran, hemipteran, lepidopteran, orthopteran
	Venezuela	White grubs, spittle bugs, aphids
M. acridum (Driver & Milner) J.F. Bisch., Rehner & Humber	Africa & Australia	Grasshoppers, locusts
Isaria fumosorosea (Wize) (formerly *Paecilomyces fumosoroseus*)	Europe & Japan	Whiteflies
	USA, Mexico & Colombia	Whiteflies, aphids, thrips
Isaria sp. (formerly *Paecilomyces* sp.)	India	Whiteflies, thrips
Conidiobolus thromboides Drechsler	Colombia, Costa Rica, Honduras	Whiteflies
	China	Aphids
	South Africa	Aphids, thrips
Nomuraea rileyi (Farl.) Samson	Columbia	Lepidopteran
Sporothrix insectorum de Hoog & H.C.Evans[b]	Brazil	Lace bug

[a] Some products (i.e. Met−52 produced by Novozymes) are recently identified as *M. brunneum*.
[b] Further species identification is required

1988; Evans 1981; Ferron et al. 1971; Glare and Milner 1991; Roberts and Hajek 1992; Tanada and Kaya 1993; Hajek and St. Leger 1994; Hajek 1997b; Boucias and Pendland 1998; Wraight and Carruthers 1999; Butt et al. 2001; Upadhyay 2003; Wraight et al. 2007; Goettel et al. 2010). Researchers recently found that some EPF, mainly *Beauveria* spp. and *Metarhizium* spp., demonstrate endophytic characteristics (Posada and Vega 2005), and have potential to suppress insect pests (Parsa et al. 2013) and plant pathogens (Ownley et al. 2008), and to promote plant growth (Elena et al. 2011). Further investigations are required to verify the role and incidence of fungal endophytic colonization in these interactions. Due to generally slow onset of mortality and incomplete efficacy, EPF when used as mycoinsecticides are not suitable for the control of insect pests in crops that have very low tolerance to damage, or when the pest pressure is high and immediate control is required.

In addition, efficacy of EPF is also restricted by environmental conditions at the site of application. Various environmental constrains, both biotic and abiotic, affect the survival, pathogenicity and persistence of fungi. Among abiotic factors, temperature, moisture, rainfall (Wraight et al. 2007), solar radiation (Braga et al. 2001), soil texture (Ferron 1971), inorganic matter, pH, aeration, and pesticides (McCoy et al. 1988; Jaros-Su et al. 1999), are the most critical components that have significant influence on field efficacies (Jaronski 2010). The optimum temperatures for infection, growth and sporulation are usually restricted to 20–30 °C, varying with species and strains (Müller-Kögler 1965; Walstad et al. 1970; Roberts and Campbell 1977; Alves et al. 1984; Hywel-Jones and Gillespie 1990; Welling et al. 1994; Ekesi et al. 1999; Milner et al. 2003), although some cold-active or heat tolerant isolates were also found to be able to grow outside the range (Glare 1992; Rath et al. 1995). Conidia or spores of many fungal species may only germinate at R.H. 90 % or above (Zimmermann 1986; Milner 1989), although it is rarely a limitation in the soil environment unless the fungi are exposed on the surface. In foliar applications, spore germination is related to the phylloplane microhabitat where, under the boundary layer of air, RH can be much higher than ambient (Jaronski 2010). In many cases, high moisture is desirable for the effective use of fungal agents, whereas dry weather conditions have been considered responsible for control failures. Biotic factors include the level of fertilizer or organic matter (Dutky 1959; Oliveira et al. 1981), plant root system, microflora and fauna populations. Parasitism by other organisms such as fungi and bacteria is one of the major factors that debilitate the control success of EPF in the field (Coles 1980; Lingg and Donaldson 1981).

Most EPF do not pose a threat to human and other vertebrates, except for immune-compromised individuals (Saik et al. 1990; Siegel and Shadduck 1990; Zimmermann 2007a, 2007b). There are a few exceptions, such as *Conidiobolus coronatus* (Costantin) Batko, *Aspergillus flavus* Link, *Paecilomyces lilacinus* (Thom) Samson, which are pathogenic to vertebrates and thus are not considered for commercialization as microbial pesticides. However, such pathogenicity may vary with strains, e.g. the strain P 251 of *P. lilacinus*, isolated in the Philippines, has been successfully developed as a safe bio-nematicide (Copping 2001). Before being registered as a microbial control agent, any EPF must undergo a series of stringent tests for potential harmful effects on mammals and other vertebrates (Laird et al. 1990; Siegel 1997; Kabaluk et al. 2010; Jaronski 2012).

4.2 Strain Improvement

Limitation in commercial use of EPF as mycoinsecticides is mostly due to slow speed of kill, low efficacy or instability in field performance (St. Leger and Wang 2009). Recent efforts in improving fungal efficacy have mainly focused on strain selection, genetic modification (St. Leger and Screen 2001), formulation (Wraight et al. 2001), and application strategies (Bateman and Chapple 2001). Because natural variation between strains of fungal species exists, strains are selected for best potential in commercialization based on virulence, host range, environmental persistence, as well as mass production and nontarget safety (Ravensberg 2011). In the process of strain selection, both biotechnological and genetic techniques may be adopted. Focusing on *M. anisopliae* and *B. bassiana*, a number of genes were found involved in fungal pathogenicity (Freimoser et al. 2005; Wang et al. 2005; Cho et al. 2006; Cho et al. 2007), e.g. cuticle degrading protease (St. Leger et al. 1992), chitinase genes (Bogo et al. 1998; Screen et al. 2001; Bagga et al. 2004; Fang et al. 2005), adhesin that assists spore attachment (Wang and St. Leger 2007), and a cell protective coat protein helping evasion the host immunity recognition (Wang and St Leger 2006). Modification of these genes may improve the virulence of fungal strains and shorten the killing time, which may be achieved by transformation systems (Bernier et al. 1989; Goettel et al. 1990; St. Leger et al. 1995; Inglis et al. 1999; Sandhu et al. 2001) and strain improvement (St. Leger et al. 1996; St. Leger 2001). For example, the overexpression of the *pr1* gene by insertion of multiple copies resulted in increased speed of kill of a host insect, despite poor sporulation ability (St. Leger 2001). Nevertheless, there are significant regulatory concerns that must be considered before a transformed fungus can be commercialized, as discussed in the regulation part of this chapter.

4.3 Mass Production

Commercial EPF can be mass-produced in two types of fermentation process, solid substrate fermentation and submerged liquid fermentation, as reviewed by Jaronski (2013).

The solid substrate fermentation generally uses organic, nutritive materials, mainly grains such as rice and barley, as substrates for the Ascomycetes (a complete list of organic substrates is presented in Bartlett and Jaronski (1988) and Jaronski (2013)). In recent years, several inorganic substrates have been evaluated. These include granules of calcined diatomaceous earth (diatomite) (Crangle (2011; Jaronski and Jackson (2012; Wikipedia (2013), and open-pored clay granules (e.g. Seramis®; Seramis (2012)). For commercial use, the conidia produced on solid substrates are dried to moisture content < 9 % w/w or $a_w \leq 0.3$ for optimal shelf life, and then separated from the dried substrate mechanically, either sieving or agitation with air collection using a cyclone dust collector (Bateman 2007; Jaronski and Jackson 2012). This technique may have limitations in massive scale-up production, due to the large amounts of solid substrates required for fermentation. See Jaronski (2013) for a fuller discussion of this topic.

More often, the submerged liquid fermentation technique can be adopted for massive scale-up production of commercial mycoinsecticides, including *I. fumosorosea* and *Lecanicillium* spp., in addition to *Metarhizium* spp. and *Beauveria* spp. (Jaronski 2013). The fermentation process in this technique requires much shorter time (hours versus days as for solid substrate fermentation). Cottonseed flour (Pharmamedia®, Archer Daniels Midland Company, Decatur, IL, USA), and casamino acids (e.g. Solulys) are the commonly used media for commercial scale fermentations. The fermented liquid culture is typically harvested with filtration, and then processed further to stabilize the fungal materials. Fungal propagules may be converted into a dry powder via simple air drying, freeze drying, or spray drying to preserve the viability of the propagules. However, it has been a major challenge to prevent the propagules, blastospores in particular, from viability loss especially during storage, although some success has been achieved by spray drying.

4.4 Improvement in Formulation and Application Technique

Mass-produced fungal propagules need to be formulated properly for being applied as biopesticides, as well as maintaining the shelf stability. EPF have been prepared for use in both liquid, such as vegetable and mineral oils (for ultra-low volume application), emulsifiable oils, and dry formulations, such as granules and wettable powder (De Faria and Wraight 2007). Fungal conidia produced with solid substrate technique can be easily formulated as wettable powders by adding dry ingredients to improve spore wetting in water for aqueous spray, as well as in oil-based formulations, because of the hydrophobic nature of the spore wall.

Formulation technology involves a specific combination and process of various ingredients, including active ingredient (propagules, e.g. conidia), adjuvant, and inert carrier (Burges 1998). Ease of use, and efficacy characteristics, such as fungal viability, storage stability, pathogenic activity, and post-application persistence, are the major considerations for product formulation. Adding oils can improve adhesion of fungal conidia to insect cuticle. For example, Leemon and Jonsson (2008) reported that the addition of 10% oil emulsion to *M. anisopliae* improved the speed of kill of the cattle tick, *Rhipicephalus microplus* (Canestrini) (formerly *Boophilus microplus*) (Ixodida: Ixodidae). Also, in an earlier study of Bateman et al. (1993), when formulated in cotton seed oil, *Metarhizium flauoviride* Gams & Roszypal demonstrated enhanced infectivity to desert locust *Schistocerca gregaria* Forsskål (Orthoptera: Acrididae) at low humidities. Proper formulation may improve conidial storage stability and shelf life. For example, hydrophobic fungal conidia tend to have good storage stability in oils (Burges 1998). A significant advantage of oil-based formulation over use of dry spores is that the oils seem to protect the spores from imbibitional damage, leading to a rapid and considerable loss in germination ability, when dry spores are suspended in water (Faria et al. 2009; Xavier-Santos et al. 2011).

Surfactants are often added in oil formulations to allow the oils to mix with water for aqueous spray applications (Santi et al. 2011). But not all surfactants or emulsification agents are compatible with fungus spores. Santos et al. (2012) tested the effect of various surfactants on conidia dispersal, germination, and colony growth, with results variable from no impact to significant inhibition of fungal growth. Choice of surfactant is critical. Jin et al. (2008) found that, varying with fungal strains, surfactants added to conidia demonstrated different hydrophilic–lipophylic balance (HLB) values for optimal wetting time and conidia suspension. Because of

Fig. 1 Directed spray application of mycoinsecticide using conventional spray equipment that directs most of the spray into the cucurbit canopy. (*Top*) Spray designed to treat two rows of cucurbits. (*Bottom*) Close-up of drop tube and nozzle arrangement. Note height of nozzles in relation to plant canopy and rearward direction of spray. (Reprint from Fig. 1 in Jaronski (2010).)

the mode of action, the fungal conidia need to have adequate surface attachment with the host insects to be effective. Due to poor permeability in soil, efficacy of the fungal products in oil-based formulations may be compromised for the control of soil-inhabiting insects that dwell deep in the soil, compared to strictly aqueous suspensions of spores. Wu (2013) demonstrated that the granular formulation of *M. anisopliae* dissolved in water appeared to have higher field efficacy than the oil-emulsified product against white grubs in turf, and the addition of a wetting agent Silwet L-77 surfactant increased the effect of control, although no statistical significances were detected.

In practical use, fungal biopesticides are normally delivered to the field by conventional spraying technique, except for products formulated as granules (*Metarhizium* spp.), which are applied as dry materials. It is extremely important to remember, however, that fungal spores require physical contact with the target insects, thus should be applied much like contact insecticides. In addition, an efficacious dose is often several hundred to several thousand spores per insect, requiring very large numbers of spores per hectare (see Jaronski (2010) for a discussion of this aspect). In many cases typical spray applications do not deliver sufficient numbers of spores to the insects for good control, for example whitefly nymphs, which are restricted to the undersides of leaves. Often ingenuity is required to modify application equipment and methods to optimize fungus spore delivery to the insect pest (Jaronski 2010) (Fig. 1).

5 Entomopathogenic Nematodes (EPNs)

EPNs are safe to plants, vertebrates and other non-target organisms (Poinar 1989; Akhurst 1990; Georgis et al. 1991; Bathon 1996). Although these nematodes are able to infect a broad range of insects in laboratory tests, they have a restricted host range in nature (Poinar 1979; Gaugler 1981, 1988). Nematodes naturally dwell in

Table 2 Nematode species that have been put into commercial production, their major target pests (Shapiro-Ilan et al. 2013), and symbiotic bacteria (Lewis and Clarke 2012)

Nematode species[a]	Symbiotic bacterium	Examples of major target pests
Steinernema carpocapsae (Weiser)	*Xenorhabdus nematophila*	Lepidoptera larvae in lawn/cryptic habitats, pecan weevil, navel orangeworm
S. feltiae (Filipjev)	*X. bovienii*	Sciarids, fungus gnats, western flower thrips
S. glaseri (Steiner)	*X. poinarii*	Black vine weevil, banana root borer, scarab grubs
S. kushidai Mamiya	*X. japonica*	Scarab grubs
S. kraussei (Steiner)	*X. bovienii*	Black vine weevil
S. riobrave Cabanillas, Poinar, and Raulston	*Xenorhabdus sp.*	Citrus root weevil, plum curculio
S. scapterisci Nguyen and Smart	*Xenorhabdus sp.*	Mole crickets
S. longicaudum Shen and Wang[b]	*Xenorhabdus sp.*	Scarab grubs
Heterorhabditis bacteriophora Poinar	*Photorhabdus luminescens laumondii*	Root weevils, billbugs, scarab grubs
H. indica Poinar, Karunakar and David	*P. luminescens akhurstii*	Citrus root weevil, sweetpotato weevil
H. marelatus Liu and Berry	*P. temperata*	Black vine weevil, scarab grubs
H. megidis Poinar, Jackson and Klein	*P. temperata*	Root weevils, scarab grubs
H. zealandica Poinar	*P. temperata*	Scarab grubs, grape root borer

[a] Information for the commercialization of insecticidal nematodes is available from the website: http://www.biocontrol.entomology.cornell.edu/pathogens/nematodes.html.
[b] Commercialized in China

soil, where they are protected from environmental extremes (e.g. desiccation, extreme temperatures and ultraviolet radiation), and thus do not normally encounter insects in other habitats, e.g. foliage-feeding lepidopteran larvae in exposed settings.

Currently, more than 85 species of EPNs have been described in the families Heterorhabditidae and Steinernematidae (Lewis and Clarke 2012). At least 13 nematode species have been put into commercial production (Shapiro-Ilan et al. 2013), for pest control in greenhouse, nurseries, turf, and other cropping systems (Table 2).

5.1 Biology

Heterorhabditis and *Steinernema* spp. possess high insecticidal effects, due to their association with the symbiotic bacteria *Photorhabdus* and *Xenorhabdus*, respectively (Forst et al. 1997). The bacteria are Gram-negative, facultative anaerobic rods in the family Enterobacteriaceae, and are found within the intestine of the infective juvenile (IJ) nematode. After entering the host, usually via insect body openings (e.g. mouth, anus, spiracles), or in some cases through the cuticle, the nematode penetrates through the gut wall, and regurgitates symbiotic bacteria into the insect hemocoel (Dowds and Peters 2002). The bacteria secret toxins to rapidly kill the host insect, and degrade insect tissues to provide nutrients for nematode development inside the host (Lewis and Clarke 2012).

In the past few decades, studies have been conducted to test the efficacies of nematode species and isolated strains against insect pests in various horticultural ecosystems including vegetables, ornamentals, orchard, greenhouse, and turf. Efficacies in insect control with EPNs have been reviewed by several authors (Nickle 1984; Kaya 1985; Begley 1990; Klein 1990; Georgis and Hague 1991; Wouts 1991; Georgis and Manweiler 1994; Shapiro-Ilan et al. 2002; Grewal et al. 2005; Georgis et al. 2006; Lacey and Shapiro-Ilan 2008; Lewis and Clarke 2012; Shapiro-Ilan et al. 2012b). Compared with many laboratory successes, field efficacy of microbial control with EPNs are often inconsistent and unsatisfactory (Georgis and Gaugler 1991; Klein 1993), due to various biotic (Kaya and Koppenhöfer 1996; Kaya 2002), such as host suitability/matching and developmental stage, inter- and intra-specific competition, nematophagous fungi, protozoan parasites, and predators; and abiotic factors (Kaya 1990; Smits 1996; Glazer 2002), such as environmental temperature, moisture, soil texture, ultraviolet light, desiccation, pH, thatch build-up, and agrichemical compatibility.

5.2 *Strain Improvement*

Superior efficacy of EPNs in biocontrol can be achieved via strain improvement. Both molecular, e.g. transgenic, and non-molecular genetic approaches, e.g. selection and hybridization, can be used (Gaugler 1987). For example, selection against scarab larvae resulted in a 72-fold increase in host-finding capability (Gaugler and Campbell 1991). Strain improvement has been also used to enhance the environmental tolerance of the insecticidal nematodes. Shapiro et al. (1997) earlier demonstrated that heat tolerance can be transferred from one *H. bacteriophora* strain to another via hybridization. In recent studies, desiccation tolerant strains of *H. bacteriophora* was screened (Mukuka et al. 2010a); major improvement in heat and desiccate tolerance was achieved by crossed-breeding and subsequent genetic selection (Mukuka et al. 2010b). Also, environmental tolerance in strains of *S. carpocapsae* was improved through hybridization (Shapiro-Ilan et al. 2005). More recently, strains of *S. feltiae* most tolerant to desiccation were selected for hybridization to further improve desiccation tolerance (Nimkingrat et al. 2013). Further advances in strain improvement are anticipated by research progress in genomics of EPNs (Bai and Grewal 2007; Ciche 2007; Bai et al. 2009).

5.3 *Mass Production*

EPNs are currently produced *in vivo* or *in vitro* (solid or liquid culture medium), as discussed in Shapiro-Ilan et al. (2012a, 2013). Production *in vivo* is based on the inoculation of insects, e.g., mealworm *Tenebrio molitor* L. (Coleoptera: Tenebrionidae), or most commonly larvae of the greater wax moth, *Galleria mellonella* L. (Lepidoptera: Pyralidae), to culture the nematodes. Mass production in this method is more difficult and expensive, due to the cost of hosts and labor. Approaches to reducing *in vivo* production costs through improved media and automation are being researched (Shapiro-Ilan et al. 2008, 2012a). Currently, production *in vitro* results in relatively higher yield with less cost. In particular, *in vitro* liquid production is considered to be most cost efficient process for mass production, and is thus more broadly used for the commercialization of nematode products. Several studies showed reduced quality or efficacy *in vitro* liquid culture than production *in vivo* or *in vitro* solid medium (Gaugler and Georgis 1991; Cottrell et al. 2011), whereas some others did not detect any significant difference (Georgis and Gaugler 1991).

Recent advances have been made in enhancing nematode quality and efficacy *in vitro* liquid production, such as optimizing media and bioprocess kinetics through modeling (Chavarria-Hernandez et al. 2006, 2010), improving inoculum and bacterial cell density (Hirao and Ehlers 2010), time of inoculation (Johnigk et al. 2004) and downstream processing (Young et al. 2002). For production both *in vivo* and *in vitro*, strain deterioration is a major concern affecting nematode efficacy and fitness, such as virulence, environmental tolerance and reproductive potential (Shapiro et al. 1996; Wang and Grewal 2002; Bai et al. 2005; Bilgrami et al. 2006). Genetic deterioration may be mitigated by taking precautions, such as minimization of serial passages, introduction of fresh genetic material, improved cryopreservation methods of stock cultures (Bai et al. 2004), or creation of homozygous inbred lines to deter deterioration of beneficial traits (Bai et al. 2005; Chaston et al. 2011; Anbesse et al. 2013).

5.4 Improvement in Formulation and Application Technique

In recent years, substantial improvements have been made in formulation and application technology to enhance the field efficacy of EPNs. For example, efficacy of nematodes in aboveground applications could be improved by mixing the nematodes with a surfactant and polymer (Schroer and Ehlers 2005). Also, Llàcer et al. (2009) reported that *S. carpocapsae* in a chitosan formulation caused high levels of control of the red palm weevil, *Rhynchophorus ferrugineus* (Olivier) (Coleoptera: Curculionidae). Shapiro-Ilan et al. (2010) found that a follow-up application of a sprayable fire gel largely improved efficacy of *S. carpocapsae* for the control of the lesser peachtree borer, *Synanthedon pictipes* (Grote & Robinson) (Lepidoptera: Sesiidae). Similar results also appeared in the study of Lacey et al. (2010), who demonstrated enhanced efficacy for the control of codling moth, *C. pomonella*, in nematode applications added with the sprayable fire gel or wood flour foam as a protective agent.

EPNs are usually delivered to the field via conventional spray, injection or irrigation methods (Georgis 1990; Bateman 1999; Koppenhöfer 2000; Wennemann et al. 2003; Reding et al. 2008). However, in situations of insufficient moisture in the

Fig. 2 The complete delivery system mounted offset to one side of the tractor. (*1*) air blower with air controller, (*2*) metering unit, (*3*) cadaver hopper, (*4*) roller chain, (*5*) nematode infected desiccated cadavers, (*6*) press-drive wheel, (*7*) packer wheel, (*8*) cadaver discharging tube, ((*9*) double disk opener). (Reprinted with permission from Zhu et al. (2011), Fig. 4.)

area of application, the insecticidal efficacy of the nematodes may be compromised, as they require a water film to reach the host and cause an infection. Under these circumstances, direct application of EPN-infected insect cadavers is considered (Creighton and Fassuliotis 1985; Jansson et al. 1993; Shapiro-Ilan et al. 2001).

Researchers found that compared with traditional delivery of aqueous suspension, application of EPN-infected cadavers could improve nematode dispersal ability (Shapiro and Glazer 1996), infectivity (Shapiro and Lewis 1999), and survival (Perez et al. 2003), and demonstrate superior efficacy in pest control (Shapiro-Ilan et al. 2003). A drawback in application of nematode-infected cadavers is that they may stick together or rupture during transportation and application process. Shapiro-Ilan et al. (2001) found that coating the cadavers with a powdery substance in starch-clay combination could alleviate the problem, without significant deleterious effects on nematode reproduction and infectivity. This technique was further advanced by partial desiccation of the cadaver to increase the shelf life by slowing down the nematode metabolism (Shapiro-Ilan et al. 2001; Perez et al. 2003). The EPN-infected cadavers need to be buried 5–7.5 cm beneath the soil surface to avoid damage from animals or exposure to sunlight. This becomes practical for field use after the development of a desiccated cadaver delivery system (as shown in Fig. 2) by Zhu et al. (2011). This system is able to deliver insect cadavers efficiently into soil in orchard, ornamental nurseries and turf, and in annual crops when plants are absent.

6 Optimal Use of Microbial Pest Control Agents

Conceptually there are three main approaches for using microbial agents to manage insect pest populations. One can introduce the microbes into a target population with the intent of permanently establishing the agent as a continuing mortality factor and thereby keeping the pest at low levels, ideally sub-economic threshold (colonization). This seems to have the case with *Entomophaga miamaiga* Humber, Shimazu & Soper and baculovirus among gypsy moth, *Lymantria dispar* L. (Lepidoptera: Erebidae), in North America (Hajek 1997a, 1999). Agricultural practices, esp. monoculture, and transient cropping can greatly affect the efficacy of this approach, however; it seems to work best in natural or perennial cropping systems. A variant of this tactic is habitat manipulation to enhance the spread of an indigenous pathogen. A successful example is the judicious use of fungicides in cotton, coupled with scouting and diagnosis of mycoses, to allow epizootics of *Neozygites fresenii* (Nowakowski) Batko to control aphids (Steinkraus et al. 1998; University of Arkansas 2013). A second approach is the inundative application of large numbers of a microbial agent into an infested crop to create an epizootic before it would normally occur to create transient insect population reduction (microbial insecticide). This tactic mimics the use of chemical insecticides, albeit with environmentally and toxicologically safer substitutes. For some microbial agents, e.g., *B. thuringiensis* and baculoviruses, inundative application has been successful. Often, however, efficacy is not satisfactory, partly because use of the microbial in this manner creates expectations of chemical-level effectiveness and ignores the many biotic and abiotic factors affecting the microbial agent. The last approach integrates a microbial agent with other tools to manage a pest population, ideally keeping the insects below an economic threshold (integrated pest management). With this approach the microbial agent is not the main tool but rather complements other tools. Therefore only moderate efficacy from the microbe may be satisfactory. For example, there is a growing body of data that indicate concurrent use of microbials, such as the EPF, with parasitoids and predators can provide satisfactory management of several greenhouse pests (Rashki et al. 2009; Brownbridge et al. 2011).

7 Regulation of Microbial Biopesticides

Similar to chemical pesticides, microbial biopesticides need to follow the regulation rules implemented by governmental agencies in each country (e.g., United States Environmental Protection Agency, abbreviated as US EPA, in the United States) for commercial production and practical use. Data requirements for the registration of microbial biocontrol agents include, (A) product analysis; (B) residues; (C) toxicology (health effects); (D) non-target environmental effects; and (E) environmental expression, in addition to possible efficacy testing (Data Requirements 2007). The purpose of regulation is to ensure that the materials do not pose any serious hazardous threat to human bodies and the non-target organisms, including vertebrates, bees, natural enemies, endangered species, etc., during production and application process. Some microbial biocontrol agents can be exempted from the registration requirements, for example, EPNs in the United States, nations in the Europe, and many other countries (Ehlers 2005; Shapiro-Ilan et al. 2013). Also, the registrant of *B. bassiana* strain HF23 in the USA was granted with waivers of data requirements for testing with freshwater aquatic invertebrates, non-target plants, non-target insects, honey bee toxicity, estuarine/marine animals and wild mammals (US EPA 2006). Despite these, registration of most microbial biocontrol agents requires risk assessment of human health and environmental safety.

Additional regulation rules may also apply to microbials imported for classical biological control. For example, in the United States, importation of non-indigenous entomopathogens need to comply the rules stated by the United States Department of Agriculture—Animal and Plant Health Inspection Service—Plant Protection and Quarantine (USDA-APHIS-PPQ) (USDA APHIS 2014). Similar to the introduction of other biocontrol agents, strict screening process is undertaken for the importation of non-indigenous microbials, as they may become established and spread to additional areas, posing potential risk to non-target organisms.

As aforementioned, numerous studies have been carried out to modify the genetic traits of entomopathogens to achieve higher pathogenicity. Commercialization of the genetically engineered strains may need to follow stricter regulation rules, as the modified microbials may have higher potential risk than their naturally occurring forms. Genetically modified microbials may have, (A) direct effects on non-target organisms via expanded lethal host range, (B) enhanced evolution of host resistance, (C) adverse effects on ecosystem, (D) unintentional development as a pest organism, (E) biological pollution/cumulative effects, and (F) possible transfer of the genetic traits to other organisms. If any of these adverse effects occur to a modified microbial, it would be almost impossible to eradicate once it becomes established (Jaronski 2012).

Conclusion

Despite the high desirability of utilizing environmentally safe biopesticides, the commercial development and broad-scale use of these microbial biocontrol agents are still largely limited, due to high costs to manufacturers and end-users, short shelf-life, and unstable and inconsistent field efficacies. In addition, some highly pathogenic microbials, like virus and some nematode species (i.e. *Steinernema scarabaei* (Koppenhöfer and Fuzy 2003; Koppenhöfer et al. 2004)), still have difficulty in massive scale production *in vitro*, although significant improvements have been achieved for the commercial production of microbial biopesticides in the past few decades. Future research should be directed to selection of more virulent strains or biotypes with broader spectra of target hosts via traditional or transgenic tools; improvement of technology in mass production to lower the costs and increase production capacity. Also, enhancement in product stability via formulations would be another direction for future research to optimize the field performance of entomopathogens. Finally, multi-categories of biopesticides with different mode of actions are desired to avoid the development of pest resistance and resurgence.

Acknowledgements We acknowledge Dr. David I. Shapiro-Ilan from the USDA-ARS (Byron, GA) for technical review of the manuscript. This work was supported by FY 2011 USDA's Pest Management Alternatives Program (PMAP), Special Research Grants Program (Award No 2011-34381-20051).

References

Akhurst RJ (1990) Safety to nontarget invertebrates of nematodes of economically important pests. In: Laird M, Lacey LA, Davidson EW (eds) Safety of microbial insecticides. CRC, Boca Raton, pp 234–238

Alves SB, Risco SH, Almeida LC (1984) Influence of photoperiod and temperature on the development and sporulation of *Metarhizium anisopliae* (Metsch.) Sorok. J Appl Entomol 97:127–129

Alves SB, Pereira RM, Lopes RB et al (2003) Use of entomopathogenic fungi in Latin America. In: Upadhyay RK (ed) Advances in microbial control of insect pests. Kluwer, New York, pp 193–211

Amiri B, Ibrahim L, Butt TM (1999) Antifeedant properties of destruxins and their use with the entomogenous fungus *Metarhizium anisopliae* for improved control of crucifer pests. Biocontrol Sci Technol 9:487–498

Amiri-Besheli B, Khambay B, Cameron S et al (2000) Inter- and intra-specific variation in destruxin production by the insect pathogenic *Metarhizium*, and its significance to pathogenesis. Mycol Res 104:447–452

Anbesse S, Sumaya NH, Dörfler AV et al (2013) Stabilisation of heat tolerance traits in *Heterorhabditis bacteriophora* through selective breeding and creation of inbred lines in liquid culture. BioControl 58:85–93

Anke H, Steiner O (2002) Insecticidal and nematicidal metabolites from fungi. In: Esser K, Lemke PA, Bennet JV, Osiewacz HD (eds) The Mycota 10. Industrial applications. Springer, Berlin, pp 109–127

Bagga S, Hu G, Screen SE et al (2004) Reconstructing the diversification of subtilisins in the pathogenic fungus *Metarhizium anisopliae*. Gene 324:159–169

Bai X, Grewal PS (2007) Identification of two down-regulated genes in entomopathogenic nematode *Heterorhabditis bacteriophora* infective juveniles upon contact with insect hemolymph. Mol Biochem Parasitol 156:162–166

Bai C, Shapiro-Ilan DI, Gaugler R et al (2004) Effect of EPN concentration on survival during cryopreservation in liquid nitrogen. J Nematol 36:281–284

Bai C, Shapiro-Ilan DI, Gaugler R et al (2005) Stabilization of beneficial traits in *Heterorhabditis bacteriophora* through creation of inbred lines. Biol Control 32:220–227

Bai X, Adams BJ, Ciche TA et al (2009) Transcriptomic analysis of the entomopathogenic nematode Heterorhabditis bacteriophora TTO1. BMC Genomics 10, art. No. 205

Bailey KL, Boyetchko SM, Längle T (2010) Social and economic drivers shaping the future of biological control: a Canadian perspective on the factors affecting the development and use of microbial biopesticides. Biol Control 52:221–229

Bartlett MC, Jaronski ST (1988) Mass production of entomogenous fungi for biological control of insects. In: Burge MN (ed) Fungi in biological control systems. Manchester University Press, Manchester, pp 61–85

Bateman RP (1999) Delivery systems and protocols for biopesticides. In: Hall FR, Menn JJ (eds) Methods in biotechnology, biopesticides: use and delivery, vol 5. Humana, Totowa, pp 509–528

Bateman R (2007) Production processes for anamorphic Fungi. IPARC, Silwood Park, Ascot, Berks. http://www.dropdata.net/mycoharvester/Mass_production.PDF. Accessed 11 Jan 2014

Bateman R, Chapple A (2001) The spray application of mycopesticide formulations. In: Butt T, Jackson C, Magan N (eds) Fungal biocontrol agents. CABI, Wallingford, pp 289–309

Bateman RP, Carey M, Moore D et al (1993) The enhanced infectivity of *Metarhizium flavoviride* in oil formulations to desert locusts at low humidities. Ann Appl Biol 122:145–152

Bathon H (1996) Impact of entomopathogenic nematodes on non-target hosts. Biocontrol Sci Technol 6:421–434

Begley JW (1990) Efficacy against insects in habitats other than soil. In: Gaugler R, Kaya HK (eds) Entomopathogenic nematodes in biological control. CRC, Boca Raton, pp 215–231

Behle R, Birthisel T (2013) Formulations of entomopathogens as bioinsecticides. In: Morales-Ramos JA, Rojas MG, Shapiro-Ilan DI (eds) Mass production of beneficial organisms: invertebrates and entomopathogens. Academic, San Diego, pp 483–517

Bernier L, Cooper RM, Charnley AK et al (1989) Transformation of the entomopathogenic fungus *Metarhizium anisopliae* to benomyl resistance. FEMS Microbiol Lett 60:261–266

Bilgrami AL, Gaugler R, Shapiro-Ilan DI et al (2006) Source of trait deterioration in entomopathogenic nematodes *Heterorhabditis bacteriophora* and *Steinernema carpocapsae* during in vivo culture. Nematology 8:397–409

Bixby A, Alm SR, Power K et al (2007) Susceptibility of four species of turfgrass-infesting scarabs (*Coleoptera: Scarabaeidae*) to *Bacillus thuringiensis* serovar *japonensis* strain Buibui. J Econ Entomol 100:1604–1610

Bogo MR, Rota CA, Pinto HJ et al (1998) A chitinase encoding gene (chit1 gene) from the entomopathogen *Metarhizium anisopliae*: isolation and characterization of genomic and full-length cDNA. Curr Microbiol 37:221–225

Bonning BC, Hoover K, Duffey S et al (1995) Production of polyhedra of the autographa californica nuclear polyhedrosis virus using the Sf21 and Tn5B1-4 cell lines and comparison with host-derived polyhedra by bioassay. J Invertebr Pathol 66:224–230

Boucias DG, Pendland JC (1998) Principles of insect pathology. Kluwer, Dordrecht

Bradley CA, Black WE, Kearns R et al (1992) Role of production technology in mycoinsecticide development. In: Leatham GF (ed) Frontiers in industrial mycology. Chapman and Hall, New York, pp 160–179

Braga GUL, Flint SD, Miller CD et al (2001) Both solar UVA and UVB radiation impair conidial curability and delay germination in the entomopathogenic fungus *Metarhizium anisopliae*. Photochem Photobiol 74:734–739

Bravo A, Likitvivatanavong S, Gill SS et al (2011) *Bacillus thuringiensis*: a story of a successful bioinsecticide. Insect Biochem Mol Biol 41:423–431

Brousseau C, Charpentier G, Bellonick S (1996) Susceptibility of spruce budworm, *Choristoneura fumiferana* Clemens, to destruxins, cyclodepsipeptidic mycotoxins of *Metarhizium anisopliae*. J Invertebr Pathol 68:180–182

Brownbridge M, Saito T, Buitenhuis R et al (2011) Developing a biologically-based IPM program for western flower thrips, *Frankliniella occidentalis*, in greenhouse floriculture. IOBC/wprs Bull 68:21–24

Buerger P, Hauxwell C, Murray D (2007) Nucleopolyhedrovirus introduction in Australia. Virol Sin 22:173–179

Burges HD (1998) Formulation of microbial biopesticides: beneficial microorganisms, nematodes and seed treatments. Kluwer, Dordrecht

Butt TM, Jackson CW, Magan N (2001) Fungi as biological agents. CABI, Wallingford

Carbonell LF, Hodge MR, Tomalski MD et al (1988) Synthesis of a gene coding for an insecticide-specific scorpion toxin neurotoxin and attempts to express it using baculovirus vectors. Gene 73:409–418

Carruthers RI, Soper RS (1987) Fungal diseases. In: Fuxa JR, Tanada Y (eds) Epizootiology of insect diseases. Wiley, New York, pp 357–416

Chaston JM, Dillman AR, Shapiro-Ilan DI et al (2011) Outcrossing and crossbreeding recovers deteriorated traits in laboratory cultured *Steinernema carpocapsae* nematodes. Int J Parasitol 41:801–809

Chavarria-Hernandez N, Espino-Garcia J-J, Sanjuan-Galindo R et al (2006) Monoxenic liquid culture of the entomopathogenic nematode *Steinernema carpocapsae* using a culture medium containing whey kinetics and modeling. J Biotechnol 125:75–84

Chavarria-Hernandez N, Ortega-Morales E, Vargas-Torres A et al (2010) Submerged monoxenic culture of the entomopathogenic nematode, *Steinernema carpocapsae* CABA01, in a mechanically agitated bioreactor: evolution of the hydrodynamic and mass transfer conditions. Biotechnol Bioprocess Eng 15:580–589

Cho EM, Liu L, Farmerie W et al (2006) EST analysis of cDNA libraries from the entomopathogenic fungus *Beauveria* (Cordyceps) *bassiana*. I. Evidence for stage-specific gene expression in aerial conidia, in vitro blastospores and submerged conidia. Microbiology 152:2843–2854

Cho EM, Kirkland BH, Holder DJ et al (2007) Phage display cDNA cloning and expression analysis of hydrophobins from the entomopathogenic fungus *Beauveria* (Cordyceps) *bassiana*. Microbiology 153:3438–3447

Ciche TA (2007) The biology and genome of Heterorhabditis bacteriophora. In: WormBook (ed) The *C. elegans* research community, WormBook. doi/10.1895/wormbook.1.135.1. http://www.wormbook.org. Accessed 20 Feb 2007

Coles RB (1980) The biology of *Cordyceps aphodii* (Sphaeriales: Clavicipitaceae). In: Crosby TK, Pottinger RP (eds) Proceedings of the 2nd Australasian conference on grassland invertebrate ecology. Palmerston North, New Zealand, pp 207–212

Copping LG (2001) The biopesticide manual, 2nd edn. British Crop Protection Council, Farnham

Cory JS, Franklin MT (2012) Evolution and the microbial control of insects. Evol Appl 5:455–469

Cottrell TE, Shapiro-Ilan DI, Horton DL et al (2011) Laboratory virulence and orchard efficacy of entomopathogenic nematodes toward the lesser peachtree borer (*Lepidoptera: Sesiidae*). Environ Entomol 104:47–53

Couch TL, Jurat-Fuentes JL (2013) Commercial production of entomopathogenic bacteria. In: Morales-Ramos JA, Rojas MG, Shapiro-Ilan DI (eds) Mass production of beneficial organisms: invertebrates and entomopathogens. Academic, San Diego, pp 415–435

Crangle RD (2011) Diatomite. 2011 minerals yearbook. U.S. Geological Survey, Washington, DC. http://minerals.usgs.gov/minerals/pubs/commodity/diatomite/myb1-2011-diato.pdf. Accessed 26 Nov 2013

Creighton CS, Fassuliotis G (1985) Heterorhabditis sp. (*Nematoda: Heterorhabditidae*): a nematode parasite isolated from the banded cucumber beetle *Diabrotica balteata*. J Nematol 17:150–153

Crickmore N, Zeigler DR, Feitelson J et al (1998) Revision of the nomenclature for the *Bacillus thuringiensis* pesticidal crystal proteins. Microbiol Mol Biol Rev 62:807–813

Crickmore N, Baum J, Bravo A et al (2013) *Bacillus thuringiensis* toxin nomenclature. http://www.btnomenclature.info/. Accessed 7 Jan 2014

Data Requirements (2007) Data requirements for biochemical and microbial pesticides. 40 CFR Part 158. http://www.epa.gov/fedrgstr/EPA-PEST/2007/October/Day-26/p20828.htm

De Faria M, Wraight S (2007) Mycoinsecticides and mycoacaricides: a comprehensive list with worldwide coverage and international classification of formulation types. Biol Control 43:237–256

Dowds BCA, Peters A (2002) Virulence mechanisms. In: Gaugler R (ed) Entomopathogenic nematology. CABI, New York, pp 79–98

Dutky SR (1959) Insect microbiology. Adv Appl Microbiol 1:175–200

Ehlers RU (2005) Forum on safety and regulation. In: Grewal PS, Ehlers RU, Shapiro-Ilan DI (eds) Nematodes as biological control agents. CABI, Wallingford, pp 107–114

Ekesi S, Maniania NK, Ampong-Nyarko K (1999) Effect of temperature on germination, radial growth and virulence of *Metarhizium anisopliae* and *Beauveria bassiana* on *Megalurothrips sjostedti*. Biocontrol Sci Technol 9:177–185

Elena GJ, Beatriz PJ, Alejandro P et al (2011) *Metarhizium anisopliae* (Metschnikoff) Sorokin promotes growth and has endophytic activity in tomato plants. Adv Biol Res 5:22–27

El-Salamouny S, Ranwala D, Shapiro M et al (2009) Tea, coffee, and cocoa as ultraviolet radiation protectants for the beet armyworm nucleopolyhedrovirus. J Econ Entomol 102:1767–1773

Evans HC (1989) Mycopathogens of insects of epigeal and aerial habitats. In: Wilding N, Collins NM, Hammond PM, Webber JF (eds) Insect-fungus interactions. Academic, London, pp 205–238

Fang W, Leng B, Xiao Y et al (2005) Cloning of *Beauveria bassiana* chitinase gene Bbchit1 and its application to improve fungal strain virulence. Appl Environ Microbiol 71:363–370

Faria M, Hajek AE, Wraight SP (2009) Imbibitional damage in conidia of the entomopathogenic fungi *Beauveria bassiana*, *Metarhizium acridum*, and *Metarhizium anisopliae*. Biol Control 51:346–354

Feng MG, Poprawski TJ, Khachatourians GG (1994) Production, formulation and application of the entomopathogenic fungus *Beauveria bassiana* for insect control: current status. Biocontrol Sci Technol 4:3–34

Ferron P (1971) Problèmes poses par la mise au point d'un procédé de lutte microbiologique contre *Melolontha melolontha* au moyen de la mycose Ã *Beauveria brongniartii* (Delacr.) Siemaszko. Phytiatrie-Phytopharmacie 10:159–168

Ferron P (1985) Fungal control. In: Kerkut GA, Gilbert LI (eds) Comprehensive insect physiology, biochemistry and pharmacology. Pergamon, Oxford, pp 313–346

Ferron P, Fargues J, Riba G (1991) Fungi as microbial insecticides against pests. In: Arora DK, Ajello L, Mukerji KG (eds) Handbook of applied mycology, vol 2: humans, animals, and insects. Dekker, New York, pp 665–706

ffrench-Constant RH, Waterfield N, Daborn P (2010) Insecticidal toxins from *Photorhabdus* and *Xenorhabdus*. In: Gilbert LI, Gill SS (eds) Insect control: biological and synthetic agents. Academic, Oxford, pp 313–327

Forst S, Dowds B, Boemare N et al (1997) *Xenorhabdus* and *Photorhabdus* spp.: bugs that kill bugs. Annu Rev Microbiol 51:47–72

Freimoser FM, Hu G, St Leger RJ (2005) Variation in gene expression patterns as the insect pathogen *Metarhizium anisopliae* adapts to different host cuticles or nutrient deprivation in vitro. Microbiology 151:361–371

Furlong MJ, Pell JK, Reddy GVP (1997) Premortality effects of *Zoophthora radicans* infection in *Plutella xylostella*. J Invertebr Pathol 70:214–220

Gaugler R (1981) Biological control potential of neoaplectanid nematodes. J Nematol 13:241–249

Gaugler R (1987) Entomogenous nematodes and their prospects for genetic improvement. In: Maramorosch K (ed) Biotechnology in invertebrate pathology and cell culture. Academic, San Diego, pp 457–484

Gaugler R (1988) Ecological considerations in the biological control of soil-inhabiting insect pests with entomopathogenic nematodes. Agric Ecosyst Environ 24:351–360

Gaugler R, Campbell JF (1991) Selection for enhanced host-finding of scarab larvae (*Coleoptera: Scarabaeidae*) in an entomopathogenic nematode. Environ Entomol 20:700–706

Gaugler R, Georgis R (1991) Culture method and efficacy of entomopathogenic nematodes (*Rhabditida: Steinernematidae* and *Heterorhabditidae*). Biol Control 1:269–274

Gaugler R, Kaya HK (1990) Entomopathogenic nematodes in biological control. CRC, Boca Raton

Georgis R (1990) Formulation and application technology. In: Gaugler R, Kaya HK (eds) Entomopathogenic nematodes in biological control. CRC, Boca Raton, pp 173–194

Georgis R, Gaugler R (1991) Predictability in biological control using entomopathogenic nematodes. J Econ Entomol 84:713–720

Georgis R, Hague NGM (1991) Nematodes as biological insecticides. Pestic Outlook 2:29–32

Georgis R, Manweiler SA (1994) Entomopathogenic nematodes: a developing biological control technology. Agric Zool Rev 6:63–94

Georgis R, Kaya HK, Gaugler R (1991) Effect of steinernematid and heterorhabditid nematodes (*Rhabditida: Steinernematidae* and *Heterorhabditidae*) on nontarget arthropods. Environ Entomol 20:815–822

Georgis R, Koppenhöfer AM, Lacey LA et al (2006) Successes and failures in the use of parasitic nematodes for pest control. Biol Control 38:103–123

Glare TR (1992) Fungal pathogens of scarabs. In: Jackson TA, Glare TR (eds) Use of pathogens in scarab pest management. Intercept, Andover, pp 63–77

Glare TR, Milner RJ (1991) Ecology of entomopathogenic fungi. In: Arora DK, Ajello L, Mukerji LG (eds) Handbook of applied mycology, vol 2: humans, animals, and insects. Marcel Dekker, New York, pp 547–612

Glazer I (2002) Survival biology. In: Gaugler R (ed) Entomopathogenic nematology. CABI, Wallingford, pp 169–187

Goettel MS, St. Leger RJ, Bhairi S et al (1990) Pathogenicity and growth of *Metarhizium anisopliae* stably transformed to benomyl resistance. Curr Genet 17:129–132

Goettel MS, Eilenberg J, Glare T (2010) Entomopathogenic fungi and their role in regulation of insect populations. In: Gilbert LI, Gill SS (eds) Insect control: biological and synthetic agents. Academic, Oxford, pp 387–431

González JM Jr, Brown BJ, Carlton BC (1982) Transfer of *Bacillus thuringiensis* plasmids coding for delta-endotoxin among strains of *B. thuringiensis* and *B. cereus*. Proc Natl Acad Sci U S A 79:6951–6955

Grewal PS, Koppenhöfer AM, Choo HY (2005) Lawn, turfgrass and pasture applications. In: Grewal PS, Ehlers R-U, Shapiro-Ilan D (eds) Nematodes as biocontrol agents. CABI, Wallingford, pp 115–146

Gröner A (1990) Safety to nontarget invertebrates of baculoviruses. In: Laird M, Lacey L, Davidson E (eds) Safety of microbial insecticides. CRC, Boca Raton, pp 135–147

Haase S, Ferrelli L, Pidre ML et al (2013) Genetic engineering of baculoviruses. In: Romanowski V (ed) Current issues in molecular virology—viral genetics and biotechnological applications. InTech, Croatia, pp 79–111

Hajek AE (1997a) Fungal and viral epizootics in gypsy moth (*Lepidoptera: Lymantriidae*) populations in Central New York. Biol Control 10:58–68

Hajek AE (1997b) Ecology of terrestrial fungal entomopathogens. Adv Microb Ecol 15:193–249

Hajek AE (1999) Pathology and epizootiology of *Entomophaga maimaiga* infections in forest Lepidoptera. Microbiol Mol Biol Rev 63:814–835

Hajek AE, St. Leger RJ (1994) Interactions between fungal pathogens and insect host. Annu Rev Entomol 39:293–322

Harrison R, Hoover K (2012) Baculoviruses and other occluded insect viruses. In: Vega FE, Kaya HK (eds) Insect pathology, 2nd edn. Academic, London, pp 73–131

Hirao A, Ehlers R-U (2010) Influence of inoculum density on population dynamics and dauer juvenile yields in liquid culture of biocontrol nematodes *Steinernema carpocapsae* and *S. feltiae* (*Nematoda: Rhabditida*). Appl Microbiol Biotechnol 85:507–515

Huber J (1986) Use of baculoviruses in pest management programs. In: Granados R, Federici B (eds) The biology of baculoviruses. Vol II: practical application for insect control. CRC, Boca Raton, pp 181–202

Hywel-Jones NL, Gillespie AT (1990) Effect of temperature on spore germination in *Metarhizium anisopliae* and *Beauveria bassiana*. Mycol Res 94:389–392

Inglis PW, Tigano MS, Valadares-Inglis MC (1999) Transformation of the entomopathogenic fungi, *Paecilomyces fumosoroseus* and *Paecilomyces lilacinus* (Deuteromycotina: Hyphomycetes) to benomyl resistance. Genet Mol Biol 22:119–123

Janmaat AF, Myers JH (2003) Rapid evolution and the cost of resistance to *Bacillus thuringiensis* in greenhouse populations of cabbage loopers, *Tricoplusia ni*. Proc R Soc Lond B270:2263–2270

Jansson RK, Lecrone SH, Gaugler R (1993) Field efficacy and persistence of entomopathogenic nematodes (Rhabditida: Steinernematidae, Heterorhabditidae) for control of sweetpotato weevil (*Coleoptera: Apionidae*) in southern Florida. J Econ Entomol 86:1055–1063

Jaronski ST (2010) Ecological factors in the inundative use of fungal entomopathogens. BioControl 55:159–185

Jaronski ST (2012) Microbial control of invertebrate pests. In: Sundh I, Wilcks A, Goettel MS (eds) Beneficial microorganisms in agriculture, food and the environment: safety assessment and regulation. CABI, Wallingford, pp 72–95

Jaronski S (2013) Mass production of entomopathogenic fungi—state of the art. In: Morales-Ramos JA, Rojas MG, Shapiro-Ilan DI (eds) Mass production of beneficial organisms: invertebrates and entomopathogens. Academic, San Diego, pp 357–413

Jaronski ST, Jackson MA (2012) Mass production of entomopathogenic Hypocreales. In: Lacey LA (ed) Manual of techniques in invertebrate pathology, 2nd edn. Academic, New York, pp 255–284

Jaros-Su J, Groden E, Zhang J (1999) Effects of selected fungicides and the timing of fungicide application in *Beauveria bassiana*-induced mortality of the Colorado potato beetle (*Coleoptera: Chrysomelidae*). Biol Control 15:259–269

Jin X, Streett DA, Dunlap CA et al (2008) Application of hydrophilic–lipophilic balance (HLB) number to optimize a compatible non-ionic surfactant for dried aerial conidia of *Beauveria bassiana*. Biol Control 46:226–233

Johnigk S-A, Ecke F, Poehling M et al (2004) Liquid culture mass production of biocontrol nematodes, *Heterorhabditis bacteriophora* (*Nematoda: Rhabditida*): improved timing of dauer juvenile inoculation. Appl Microbiol Biotechnol 64:651–658

Johnson VW, Pearson JF, Jackson TA (2001) Formulation of *Serratia entomophila* for biological control of grass grub. New Zealand plant protection, Proceedings of a conference, vol 54. Palmerston North, New Zealand, 14–16 Aug 2001, pp 125–127

Kabaluk JT, Svircev AM, Goettel MS et al (eds) (2010) The use and regulation of microbial pesticides in representative jurisdictions worldwide. IOBC Global, p 99. http://www.iobc-global.org/publications.html

Kaya HK (1985) Entomogenous nematodes for insect control in IPM systems. In: Hoy MA, Herzog DC (eds) Biological control in agricultural IPM systems. Academic, Orlando, pp 283–302

Kaya HK (1990) Soil ecology. In: Gaugler R, Kaya HK (eds) Entomopathogenic nematodes in biological control. CRC, Boca Raton, pp 93–115

Kaya HK (2002) Natural enemies and other antagonists. In: Gaugler R (ed) Entomopathogenic nematology. CABI, Wallingford, pp 189–203

Kaya HK, Koppenhöfer AM (1996) Effects of microbial and other antagonistic organism and competition on entomopathogenic nematodes. Biocontrol Sci Technol 6:357–371

Khan S, Guo L, Maimaiti Y et al (2012) Entomopathogenic fungi as microbial biocontrol agent. Mol Plant Breed 3:63–79

Kim HS, Jung MH, Ahn S et al (2002) Structure elucidation of a new cyclic hexadepsipeptide from *Beauveria felina*. J Antibiot 55:598–601

Kirsch K, Schmutterer H (1988) Low efficacy of a *Bacillusthuringiensis* (Berl) formulation in controlling the diamondback moth, *Plutella xylostella* (L), in the Philippines. J Appl Entomol 105:249–255

Klein MG (1990) Efficacy against soil-inhabiting insect pests. In: Gaugler R, Kaya HK (eds) Entomopathogenic nematodes in biological control. CRC, Boca Raton, pp 195–214

Klein MG (1993) Biological control of scarabs with entomopathogenic nematodes. In: Bedding R, Akhurst R, Kaya H (eds) Nematodes and the biological control of insect pests. CSIRO, East Melbourne, pp 49–58

Koppenhöfer AM (2000) Nematodes. In: Lacey LA, Kaya HK (eds) Field manual of techniques in invertebrate pathology. Kluwer, Dordrecht, pp 283–301

Koppenhöfer AM, Fuzy EM (2003) *Steinernema scarabaei* for the control of white grubs. Biol Control 28:47–59

Koppenhöfer AM, Fuzy EM, Crocker R et al (2004) Pathogenicity of *Heterorhabditis bacteriophora*, *Steinernema glaseri*, and *S. scarabaei* (Rhabditida: Heterorhabditidae, Steinernematidae) against 12 white grub species (Coleoptera: Scarabaeidae). Biocontrol Sci Technol 14:87–92

Krasnoff SB, Gibson DM, Belofsky GN et al (1996) New destruxins from the entomopathogenic fungus *Aschersonia* sp. J Nat Prod 59:485–489

Lacey LA, Shapiro-Ilan DI (2008) Microbial control of insect pests in temperate orchard systems: potential for incorporation into IPM. Annu Rev Entomol 53:121–144

Lacey LA, Shapiro-Ilan DI, Glenn GM (2010) Post-application of anti-desiccant agents improves efficacy of entomopathogenic nematodes in formulated host cadavers or aqueous suspension against diapausing codling moth larvae (Lepidoptera: Tortricidae). Biocontrol Sci Technol 20:909–921

Laird M, Lacey LA, Davidson EW (1990) Safety of microbial insecticides. CRC, Boca Raton

Lasa R, Williams T, Caballero P (2008) Insecticidal properties and microbial contaminants in a *Spodoptera exigua* multiple nucleopolyhedrovirus (Baculoviridae) formulation stored at different temperatures. J Econ Entomol 101:42–49
Leemon DM, Jonsson NN (2008) Laboratory studies on Australian isolates of *Metarhizium anisopliae* as a biopesticide for the cattle tick *Boophilus microplus*. J Invertebr Pathol 97:40–49
Lewis EE, Clarke DJ (2012) Nematode parasites and entomopathogens. In: Vega FE, Kaya HK (eds) Insect pathology, 2nd edn. Academic, San Diego, pp 395–443
Lingg AJ, Donaldson MD (1981) Biotic and abiotic factors affecting stability of *Beauveria bassiana* conidia in soil. J Invertebr Pathol 38:191–200
Liu CM, Huang SS, Tzeng YM (2004) Analysis of destruxins produced from *Metarhizium anisopliae* by capillary electrophoresis. J Chromatogr Sci 42:140–144
Llàcer E, Martinez de Altube MM, Jacas JA (2009) Evaluation of the efficacy of *Steinernema carpocapsae* in a chitosan formulation against the red palm weevil, *Rhynchophorus ferrugineus*, in *Phoenix canariensis*. BioControl 54:559–565
Lomer CJ, Bateman RP, Johnson DL et al (2001) Biological control of locusts and grasshoppers. Annu Rev Entomol 46:667–702
Lord JC (2005) From Metchnikoff to Monsanto and beyond: the path of microbial control. J Invertebr Pathol 89:19–29
Maeda S (1989) Increased insecticidal effect by a recombinant baculovirus carrying a synthetic diuretic hormone gene. Biochem Biophys Res Commun 165:1177–1183
McCoy CW, Samson RA, Boucias DG (1988) Entomogenous fungi. In: Ignoffo CM (ed) CRC handbook of natural pesticides. Microbial insecticides, part A. Entomogenous protozoa and fungi, vol 5. CRC, Boca Raton, pp 151–236
McGaughey WH (1985) Insect resistance to the biological insecticide *Bacillus thuringiensis*. Science 229:193–195
Miller LK (1997) Introduction to the Baculoviruses. In: Miller LK (ed) The Baculoviruses. Plenum, New York, pp 1–6
Milner RJ (1989) Ecological considerations on the use of *Metarhizium* for control of soil-dwelling pests. In: Robertson LN, Allsopp PG (eds) Proceedings of a soil-invertebrate workshop. QDPI Indooroopilly, Queensland, pp 10–13
Milner RJ, Lozano LB, Driver F et al (2003) A comparative study of two Mexican isolates with an Australian isolate of *Metarhizium anisopliae* var. *acridum*—strain characterisation, temperature profile and virulence for wingless grasshopper, *Phaulacridium vittatum*. BioControl 48:335–348
Moscardi F, Morales L, Santos B (2002) The successful use of AgMNPV for the control of velvet bean caterpillar, *Anticarsia gemmatalis*, in soybean in Brazil. Proceedings of the VIII international on invertebrate pathology and microbial control and XXXV annual meeting of the Society for Invertebrate Pathology. Foz do Iguassu, Brazil, pp 86–91
Moscardi F, Souza MLd, Castro MEBd et al (2011) Baculovirus pesticides: present state and future perspectives. In: Ahmad I, Ahmad F, Pichtel J (eds) Microbes and microbial technology: agricultural and environmental applications. Springer, New York, pp 415–445
Mukuka J, Strauch O, Ehlers R-U (2010a) Variability in desiccation tolerance among different strains of the entomopathogenic nematode *Heterorhabditis* spp. Nematology 12:711–720
Mukuka J, Strauch O, Hoppe C et al (2010b) Major improvement of heat and desiccation tolerance in *Heterorhabditis bacteriophora* through cross-breeding of tolerant strains and successive genetic selection. BioControl 55:511–521
Müller-Kögler E (1965) Pilzkrankheiten bei Insekten. Anwendung zur biologischen Schädlingsbekämpfung und Grundlagen der Insektenmykologie. P. Parey, Berlin, p 444
Nickle WR (1984) Plant and insect nematodes. Marcel Dekker, New York
Nimkingrat P, Strauch O, Ehlers RU (2013) Hybridisation and genetic selection for improving desiccation tolerance of the entomopathogenic nematode *Steinernema feltiae*. Biocontrol Sci Technol 23:348–361
Oliveira DP, Chaves BM, Loures EG (1981) Estudo comparativo da sobrevivencia de *Metarhizium anisopliae* (Metsch.) Sorokin em diferentes tipos de solo. Rev Theobroma 11:233–239

O'Reilly DR, Miller LK (1991) Improvement of a baculovirus pesticide by deletion of the egt gene. Biotechnology 9:1086–1089

Ownley BH, Griffin MR, Klingeman WE et al (2008) *Beauveria bassiana*: endophytic colonization and plant disease control. J Invertebr Pathol 98:267–270

Parsa S, Ortiz V, Vega FE (2013) Establishing fungal entomopathogens as endophytes: towards endophytic biological control. J Vis Exp 74:e50360

Pemsel M, Schwab S, Scheurer A et al (2010) Advanced PGSS process for the encapsulation of the biopesticide *Cydia pomonella* granulovirus. J Supercrit Fluids 53:174–178

Perez EE, Lewis EE, Shapiro-Ilan DI (2003) Impact of host cadaver on survival and infectivity of entomopathogenic nematodes (Rhabditida: Steinernematidae and Heterorhabditidae) under desiccating conditions. J Invertebr Pathol 82:111–118

Poinar GO Jr (1979) Nematodes for biological control of insects. CRC, Boca Raton

Poinar GO Jr (1989) Non-insect hosts for the entomogenous rhabditoid nematodes *Neoaplectana* (Steinernematidae) and *Heterorhabditis* (Heterorhabditidae). Rev Nématol 12:423–428

Posada F, Vega FE (2005) Establishment of the fungal entomopathogen *Beauveria bassiana* (Ascomycota: Hypocreales) as an endophyte in cocoa seedlings (*Theobroma cacao*). Mycologia 97:1195–1200

Rajagopal R, Mohan S, Bhatnagar RK (2006) Direct infection of *Spodoptera litura* by *Photorhabdus luminescens* encapsulated in alginate beads. J Invertebr Pathol 93:50–53

Rashki M, Kharazi-pakdel A, Allahyari H et al (2009) Interactions among the entomopathogenic fungus, *Beauveria bassiana* (Ascomycota: Hypocreales), the parasitoid, *Aphidius matricariae* (Hymenoptera: Braconidae), and its host, *Myzus persicae* (Homoptera: Aphididae). Biol Control 50:324–328

Rath AC, Anderson GC, Worledge D et al (1995) The effect of low temperatures on the virulence of *Metarhizium anisopliae* (DAT F-001) to the subterranean scarab, *Adoryphorus couloni*. J Invertebr Pathol 65:186–192

Ravensberg WJ (2011) Critical factors in the successful commercialization of microbial pest control products. In: Ravensberg WJ (ed) A roadmap to the successful development and commercialization of microbial pest control products for control of arthropods (Progress in biological control). Springer, Dordrecht, pp 295–356

Raymond B, Johnston PR, Nielsen-LeRoux C et al (2010) *Bacillus thuringiensis*: an impotent pathogen? Trends Microbiol 18:189–194

Reddy GVP, Furlong MJ, Pell JK et al (1998) *Zoophthora radicans* infection inhibits the response to and production of sex pheromone in the diamondback moth (Note). J Invertebr Pathol 72:167–169

Reding ME, Zhu H, Derksen RC (2008) Drip-chemigation with imidacloprid and nematodes for control of scarab larvae in nursery crops. J Environ Hort 26:93–100

Reid S, Chan L, van Oers MM (2013) Production of entomopathogenic viruses. In: Morales-Ramos JA, Rojas MG, Shapiro-Ilan DI (eds) Mass production of beneficial organisms: invertebrates and entomopathogens. Academic, San Diego, pp 437–482

Roberts DW, Campbell AS (1977) Stability of entomopathogenic fungi. Misc Publ Entomol Soc Am 10:19–76

Roberts DW, Hajek AE (1992) Entomopathogenic fungi as bioinsecticides. In: Leatham GF (ed) Frontiers in industrial mycology. Chapman and Hall, New York, pp 144–159

Robert PH, Riba G (1989) Toxic and repulsive effect of spray *per os* and systemic application of destruxin E to aphids. Mycopathologia 108:170–183

Ruiu L (2013) *Brevibacillus laterosporus*, a pathogen of invertebrates and a broad-spectrum antimicrobial species. Insects 4:476–492

Saik JE, Lacey LA, Lacey CM (1990) Safety of microbial insecticides to vertebrates—domestic animals and wildlife. In: Laird M, Lacey LA, Davidson EW (eds) Safety of microbial insecticides. CRC, Boca Raton, pp 115–132

Salamouny SE, Shapiro M, Ling KS et al (2009) Black tea and lignin as ultraviolet protectants for the beet armyworm nucleopolyhedrovirus. J Entomol Sci 44:50–58

Samson RA, Evans HC, Latgé JP (1988) Atlas of Entomopathogenic Fungi. Springer, Berlin

Sandhu SS, Kinghorn JR, Rajak RC et al (2001) Transformation system of *Beauveria bassiana* and *Metarhizium anisopliae* using nitrate reductase gene of *Aspergillus nidulans*. Indian J Exp Biol 39:650–653

Sansinenea E, Vázquez C, Ortiz A (2010) Genetic manipulation in *Bacillus thuringiensis* for strain improvement. Biotechnol Lett 32:1549–1557

Santi L, Silva LADe, Silva WOBd et al (2011) Virulence of the entomopathogenic fungus *Metarhizium anisopliae* using soybean oil formulation for control of the cotton stainer bug, *Dysdercus peruvianus*. World J Microbiol Biotechnol 27:2297–2303

Santos PdS, Silva, Monteiro AC et al (2012) Selection of surfactant compounds to enhance the dispersion of *Beauveria bassiana*. Biocontrol Sci Technol 22:281–292

Schroer S, Ehlers R-U (2005) Foliar application of the entomopathogenic nematode *Steinernema carpocapsae* for biological control of diamondback moth larvae (*Plutella xylostella*). Biol Control 33:81–86

Screen SE, Hu G, St. Leger RJ (2001) Transformants of *Metarhizium anisopliae* sf. *anisopliae* overexpressing chitinase from *Metarhizium anisopliae* sf. *acridum* show early induction of native chitinase but are not altered in pathogenicity to *Manduca sexta*. J Invertebr Pathol 78:260–266

Seramis GmbH (2012) Seramis special growing granules. http://www.seramisuk.co.uk/technical-data.html. Accessed 23 Dec 2012

Shah PA, Goettel MS (1999) Directory of microbial control products and services. Society for Invertebrate Pathology, Gainesville

Shapiro M, Argauer R (1997) Components of the stilbene optical brightener tinopal LPW as enhancers for the gypsy moth (Lepidoptera: Lymantriidae) baculovirus. J Econ Entomol 90:899–904

Shapiro DI, Glazer I (1996) Comparison of entomopathogenic nematode dispersal from infected hosts versus aqueous suspension. Environ Entomol 25:1455–1461

Shapiro DI, Lewis EE (1999) Comparison of entomopathogenic nematode infectivity from infected hosts versus aqueous suspension. Environ Entomol 28:907–911

Shapiro DI, Glazer I, Segal D (1996) Trait stability and fitness of the heat tolerant entomopathogenic nematode *Heterorhabditis bacteriophora* IS5 strain. Biol Control 6:238–244

Shapiro DI, Glazer I, Segal D (1997) Genetic improvement of heat tolerance in *Heterorhabditis bacteriophora* through hybridization. Biol Control 8:153–159

Shapiro-Ilan DI, Lewis EE, Behle RW et al (2001) Formulation of entomopathogenic nematode-infected cadavers. J Invertebr Pathol 78:17–23

Shapiro-Ilan DI, Gouge DH, Koppenhöfer AM (2002) Factors affecting commercial success: case studies in cotton, turf and citrus. In: Gaugler R (ed) Entomopathogenic nematology. CABI, Wallingford, pp 333–355

Shapiro-Ilan DI, Lewis EE, Son Y et al (2003) Superior efficacy observed in entomopathogenic nematodes applied in infected-host cadavers compared with application in aqueous suspension. J Invertebr Pathol 83:270–272

Shapiro-Ilan DI, Stuart RJ, McCoy CW (2005) Targeted improvement of *Steinernema carpocapsae* for control of the pecan weevil, *Curculio caryae* (Horn) (Coleoptera: Curculionidae) through hybridization and bacterial transfer. Biol Control 34:215–221

Shapiro-Ilan DI, Rojas MG, Morales-Ramos JA et al (2008) Effects of host nutrition on virulence and fitness of entomopathogenic nematodes: lipid and protein based supplements in *Tenebrio molitor* diets. J Nematol 40:13–19

Shapiro-Ilan DI, Cottrell TE, Mizell RF et al (2010) Efficacy of *Steinernema carpocapsae* for control of the lesser peachtree borer, *Synanthedon pictipes*: improved aboveground suppression with a novel gel application. Biol Control 54:23–28

Shapiro-Ilan DI, Han R, Dolinksi C (2012a) Entomopathogenic nematode production and application technology. J Nematol 44:206–217

Shapiro-Ilan DI, Bruck DJ, Lacey LA (2012b) Principles of epizootiology and microbial control. In: Vega FE, Kaya HK (eds) Insect pathology, 2nd edn. Academic, San Diego, pp 29–72

Shapiro-Ilan DI, Han R, Qiu X (2013) Production of entomopathogenic nematodes. In: Morales-Ramos JA, Rojas MG, Shapiro-Ilan DI (eds) Mass production of beneficial organisms: invertebrates and entomopathogens. Academic, San Diego, pp 321–355

Siegel JP (1997) Testing the pathogenicity and infectivity of entomopathogens to mammals. In: Lacey LA (ed) Manual of techniques in insect pathology. Academic, San Diego, pp 325–336

Siegel JP, Shadduck JA (1990) Safety of microbial insecticides to vertebrates—humans. In: Laird M, Lacey LA, Davidson EW (eds) Safety of microbial insecticides. CRC, Boca Raton, pp 101–114

Smits PH (1996) Post-application persistence of entomopathogenic nematodes. Biocontrol Sci Technol 6:379–387

Snow FH (1893) Contagious diseases of the chinch bug. Second Ann. Rpt. Dir. Kans. Univ. Exp. Sta. for the year 1892

St. Leger RJ (2001) Notification of intent to release a transgenic strain of *Metarhizium anisopliae* (document submitted by the University of Maryland as required by FIFRA). http://www.epa.gov/pesticides/biopesticides/other-docs/release_notification.htm. Accessed 16 Aug 2001

St. Leger RJ, Screen S (2001) Prospects for strain improvement of fungal pathogens of insects and weeds. In: Butt TM, Jackson CW, Magan N (eds) Fungi as biocontrol agents. CABI, Wallingford, pp 219–237

St. Leger RJ, Wang C (2009) Entomopathogenic fungi and the genomic era. In: Stock SP, Vandenberg J, Glazer I, Boemare N (eds) Insect pathogens: molecular approaches and techniques. CABI, Wallingford, pp 366–400

St. Leger RJ, Frank DC, Roberts DW et al (1992) Molecular cloning and regulatory analysis of the cuticle-degrading-protease structural gene from the entomopathogenic fungus *Metarhizium anisopliae*. Eur J Biochem 204:991–1001

St. Leger RJ, Shimizu S, Joshi L et al (1995) Co-transformation of *Metarhizium anisopliae* by electroporation or using the gene gun to produce stable GUS transformants. FEMS Microbiol Lett 131:289–294

St. Leger RJ, Joshi L, Bidochka MJ et al (1996) Construction of an improved mycoinsecticide overexpressing a toxic protease. Proc Natl Acad Sci U S A 93:6349–6354

Steinkraus DC, Boys GO, Bagwell RD et al (1998) Expansion of extension-based aphid fungus sampling service to Louisiana and Mississippi. Proc Beltwide Cotton Conf 2:1239–1242

Tabashnik BE (1994) Evolution of resistance to *Bacillus thuringiensis*. Annu Rev Entomol 39:47–79

Tabashnik BE, Gassmann AJ, Crowder DW et al (2008) Insect resistance to *Bt* crops: evidence versus theory. Nat Biotechnol 26:99–202

Tanada Y, Kaya HK (1993) Insect pathology. Academic, San Diego

Teera-Arunsiri A, Suphantharika M, Ketunuti U (2003) Preparation of spray-dried wettable powder formulations of *Bacillus thuringiensis*-based biopesticides. J Econ Entomol 96:292–299

Thomsen L, Eilenberg J (2000) Time-concentration mortality of *Pieris brassicae* (Lepidoptera: Pieridae) and *Agrotis segetum* (Lepidoptera: Noctuidae) larvae from different destruxins. Environ Entomol 29:1041–1047

Townsend RJ, Ferguson CM, Proffitt JR et al (2004) Establishment of *Serratia entomophila* after application of a new formulation for grass grub control. New Zealand Plant Protection, vol 57. Proceedings of a conference, Hamilton, New Zealand, 10–12 Aug 2004, pp 310–313

University of Arkansas (2013) Cotton aphid fungus sampling service. http://www.uark.edu/misc/aphid/project/project.htm. Accessed 6 Jan 2014

Upadhyay RK (2003) Advances in microbial control of insect pests. Kluwer, New York

USDA APHIS (2014) Biological control organism permits. http://www.aphis.usda.gov/plant_health/permits/organism/biological_control/index.shtml. Accessed 13 Jan 2014

US EPA (2006) Biopesticide registration action document, *Beauveria bassiana* I-IF23 (PC Code 090305). US Environmental Protection Agency, Washington, DC. http://www.epa.gov/oppbppd1/biopesticides/ingredients/tech_docs/brad_090305.pdf. Accessed 13 Jan 2014

US EPA (2012) *Chromobacterium subtsugae* strain PRAA4-1T (016329) Fact Sheet. http://www.epa.gov/pesticides/chem_search/reg_actions/registration/fs_PC-016329_20-Aug-12.pdf. Accessed 10 Jan 2014 van Frankenhuyzen K (2013) Cross-order and cross-phylum activity of *Bacillus thuringiensis* pesticidal proteins. J Invertebr Pathol 114:76–85

Walstad JD, Anderson RF, Stambaugh WJ (1970) Effects of environmental conditions on two species of muscardine fungi (*Beauveria bassiana* and *Metarhizium anisopliae*). J Invertebr Pathol 16:221–226

Wang X, Grewal PS (2002) Rapid genetic deterioration of environmental tolerance and reproductive potential of an entomopathogenic nematode during laboratory maintenance. Biol Control 23:71–78

Wang C, RJ St. Leger (2006) A collagenous protective coat enables *Metarhizium anisopliae* to evade insect immune responses. Proc Natl Acad Sci U S A 103:6647–6652

Wang C, RJ St. Leger (2007) The *Metarhizium anisopliae* perilipin homolog MPL1 regulates lipid metabolism, appressorial turgor pressure, and virulence. J Biol Chem 282:21110–21115

Wang C, Hu G, RJ St. Leger (2005) Differential gene expression by *Metarhizium anisopliae* growing in root exudate and host (*Manduca sexta*) cuticle or hemolymph reveals mechanisms of physiological adaptation. Fungal Genet Biol 42:704–718

Welling M, Nachtigall G, Zimmermann G (1994) *Metarhizium* spp. isolates from Madagascar: morphology and effect of high temperature on growth and infectivity to the migratory locust, *Locusta migratoria*. Entomophaga 39:351–361

Wennemann L, Cone WW, Wright LC et al (2003) Distribution patterns of entomopathogenic nematodes applied through drip irrigation systems. J Econ Entomol 96:287–291

Wikipedia (2013) Diatomaceous earth. Wikipedia, the free encyclopedia. http://en.wikipedia.org/w/index.php?title=Diatomaceous_earth&oldid=528917162. Accessed 26 Nov 2013

Wouts WM (1991) *Steinernema* (*Neoaplectana*) and *Heterorhabditis* species. In: Nickle WR (ed) Manual of agricultural nematology. Marcel Dekker, New York, pp 855–897

Wraight SP, Carruthers RI (1999) Production, delivery, and use of mycoinsecticides for control of insect pests of field crops. In: Hall FR, Menn JJ (eds) Biopesticides: use and delivery. Humana, Totowa, pp 233–269

Wraight SP, Jackson MA, de Lock SL (2001) Production, stabilization and formulation of fungal biocontrol agents. In: Butt TM, Jackson CW, Magan N (eds) Fungi as biocontrol agents. CABI, Wallingford, pp 253–287

Wraight SP, Inglis GD, Goettel MS (2007) Fungi. In: Lacey LA, Kaya HK (eds) Field manual of techniques in invertebrate pathology, 2nd edn. Springer, Dordrecht, pp 223–248

Wu S (2013) Efficacy of entomopathogenic nematodes and entomopathogenic fungi against masked chafer white grubs, *Cyclocephala* spp. (Coleoptera: Scarabaeidae). Ph. D. Dissertation, Virginia Polytechnic Institute and State University Blacksburg, VA

Xavier-Santos S, Lopes RB, Faria M (2011) Emulsifiable oils protect *Metarhizium robertsii* and *Metarhizium pingshaense* conidia from imbibitional damage. Biol Control 59:261–267

Young JM, Dunnill P, Pearce JD (2002) Separation characteristics of liquid nematode cultures and the design of recovery operations. Biotechnol Prog 18:29–35

Zhu H, Grewal PS, Reding ME (2011) Development of a desiccated cadaver delivery system to apply entomopathogenic nematodes for control of soil pests. Appl Eng Agric 27:317–324

Zimmermann G (1986) Insect pathogenic fungi as pest control agents. In: Franz JM (ed) Progress in zoology: biological plant and health protection, vol 32. Gustav Fischer, New York, pp 217–231

Zimmermann G (2007a) Review on safety of the entomopathogenic fungus *Metarhizium anisopliae*. Biocontrol Sci Technol 17:879–920

Zimmermann G (2007b) Review on safety of the entomopathogenic fungus *Beauveria bassiana* and *Beauveria brongniartii*. Biocontrol Sci Technol 17:553–596

Plant Breeding: A Tool for Achieving Food Sufficiency

M. K. Osei, M. D. Asante, A. Agyeman, M. A. Adebayo and H. Adu-Dapaah

1 Introduction

The UN predicts that 9.2 billion people are likely to occupy our planet by 2050. This is a population increase of 50 %, but food production will need to double as more people adopt a western diet. The predicted changes in rainfall and temperature, and an increase in extreme weather events (floods and droughts) all pose significant risks to agriculture. For example, 4 million tonnes of rice are lost annually due to floods in India and Bangladesh alone. The global distribution of certain crop pests and diseases is already expanding. For example, wheat stripe rust appeared in South Africa for the first time in 1996 and in Western Australia in 2002. The notorious tobacco whitefly has also steadily spread northwards over the past two decades from the tropics into temperate regions, infesting and spreading viruses to horticultural crops such as tomatoes, cucumbers, and beans. In addition, one of Europe's most destructive insect pests of brassica crops, the diamond-back moth, is now regularly appearing on UK crops.

Food insecurity exists when people lack sustainable physical or economic access to enough safe, nutritious, and socially acceptable food for a healthy and productive life. Food insecurity is part of a continuum that includes hunger (food deprivation), malnutrition (deficiencies, imbalances, or excesses of nutrients), and famine. There are many complex reasons for which an individual becomes food insecure. Food insecurity can result from declines in agricultural production as a result of declining soil fertility and water stresses. Sub-Saharan Africa's (SSA) dependency on rain-fed agriculture makes it vulnerable to climate change and variability which is a major threat to food security. Biotic (pests and diseases) and abiotic stresses (e.g., droughts, floods) result in major crop losses and consequent increases in food prices. Changes in climate and increases in extreme weather events, such as floods

M. K. Osei (✉) · M. D. Asante · A. Agyeman · H. Adu-Dapaah
CSIR-Crops Research Institute, P.O.BOX 3785, Kumasi, Ghana
e-mail: oranigh@hotmail.com

M. A. Adebayo
Ladoke Akintola University of Technology, Ogbomoso, Nigeria

D. Nandwani (ed.), *Sustainable Horticultural Systems*, Sustainable Development and Biodiversity 2, DOI 10.1007/978-3-319-06904-3_11

and drought, have become major challenges to stability of food supply and people's livelihoods sub-Saharan Africa. These weather events have made it more difficult for farmers to earn a stable income. In addition, climate change impacts have resulted in an increase of pest and diseases, which in turn has led to the misuse and abuse of pesticides with their negative effects on human health and the environment. Another major challenge to achieving food security in sub-Saharan Africa is the high rate of population growth, which increases the amount of food needed to adequately feed people in the sub-region

By 2020, the number of food insecure people in Sub-Saharan Africa (SSA) is projected to exceed 500 million out of a total population of roughly 1 billion. In other words, without any significant increase in investment or change in historical trends of major indicators, more than half of the region's population will consume less than the nutritional target of 2100 calories per day per person. Contrary to SSA's projections of future food insecurity, the number of food-insecure people in Asia is projected to decline from 433 million in 2010 to 320 million in 2020. The total number of stunted children is also expected to reach 64.2 million in 2020.

This chapter will focus on the contribution of plant breeding to a more sustainable food sufficiency.

2 Food Security: Fact or Myth

2.1 Food Insecurity

Food insecurity exists when people lack sustainable physical or economic access to enough safe, nutritious, and socially acceptable food for a healthy and productive life. The U.S. Department of Agriculture (USDA) also defines food insecurity as meaning "consistent access to adequate food is limited by a lack of money and other resources at times during the year." Food insecurity can also accurately be described as "a financial juggling act, where sometimes the food ball gets dropped". Rapid population growth which is usually found in poor African and other third world countries puts people at increased risk of food crises (e.g., the population of Niger increased from 2.5 to 15 million from 1950 to 2010). According to some estimations done by FAO, Africa will produce enough food for only about a quarter of their population by 2025 if the current growth rate continues. Certain groups of people are particularly vulnerable to food insecurity, including women (especially low income pregnant and lactating women), victims of conflict, the ill, migrant workers, low-income urban dwellers, the elderly, and children under five.

The consequences of food insecurity relates to significant environmental and socio-economic problems such as extreme poverty and malnutrition. Poverty is unmistakably the driving factor in the lack of resources to purchase or otherwise procure food and many people cannot afford food because of poverty. Poverty, combined with other socioeconomic and political problems, creates the bulk of food insecurity around the world. Malnutrition is also a direct consequence of food insecurity and the underlying cause of death of more than 2.6 million children each year, a third of under-five deaths, and a third of total child deaths worldwide. Nutritional consequences of insufficient food or under-nutrition include protein energy malnutrition, anemia, vitamin A, iodine and iron deficiencies. Iron deficiency contributes to the deaths of young women during pregnancy and childbirth. It is a leading cause of anemia. African countries have among the highest risk of iron deficiency in the world. Malnutrition results in high rates of infant and maternal mortality coupled with disease and disability. The aggregate costs of food and nutrition insecurity in Africa impose a heavy burden on efforts to foster sustained economic growth and improve general welfare.

2.2 *Evidence of Food Insecurity in Africa*

In sub-Saharan Africa, nearly 240 million people or one out of four people lack adequate food for a healthy and active life. An estimated 1 million children under the age of five are at risk of severe acute malnutrition in the Sahel region of West Africa (July 2012 executive brief of the Food and Agriculture Organization of the United Nations) (FAO). One in three developing country preschoolers—178 million children under the age of 5—suffers from stunting as a result of chronic malnutrition. Eighty percent of these children live in just 20 countries in Africa and the Asia Pacific region. The absolute number of stunted pre-school children has actually increased by more than 14.5 to 60 million, between 1990 and 2010. Ethiopia, Nigeria and the Democratic Republic of the Congo account for 40 % of all the stunted preschoolers in Africa. About 30 million children in sub-Saharan Africa are underweight and women in the region have on average 5.1 children, which is almost twice the world average of 2.5 children.

Changes in climate and increases in extreme weather events, such as floods and drought, have become major challenges to stability of food supply and people's livelihoods in sub-Saharan Africa. These weather events have made it more difficult for farmers to earn a stable income. In addition, climate change impacts have resulted in an increase of pest and diseases, which in turn has led to the misuse and abuse of pesticides with their negative effects on human health and the environment. Another major challenge to achieving food security in sub-Saharan Africa is the high rate of population growth, which increases the amount of food needed to adequately feed people in the sub-region.

2.3 *Projections of Future Food Insecurity in Africa*

By 2020, the number of food insecure people in Sub-Saharan Africa (SSA) is projected to exceed 500 million out of a total population of roughly 1 billion. In other words, without any significant increase in investment or change in historical trends of major indicators, more than half of the region's population will consume less than the nutritional target of 2100 calories per day per person. Contrary to SSA's projections of future food insecurity, the number of food-insecure people in Asia is projected to decline from 433 million in 2010 to 320 million in 2020. The total number of stunted children is also expected to reach 64.2 million in 2020.

FAO's recent estimates indicate that food insecurity has been increasing in SSA and is a source of growing concern to African governments. Strategies to address food insecurity must therefore aim to increase agricultural productivity in the developing world in order to tackle poverty. It must also provide long-term improvements in crop yields to keep up with demand as the world's population grows. Increasing availability, access, utilization and stability of nutritious and healthy food in order to minimize the impacts of food insecurity includes breeding of horticultural crops.

In short, no matter what the arguments are, there are an estimated 870 million chronically undernourished people in the world, 98% of who live in developing countries (FAO 2012). There is a strong correlation between the level of development and food security (Food Security Network 2012). Food security is dependent on a strong and sustainable food system which comprise of production, processing, distribution, marketing, acquisition, and consumption of food. These are lacking in many developing countries making them food insecure. Food insecurity is a fact and not a myth. Food insecure countries/ regions must put proper structures in place to overcome undernourishment. The best way to overcome undernourishment is to produce sufficient quantities of food at places where it is easily accessible and affordable to those who need it. The backbone of food production is the development of crop varieties and sustainable food production systems. The breeding of horticultural crops is especially crucial because vegetables and fruits are the best suited to provide vitamins that helps to overcome "hidden hunger" in addition to hunger itself.

3 Overview of Production Capacity of Horticultural Crops

Horticulture is the science, technology and business involved in intensive plant cultivation for human use. It is practiced from the individual level in a garden up to the activities of a multinational corporation. It is very diverse in its activities, incorporating plants for food (fruits, vegetables, mushrooms, culinary herbs) and non-food crops (flowers, trees and shrubs, turf-grass, hops, medicinal herbs). It also includes related services in plant conservation, landscape restoration, landscape and garden design/construction/maintenance, arboriculture, horticultural therapy, and much more. This range of food, medicinal, environmental, and social products and services are all fundamental to developing and maintaining human health and well-being. Horticultural produce and processed products from the developing world are becoming increasingly popular both in domestic and international markets.

Sustainable horticultural production requires an integrated consideration of inputs (germplasm, water, fertilizer, pesticides and growth regulators), as well as potential consequences of these inputs to man and his environment (salinization, contamination of water supplies, loss of soil structure and fertility). The value of horticultural crops is seasonal, and systematic horticultural production can staggered during the season to reduce market fluctuations and increase farmers' incomes. In the developing world, the development of horticultural crops offers the opportunities to meet domestic food needs and diversify incomes. In addition, horticulture affords excellent opportunities for improvement of human health, and farmer household economic and social advancement. However, less than 50% of the value of harvested horticultural crops gets to the ultimate consumer and the rest is lost between the farms and through the market and the consumer. This scenario creates a huge opportunity to explore and adopt modern production techniques and postharvest technologies that are cost-effective, appropriate, scalable, innovative, and environmentally friendly. These methods and technologies would ensure congenial production environment and good postharvest handling, improve profits for producers and marketers as well as provide commercial opportunities for local manufacturers.

Globally competitive horticultural production will be largely influenced by the impact of climate change in the foreseeable future. Hence, new opportunities for strengthening the horticultural sector will emerge, not necessarily from increased land area for production but from the development of horticultural marketing, plant breeding, agronomic management techniques, horticultural engineering in the fields of mechanization, robotics, traceability and food safety Milbank (2007). Quality seed and planting stock of high yielding cultivars represent a package of genetic technology that is critical to a sound horticulture supply chain. Result-oriented research efforts must be directed towards the development of suitably adapted modern and traditional varieties of horticultural crops for enhanced productivity. The rich endemic horticultural diversity in the developing regions should be exploited for the development and advancement of the indigenous and traditional crops in local, regional and export markets. Advances in horticulture require an atmosphere of sound legislative and policy framework, social and political will, adequate local and regional infrastructure. Additionally, institutions whose primary concerns are capacity building, management instruments, and monitoring and evaluation should be established and strengthened in developing countries. Horticultural crops diversity diets and increase dietary consumption of essential vitamins, micronutrients, antioxidants, and fibres. For horticultural produce to be admitted into these modern market outlets, they must satisfy certain stringent quality requirements. These supermarkets are opened in the developing countries by big companies that are owned by entrepreneurs from the developed countries and their organization and operational strategies are in sharp contrast to the underdeveloped poorly structured traditional local and regional market outlets existing in developing countries. This development poses a huge challenge to the small scale horticultural crop producers who are grossly deficient in resource base and are lacking in skills to access and interpret market information, who operate without adequate financial, human or social capital to develop the linkages needed to succeed in the market. Current trends in globalization and fast dissemination of research knowledge will continue to reduce the gap between the advanced market structures of the developed world and the poor market facilities existing in the developing countries. Poor postharvest management, poor quality standards and food safety protocols, and quality deterioration, high contamination risks, and short shelf life of horticultural produce result in rejection of produce by consumers. These stem from lack of access to critical postharvest knowledge, technology, and infrastructure by the small scale horticulture producers mostly in the developing world. Research and development of appropriate postharvest technologies for small and medium-scale horticultural crop farmers, value-added processing techniques, food safety protocols and quality standards for horticultural commodities can help to reduce postharvest losses, improve food safety, and contribute to increased producer incomes and the subsequent development of rural economies.

4 Current Horticultural Systems and Potential for Contributing to Food Security

Global production and exports of horticultural crops are rising steadily (Spore 2007). The overall challenge is to achieve sufficient horticultural production and fulfill the growing world demand to facilitate the socioeconomic development of farmers in developing countries, while preserving the environment and reducing risks for human health and ecosystems. However, yield increases have been smaller than area growth and have been negligible or even negative in the least developed countries. While experience shows that horticulture can offer good opportunities for poverty reduction because it increases income and generates employment, care must be taken that small and poor farmers are not excluded from the opportunities in these market sectors. Consequently, we argue that development agencies must put more emphasis on horticultural research and development, especially in the following priority areas: genetic improvement, safe production systems, commercial seed production, postharvest facilities, and the urban/peri-urban environment.

5 Horticulture's Dwindling Fortunes: Plant Breeding to the Rescue

When people invented agricultural technology about 10,000 years ago, human cultures and several other life forms were largely altered by the new inventions. The agricultural evolution heralded the era of changing plant forms; with genetics, and later plant breeding, serving as agents of change. Prior to the evolution, plants

Fig. 1 Wild (*left*) and Modern (*right*) tomato. (*Photo by Thomas Bruce, UC, Davies*)

existed in unattractive wild forms with enormous inherent yet-to-be-tapped genetic benefits. Humans' quest for food, feed, fuel, and fibre, drew them to the jungles in search of resources. We later realized that so many of those wild plant forms possessed the traits and attributes that will support all life forms for all generations. However, the gap between wild plants and the domesticated plant types with which we are accustomed today is remarkably huge; widest and most dramatic in many cereal, fruit, and vegetable crops, where the ancestors may be almost indiscernible. Out of the existing huge number of plant species, only about 500 were known to have been domesticated worldwide for the use of man. Notable among the earliest domesticated and "genetically" transformed wild plant types were horticultural crops like cabbage (*Brassica oleracea*), tomato (formerly *Lycopersicon esculentum*, now *S. lycopersicum*), and others that are grouped under the heading of 'brassicas': cauliflower, brussels sprouts, etc. Tomato, for instance, had a huge transformation from its wild ancestor into its current food form. It is probably hard for anyone to believe that today's beautiful and juicy tomato vegetable was once considered lethally poisonous! Wild tomato versions were small, cherry-like yellow fruits which were later "genetically" transformed into big juicy red fruits through several generations of natural selection and breeding (Fig. 1).

Prior to the time that humans began making conscious efforts at developing improved plant types through the application of the principles of genetics and breeding, plants growing in the wild were, and still are, subject to evolutionary force demanding what Charles Darwin, in the nineteenth century, dubbed as "the survival of the fittest." He formulated the idea of evolution which suggests that plants and animals are in constant competition, and therefore the fittest or the best adapted survives and transmits its genes to the next generation while those that are less adapted dies or fails to reproduce and so does not pass its genes to any progeny (Kingsbury 2009). Hence, the force of natural selection purges plants and animals of appreciable chunk of genetic materials and consequently, genetic biodiversity; yet the future of mankind depend on preservation of the genetic diversity of the world's food crops for future generations. Later, the evolutionary process came heavily under human control through agriculture and breeding. In the practice, decisions are taken by the agriculturist (farmer or scientist) on whom plants are "fit" based on certain defined criteria that meet human needs, and which ones are unfit and regarded as weeds. Systematic or artificial selection and breeding advanced as humans continue to exert evolutionary pressure on plants by intentionally favoring some forms or types of a particular plant species and deliberately discriminating against others. The unselected types are discarded and consequently lose their genetic constituents, particularly the unique genetic codes that are not shared with the selected types. For instance, the rural farmer exerts evolutionary pressure on the wild tomato (*Lycopersicon lycopersicum*) by selecting for the "fitter" types having bigger and juicier fruits. Modern scientists do the same, except that they apply the knowledge of heredity to direct their actions in order to achieve desired results. Over time, the genetic base of the transformed plant types become dramatically narrow even though these humanly guided evolutionary processes of selection and breeding have brought enormous improvements to the wild plants, turning them to invaluable food, feed, and fibre resources for our continued existence. According to Kingsbury (2009), selection involves picking the best *and discarding the rest* from each generation; "breeding" is about a more active and directed engagement of crossing genetically different plant genotypes for achieving some desired resultant progenies.

Horticultural crops such as fruit and vegetable crops attracted huge genetic and breeding research attention that birthed the various varieties that we relish today. Just like in tomato, primitive plant breeders selected for uniformity in plant architecture, fruit size, color and taste while modern breeders followed up with additional selections for other genetic improvements. In doing this over several years, "desirable" genes were accumulated in the new crop types and the "undesirable" ones discarded. Unfortunately in so many occasions, the discarded genes were never conserved, never to be regained again. New elegant genetically uniform horticultural crop types emerged, having narrow genetic base and enormous genetic vulnerability to new strains of disease-causing pathogens and insect pests, and other abiotic stress factors.

The importance of maintaining a robust genetic diversity for any crop was underscored by the outbreak of southern corn blight (*Helminthosporium maydis*) in the late 1960s particularly in the United States of America. The incident created a nationwide panic in the corn industry of the US because most corn fields succumbed to the "new" disease, forcing the national yield down by 40 %. One of the factors that fuelled the disaster was lack of resistance genes in the "narrow" genetic backgrounds of the hybrid corn varieties that farmers were planting at the time. Corn farmers could choose from over 600 different corn cultivars in 1948 but two decades later, only six hybrid maize varieties produced over 70 % of total US corn yield, and only two varieties produced 96 % of the peas. Naturally, the plant hosts and their associated pathogens maintain a constant state of equilibrium which, however, shifts in response to evolution of either or both. Hence, a crop variety that is considered as resistant to a pathogen (resistance depicts a state in which a host plant responds to the presence of a pathogen by certain mechanisms that reduce or restrict the establishment, growth, and reproduction of the pathogen) suddenly succumbs to a new or exotic variant/strain of a particular pathogen with which it has always been associated. The exotic strain of the pathogen then produces its progeny that, through natural selection, is able to expose the vulnerabilities of the genetically uniform host plant, sometimes to a point of total devastation. The same new pathogen variant may not be able to affect a genetically diverse landrace population of the same crop variety that much.

New strains of the late blight of tomatoes and potatoes caused by *Phytophthora infestans* re-emerged in the US and Canada in the late 1980s and early 1990s and caused widespread devastation (Fry and Godwin 1997) to a degree that is similitude of what caused the famine in Ireland. The devastation which attracted national press attention was attributed to different exotic strains of the pathogen that spread over the world.

In the course of breeding for agronomic and economically important traits in horticultural crops, we see plant breeders achieve a horticultural crop variety having a narrow genetic base as a result of rigorous selections for genetic and morphological uniformities. Hence, the problem of narrow genetic diversity in crops generally has been partly created by plant breeding. Major crops that are bred for high productivity are reaching yield ceilings as a result of dwindling genetic variability. Yield (genetic yield potential or production) here means the harvestable biomass that one may achieve by growing a particular genetic material (Ribou et al. 2013).

Generally, improved agronomic crop management that entails fertilization, integrated pest and disease control, environmental-friendly and timely weed control, improved water availability through modern irrigation systems has resulted in an upsurge in realized yields in agriculture over time; increased genetic yield potential has not made substantial contribution. The yields of many conventional horticultural crops have plateau, so much that breeders no longer achieve ground-breaking success of a "bump" from their efforts at breeding for yield increase. The genetic implication is that the individuals in the population of an elite crop germplasm now have so many genes in common and consequently differ in too few gene loci. Since yield is a complex polygenic trait, crossing individuals that share many identical loci may not produce the kind of complementarities that is needed at the loci that control yield in other to achieve heterosis. "Heterosis is the phenotypic superiority of a hybrid over its parents in growth rate, reproductive success, and yield" (Lippman and Zamir 2007). Interestingly, the solution to the problem of narrow genetic base with the attendant yield ceiling or yield plateau lies with plant breeding. Plant breeders are always eager to diversify their base genetic resources through introduction or introgression, mutation, hybridization, and through molecular breeding. Biodiversity is a key agricultural resource that is needed for improvement. Breeders look for novel alleles in introduced germplasm in order to improve their local adapted genetic materials. Relative wild and weedy species of agricultural crops are collected and conserved on the anticipation that they serve as repositories of favorable novel alleles that can be introgressed into conventional species for genetic improvement and diversification. Mutagenic agents such as radioactive substances are employed to create useful variations; through the rate of success is usually very slim. Genetically diverse individuals in a population are often systematically crossed in order to achieve hybrids that share the characteristics of the parents. During the last 20 years, molecular markers have entered the scene of genetic improvement in a wide range of horticultural crops. Among the major traits targeted for improvement in horticultural breeding programs are disease resistance, fruit yield and quality, tree shape, floral characteristics, cold hardiness, and dormancy.

Today, molecular markers are being used for germplasm characterization, genetic mapping, gene tagging, and gene introgression from exotic species. The most widespread use of markers has been in germplasm characterization. Although genetic maps have been developed for most of our important fruit and vegetable species, and a number of horticultural important gene loci have been tagged, only a few instances of application of molecular markers for progeny selection have been reported. New, easy to perform allele testing methods are needed to bridge this large gap between marker development and application. Research on tomato provides an elegant example for simultaneous discovery and introgression of quantitative trait loci from wild species with the aid of markers. Molecular markers and genetic maps have not been widely applied to ornamental, medicinal and other horticultural species and only a handful of such species is used in marker development and testing.

6 Importance of Biodiversity in Plant Breeding

Plant breeding as it is today is the combined effort of farmers and plant breeders spanning a period of over 10,000 years. Before this period, farmers got their food from hunting wild plants and animals. Farmers probably started farming because population increase put a strain on the food they were gathering from hunting. The emergence of farming came plant breeding because farmers had to select plants based on certain desirable traits and the selected plant were used for the next planting. Farmers were therefore the first breeders even though they might not have understood the science of breeding as we know it today.

Plant breeding involves the selection of plants with desirable traits. These desirable traits can be classified broadly as yield potential, resistance/tolerance to biotic stresses, resistance/tolerance to abiotic stresses and quality which includes appearance, taste, cooking or processing character and nutritional value. Plant breeders, therefore, need plant populations that have wide variation/diversity for all kinds of traits to make selection possible and to create new genotypes. Biodiversity is therefore the foundation of plant breeding. There cannot be breeding for a trait if variants of that trait do not exist. This makes it crucial to preserve a set of genotypes (germplasm) that are representative of the variation within a species. Even though genetic diversity is the greatest resource of the plant breeder, the breeding and promotion of modern varieties is a threat to the maintenance of genetic diversity. This is because seed companies promote only a few uniform varieties with the desired traits and this could lead to genetic erosion. For example, 52 % of vegetables grown worldwide receive breeding attention; of these only 17 % are found in large scale breeding programmes (Dias 2011). Vegetable biodiversity is reported to be eroding at a rate of 1.5–2.0 % per annum (Dias 2011). The challenge of genetic erosion is mitigated by international efforts at conserving large amounts of germplasm of various crops in gene banks around the world.

7 Methods of Breeding Horticulture Crops

The methods used by breeders to develop improved varieties of various crops depend on the reproductive system of that crop. Horticultural crops are either asexually propagated or sexually propagated (self-pollinated or crop pollinated). For self-pollinated crops, breeders exploit their homozygous nature by focusing on selecting for individuals within a population. Breeders who work on cross-pollinated crops on the other hand, focus on populations instead of individual plants. In this paper we will discuss breeding methods that are applicable to sexually and asexually propagated crops, followed by methods that are usually applied for self-pollinated and cross-pollinated crops.

7.1 Methods for Breeding Asexually Propagated, Self-pollinated and Cross Pollinated Crops

Introduction As mentioned above plant breeders depend on the diversity amongst genotypes in order to make effective selection of improved varieties. A breeder would normally access advance breeding lines from state, national or international breeding institutions and gene banks. These can be tested in the local environment and released as varieties per se based a performance criteria determined by the breeder.

Clonal Selection A clone is an asexually propagated population of genetically similar genotypes. Most clonally propagated horticultural plants are mostly perennials. Selection of clones of natural and induced variants is an important breeding procedure for horticultural crops. The variants in a clone arise by mutation which results in bud sports, chimeras, or genetic mosaics (Sleper and Poehlman 2006). These mutations are particularly useful in developing new clones of ornamental crops. A high mutation rate has also been reported in sugarcane genotypes maintained through tissues culture (Sleper and Poehlman 2006).

In vitro Breeding In this method, tissue culture techniques are used to create new crops. The phenomenon of Somaclonal variation gives breeders an opportunity to select for new varieties. Somaclonal variation is the changes that occur in somatic cells during the mitotic divisions of a single clonal plant that result in variation in derived clonal individuals (Rodriguez-Enriquez et al. 2011). This method has been used to develop new varieties of horticultural crops such as strawberries (Biswas et al. 2009). The method involves the induction of Somaclonal variation using various tissue culture techniques, screening Somaclonal materials obtained, selection and propagation of promising variants in the nursery, field evaluation of selected variants for up to three generations and selection of new lines based on field performance (Biswas et al. 2009).

Mutation Breeding Mutations are changes in nucleotide sequences of a gene which occur either spontaneously or are artificially induced using mutagenic agents. Selection for spontaneous mutations has played a major role in the development of horticultural crops. For example, many varieties of ornamental plants as well as certain desirable fruit traits of apples, bananas and pineapples have been obtained from spontaneous mutations (Sparrow 1979). Over the last 60 years, induced mutations have been successfully used by breeders to create more than 3000 mutant varieties in about 180 plant species including horticultural plants (Shu et al. 2007). More recently, horticultural crops that are developed through mutation breeding use molecular tools to make mutation breeding more targeted (Shu 2009; Wilde et al. 2012). Molecular mutation breeding has been used to develop traits such as longer shelf-life (tomato, melon), improved starch quality (potato), and virus-resistance (peppers, tomato) which are being used in commercial varieties (Wilde et al. 2012). Over 16 horticultural crops have been screened for natural or induced allelic diversity in more than 100 candidate genes (Wilde et al. 2012). Polymorphisms in these candidate genes can facilitate the development of cultivars with value-added traits. With the recent advances in genomics and the drastic reduction in the cost of genotyping, breeders can now use mutation techniques in a more efficient and precise manner to develop new varieties.

Marker-assisted Breeding In marker-assisted breeding, genetic markers associated with traits or quantitative trait loci (QTL or genes) are used to facilitate the selection of the desired phenotype. Selection is thus done indirectly using the DNA markers. Marker-assisted selection requires the development of genome linkage map, location of the QTL/gene of interest and the identification/development of molecular markers for breeding. Molecular markers are also used to select the desired genetic background of a cultivar especially during backcross breeding. In recent times, the genomes of many horticultural crops have been sequenced and the cost of genotyping is declining rapidly. This has made the integration of molecular markers into the breeding of horticultural crops very feasible.

7.2 Methods for Breeding Self-pollinated Crop

7.2.1 Methods with Selection Only

Mass Selection The method involves the selection of similar plants on the basis of phenotype. Seeds from the selected plants are bulked into composite for the next generation without progeny testing (Fig. 2).

It is an effective method for the improvement of landraces. The aim is not to produce a single line but simply to eliminate defective or unproductive components of a landrace, so that the proportion of productive plants is increased and the overall yield is raised. Cultivars developed from mass selection are normally uniform for highly heritable traits such as maturity and colour. Differences in quantitative traits

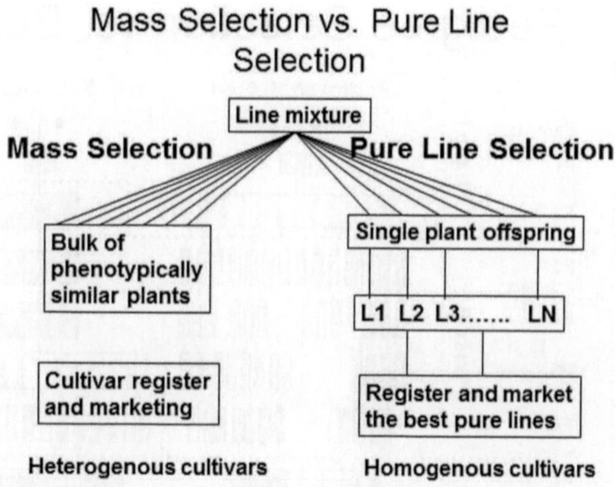

Fig. 2 Schematic presentation of mass and pure line breeding methods

such yield and size may still be present. Mass selection has been used successfully to develop improve varieties of spinach, beetroot, cole crops, bulb crops and sweet-potato (AVRDC 2000).

Pure Line Selection Pure line selection involves the isolation of pure lines from a mixed population (Fig. 2).

Pure line selection has been practiced by farmers for a long a time. With this method, an atypical plant in the field is selected and multiplied as a new cultivar. The new cultivar is developed from a single plant. Progeny testing is crucial in pure line selection in order to determine the behaviour of the selected plant. The superiority of the selected line compared to the mixed population is also evaluated. If superiority is proven, the population is increased, named and released as a variety.

7.2.2 Methods with Hybridization Followed by Selection

Methods that involve hybridization followed by selection include bulk-population breeding, pedigree selection, single-seed descent, backcrossing and double haploid method.

Bulk Breeding In bulk breeding, variation is created through crossing (hybridization). Seeds harvested in the F1 through F4 generations are bulked without selection. Selection is delayed until advanced generations (F5–F8) when most segregation has stopped (Fig. 3).

Pedigree Method Pedigree selection is used to combine the desirable traits of two or more genotypes in a new improved variety or inbred line. Its main objective is to isolate superior, recombinant, homozygous lines. Selection is performed from the F2 generation onward until genetic purity is reached (Fig. 3).

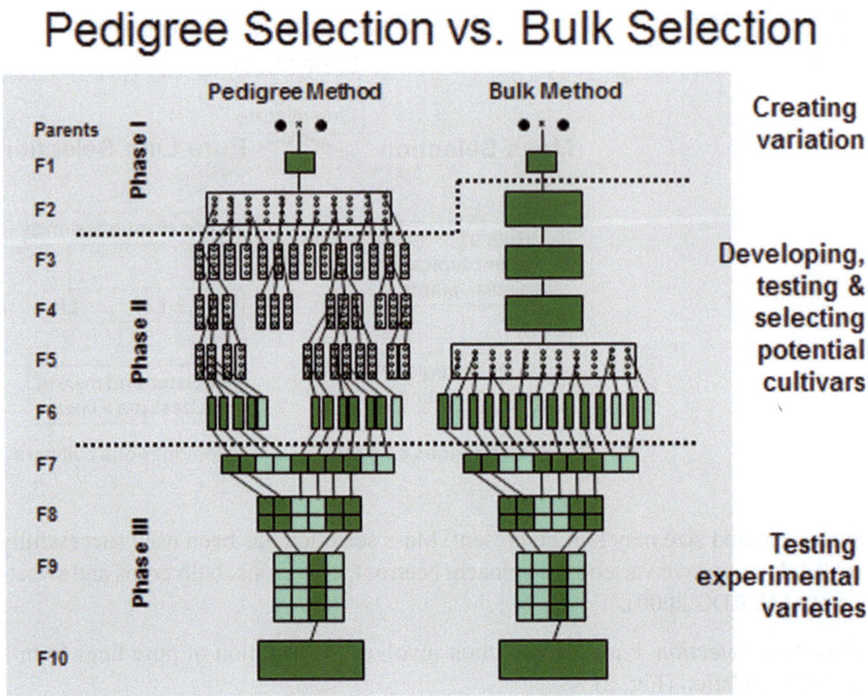

Fig. 3 Schematic presentation of pedigree and bulk breeding methods

Single-Seed Descent Single Seed Descent procedure is based on advancing generations through single seed from each plant in each generation. This separates the inbreeding phase from the selection phase. Selection is done at F6 generation. The progeny of the selected plants are tested in replicated trials and recommended for release as new varieties.

Backcross Method Backcrossing is a form of recurrent hybridization by which a desirable allele for a character is substituted for an alternative, undesirable allele. After each back cross, hybrid plants are identified with the gene under consideration and are back crossed again with the recurrent parent. Backcrossing to the recurrent parent is continued for 6–8 generations. After backcrossed plants are back to type, they are selfed to produce homozygous plants.

Double Haploid Method In double haploid breeding, crosses are made between selected parents. The F1 plants are cultured and chromosome number of haploid plants (anthers) is doubled with colchicines to produce double-haploid plants. Progenies of double-haploid plants are evaluated in the field. Superior lines are recommended for release. The double-haploid method has the advantage of achieving homozygosity in one generation (Fig. 4).

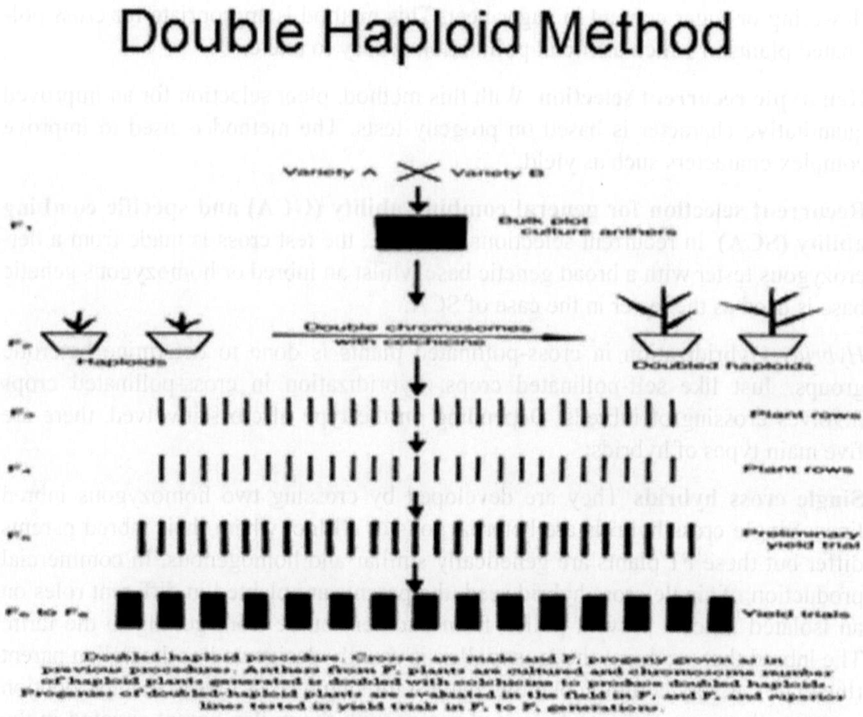

Fig. 4 Schematic presentation of double haploid

7.3 *Breeding Cross-pollinated Crops*

In cross pollinated crops, the breeder focuses more on populations than individual plants and on quantitative analysis rather than qualitative traits.

Recurrent Selection Recurrent selection is a breeding system used to increase the frequency of desired alleles of a population by repeated cycles of selection (Sleper and Poehlman 2006). In recurrent selection, a source population superior for the character being improved is identified by the breeder. Superior genotypes selected from the source population are intermated to produce new gene combinations leading to an improved expression of the character of interest. Superior plants for the character of interest are propagated from selfed seed and intermated. The cycle of selection of superior plants and intermating continues until the desired goal is achieved. Recurrent selection may be repeated as long as superior genotypes are being generated.

Types of recurrent selection include;

Phenotypic recurrent selection Selection to improve a quantitative character based on visual observation or physical measurement of a character such as early flowering or sugar content in sugar beet. This method is appropriate for cross pollinated plants in which artificial pollination is easy to make.

Genotypic recurrent selection With this method, plant selection for an improved quantitative character is based on progeny tests. The method is used to improve complex characters such as yield.

Recurrent selection for general combing ability (GCA) and specific combing ability (SCA) In recurrent selections for GCA, the test cross is made from a heterozygous tester with a broad genetic base whilst an inbred or homozygous genetic base is used as the tester in the case of SCA.

Hybrids Hybridization in cross-pollinated plants is done to determine heterotic groups. Just like self-pollinated crops, hybridization in cross-pollinated crops involves crossing of inbreds. Depending on the type of cross involved, there are five main types of hybrids;

Single cross hybrids They are developed by crossing two homozygous inbred lines. Single cross hybrids are heterozygous at all loci where their inbred parents differ but these F1 plants are genetically similar and homogenous. In commercial production of single cross hybrid seed, the parents are planted in different roles on an isolated field to prevent pollen from another source from getting to the farm. The inbred that produces the most pollen is usually designated as the pollen parent (the male). A very common planting pattern for single cross hybrid seed production is one pollen parent row to four seed parent with the pollen parent situated in the middle (Sleper and Poehlman 2006).

Double cross hybrids These are hybrids developed from crosses between two single crosses. Each single cross comes from a cross between two independent inbreds.

Top cross hybrids It is a hybrid obtained from a cross between inbred line or single cross and a genetically mixed population.

Synthetic hybrids A synthetic variety is a random mating population produced by crossing a group of pre-tested inbreds usually numbering 5–10. Ideally, the lines included must have high general combining ability. Synthetic varieties are maintained by open-pollination in isolation. Synthetic varieties have been used widely in crops where pollination is difficult to control e.g. forage crops. For horticulture crops, synthetics have been produced for cabbage, cauliflower, turnip, radish and beet (AVRDC 2000).

Composite varieties A composite variety is developed by mixing seed of two or more genotypes or by intermating clones in isolation (Sleper and Poehlman 2006). In forming the cultivar equal quantities of seed from each component strain or clone are blended to form the breeder seed. After the initial constitution, the composite variety is maintained by open-pollination and it is not reconstituted as originally produced. This differentiates composite varieties from synthetics, where the original parents are maintained for reconstitution of the variety.

8 On-going Breeding Horticultural Programmes to Address Food Insecurity

Horticultural breeding plays a key role in addressing food and nutritional insecurity for those most vulnerable to stresses imposed by resource scarcity and climate change. Horticultural crops comprises of fruit and nut crops, vegetables, food legumes, roots and tubers, medicinal, aromatic and ornamental plants. These contribute greatly both to satisfying food security and to integrate and diversify dietary needs of mankind.

Horticultural breeding programmes in most African countries focuses on producing improved crop varieties aimed at producing disease resistant lines and increasing crop yields and nutritional qualities and other traits of commercial value. For instance, an AVRDC project in West and Central Africa has distributed more than 700 kg of improved varieties of the African eggplant and other indigenous vegetable seeds to local farmers aimed at increasing crop yields by over 50 %. Particularly, the Gambia, Mali, Senegal and Togo have formed a new Regional Network that will increase vegetable production, marketing and consumption in West Africa. This Network offers farmers significant benefits in terms of yield improvement, agronomic performance and consistency of end-use quality.

In Kenya, the introduction of a hybrid (F1) watermelon variety known as Sukari yielded 300 ton of the fruit per hectare, which was more than 150 times the harvest of the local variety, which yielded on average 12 ton of fruit per hectare. Watermelon is an excellent source of vitamin C, potassium and a very good source of vitamin A, notably through its concentration of beta-carotene. Pink watermelon is also a source of the potent carotenoid antioxidant, lycopene. Watermelon has moved up to the front of the line in recent research studies on high-lycopene foods. Thus the increase in water melon production will impact significantly on nutrition and food security. Another example is the production of East Africa Highland banana, which is an important food and income crop in East Africa. The production of the cooking banana also known as the "matoke" in the region is poised to get a boost with the distribution of the first-ever, high-yielding, and disease-resistant hybrid varieties. The 26 hybrid varieties dubbed NARITAs (NARO_IITA) are a result of over 20 years of joint breeding efforts between the National Agricultural Research Organization (NARO) of Uganda and the International Institute of Tropical Agriculture (IITA). With nearly 60 % higher yield than the local "matoke", these hybrids are also resistant to black Sigatoka, an important fungal disease of the crop worldwide that affects the leaves and leads to losses of 30–80 %. These hybrid bananas spell good news for millions of smallholder farmers growing the crop and have great potential in efforts to boost the region's food security and poverty reduction efforts.

Horticultural programmes also breed for health to sustain lives. In Eastern and Southern Africa, there is an on-going sweetpotato breeding programme by the International Potato Center (CIP) aimed at increasing the vitamin A content of the crop. Vitamin A is commonly known as the vitamin needed for good eyesight. It is also vital for regulating genes, maintaining healthy skin, supporting the immune system and producing red blood cells. Vitamin A deficiency primarily causes impaired vision and increases susceptibility to infectious diseases. The CIP focuses on producing orange-fleshed sweetpotato varieties, which have high ß-carotene contents. These orange-fleshed sweetpotato varieties also serve as cheap and complementary sources of vitamin A to the rural and urban poor families. It is important to note that Sweetpotato is an important household food security crop in Eastern and Southern Africa, particularly in densely populated, intensively cultivated mid-elevation farming areas because it is a sustainable and affordable source of micronutrients in diets and can therefore help to address malnutrition. Although the crop is mainly cultivated for domestic consumption, both fresh roots and leaves are increasingly gaining market potential for cash income.

Harvest Plus disseminated orange-fleshed sweet potato to more than 24,000 households in Mozambique and Uganda to see if more vitamin A could be provide through food. As a result of this, the orange fleshed sweetpotato contributed 78 % of the total vitamin A intake for children aged between 6 and 35 months in Mozambique and 53 % in Uganda.

Other breeding programmes in Sub-Saharan Africa have focused on the production of drought-tolerant horticultural crop varieties. This is because Sub-Saharan Africa already has a highly variable and unpredictable climate and is acutely vulnerable to droughts. It is estimated that over 30 % of the Sub-Saharan countries live in drought prone areas. Large parts of the region are projected to become drier, increasing the number of people at risk of hunger and poverty by tens of millions with the advent of climate change.

In Sub-Saharan Africa, farmers who used improved legume varieties by ICRISAT have found that the new varieties have brought a dramatic increase in yields in recent years, with benefits for their income and the land. New cultivars of chickpea that combine early maturity and disease resistance, especially the large-seeded Kabuli, have been rapidly adopted in Ethiopia, Tanzania, Sudan and Kenya, leading to a 50 % increase in cultivated acreage and a doubling of output over a 30 year period from 1977. Chickpea is nutritionally rich, an excellent source of high-quality protein with a wide range of essential amino acids. It is high in dietary fiber and low in fat, most of it polyunsaturated. It also provides dietary phosphorus, magnesium, iron and zinc. Chickpea exports from Eastern Africa increased commensurately creating new income generation opportunities for farmers, proving that getting improved seed varieties into the hands of smallholder farmers does increase food production in the developing world.

Participatory Plant Breeding (PPB) programmes for horticultural crops in Sub-Sahara Africa have contributed to all three aspects of food security (ie., availability, access and utilization). PPB is the development of a plant breeding program in collaboration between breeders and farmers, marketers, processors, consumers, and policy makers (food security, health and nutrition, employment). PPB is also breeding that involves close farmer-researcher collaboration to bring about plant genetic improvement within a species. Several studies suggest that farmers can increase their revenue or make cost savings through the implementation of PPB. Informal rural seed networks combined with PPB measures play an important role in assuring access to food, particularly for the poorest members of farming communities. Owing to its decentralized organization, PPB programs generate many different varieties for different production conditions and purposes. The approach tends to enhance food diversity and to maintain traditional knowledge that can help counteract food shortages and malnutrition. Thus, the unique potential of PPB for improving food security is the impact it can make in those very situations where people are afflicted by food insecurity.

In Kenya and Tanzania CABI has tested three farmer-led seed production models (contract model, research model and quality declared seed model) to promote better access to high quality African indigenous vegetable (AIV) seeds among African smallholders. This has helped relevant stakeholders develop the necessary skills to establish and manage seed production and marketing enterprises. Under the contract model, the quantity and quality of seed produced by farmers participating in the project increased significantly: Amaranthus increased from 2134 to 5918 kg; African nightshade from 3832 to 27,997 kg; Jute mallow from 1770 to 17,706 kg; and Crotalaria from 6669 to 24,253 kg. As a result, farmers earned an average of US$ 4500 per year. East African farmers have been given a chance through participatory plant breeding to influence final production of preferred varieties of bananas, beans and pumpkins.

According to breeding experts this will increase the uptake of such new crop varieties because farmers input on preferred characteristics have been taken into account. In recent years, tissue culture and the application of molecular approaches have been used in breeding of horticultural crops. As a result, manipulation of generation cycles for a faster breeding of horticultural crops through in vitro culture systems has been done. This technique is expected to improve yields and breed disease resistant crop varieties. In a joint FAO/IAEA plant breeding programme, the duration of generation cycles in the food legume Bambara groundnut (*Vigna subterranean*) was significantly shortened through in vitro and in vivo methods. More than four generations per year as compared to 1 or <2 generations in the control field/greenhouse system was obtained. Bambara groundnut is a rich source of protein (16–25 %) in sub-Saharan Africa and its seeds are valued both for their nutritional and economic importance. The seeds command a high market price, with demand far outweighing supply in many areas (Coudert 1982).

Horticultural breeding plays a key role in addressing food insecurity in SSA. Horticultural breeding makes food available through crop improvement programmes. Increases in horticultural crop yields are realized through improved seeds, integrated pest management etc. Food is regularly and periodically available and affordable and food prices are stable. Horticultural crop production creates jobs by providing opportunities for diversification of income. This enables a household to acquire adequate amounts of food or to have access to adequate food at all times. Biofortification procedures addresses food insecurity by fighting micronutrient deficiency through horticultural breeding and this could contribute dramatically to improve health and reduce childhood and women mortality in many countries.

9 Widening the Horticultural Space through Genetically Modified Crops

Challenges to Horticultural Crop Production in Sub-Saharan Africa Horticultural crop production in Sub-Saharan Africa is saddled with some constraints that prevent it from attaining potential. These constraints include the establishment of farms with low yielding and poor quality local varieties due to inadequate agronomic and/or technological knowledge on production. Another major constraint is the lack of an efficient production, multiplication, and distribution system of seed to meet local demand. For example, in vegetable production, locally produced seeds by both individuals and communities are often infected by seed-borne or seed transmitted viruses, fungi, bacteria and insects, and are genetically diverse (João 2010), thus reducing the quality of the seeds and the subsequent crops. Horticultural crop production is also vulnerable to seasonal and inter-annual variation of rainfall. Over reliance on rain fed production therefore, results in more than 60% vegetable yield losses.

Plant Breeding as a Means of Improving Agricultural Productivity—Conventional Methods and Their Limitations About 70% of species used as vegetable are gathered from the wild according to an inventory study in Benin (Achigan-Dako et al. 2011). The situation is similar to what pertains in other sub-Saharan African countries and thus there is the need for a strong horticultural domestication programme under pinned by breeding and crop improvement. Plant breeding programmes has over the years significantly increased the productivity and quality of the plants we grow for food, feed and fiber. The selection for characters such as faster growth, higher yields, pest and disease resistance, larger seeds, or sweeter fruits has dramatically changed domesticated plant species compared to their wild relatives.

Plant breeding can be accomplished through many different techniques ranging from simply selecting plants with desirable characteristics for propagation (conventional breeding), to more complex molecular techniques. Conventional plant breeding is a genetic modification of plants through sexual crosses using parents selected for desirable traits. This breeding method results in open pollinated varieties (OP) or hybrid varieties and these have had tremendous impact on agricultural productivity over the last decades. Horticultural farmers who use improved varieties have witnessed increases in yields and quality of their crops, increased tolerance to insect pests, resistance to viruses, bacteria and fungi, etc. However, conventional plant breeding procedures also have their limitations. For instance, it has not been successful in introducing high levels of resistance to some bacterial pathogens into acceptable crop varieties. Breeding can also only be done between two plants that can sexually mate with each other. This limits the new traits that can be added to those that already existing in a particular species. In addition, when plants are crossed, many traits are transferred along with the trait/s of interest including those traits that have undesirable effects on yield potential.

Use of Genetically Modified Crops Recombinant DNA techniques offer a great potential to widen the gene pool that is available to plant breeders and help them to develop better crops. The technology allows DNA from any source to be transferred to specific crops. It involves identification and isolation of the gene(s) of from theoretically any organism. The gene is altered and introduced into the crop of interest. The new crop is referred to as genetically modified crop (GM crops). The GM technique offer opportunities to accelerate the efficiency and extent of further crop improvement by the transfer of genes conferring resistance to pests, diseases, herbicides and environmental stress, as well as quality traits such as improved post-harvest storage, flavour, nutritional content and colour. GM crops could thus help plant breeders to respond better to the needs of farmers and consumers. Consumers needs for fruits and vegetables with reduced blemishes from pests/diseases and reduced pesticide residues could be better responded to through the development of GM horticultural crops. GM horticultural crops that have been commercialized include capsicum, papaya, petunia, plum, potato, rock melon, rose, squash and tomatoes. These crops were mainly modified for disease resistance or quality traits including colour and long shelve-life (Fitzgerald 2011). Many more GM horticultural crops are in the pipeline.

However, because of the unconventional methods involved in the breeding of GM crops there is a heavy regulation of these crops. In order to take fully advantage of this technology in sub-Saharan Africa, biosafety laws and regulations must be put in place and more personal has to be trained to use the technology.

Conclusion

Food self-sufficiency cannot be achieved without plant breeding. Over the years, the horticultural industry has benefited immensely from the breeding of improved varieties. Thousands of varieties of horticultural crops have been bred for various characters including higher yields, tolerance to biotic and abiotic stresses and end-user traits. However, many challenges still exist especially in the area of changing consumer preferences and emerging pest and diseases. The challenge with biotic and abiotic stresses is expected to become worse with climate change. Plant breeders in sub-Saharan Africa must therefore combine conventional and biotechnology approaches in order to develop varieties that will meet current and future challenges.

References

Achigan-Dako EG, N'Danikou S, Assogba-Komlan F, Ambrose-Oji B, Ahanchede A Pasquini MW (2011) Diversity, geographical and consumption patterns of traditional vegetables in sociolinguistic communities in Benin: implications for domestication and utilization. Econ Bot. doi:10.1007/s12231-011-9153-4

AVRDC Africa Regional Program (2000) Varietal evaluation and seed production of vegetable crops held at AVRDC African Regional Program, Arusha, Tanzania, 29 Sept to 05 Oct 1997 Asian Vegetable Research and Development Center-Africa Regional Program, Arusha, Tanzania

Biswas MK, Dutt M, Roy UK, Islam R, Hossain M (2009) Development and evaluation of in vitro Somaclonal variation in strawberry for improved horticultural traits. Sci Horticult 122:409–416
Coudert MJ (1982) Market openings in West Africa for Cowpeas and Bambara groundnuts. Int Trade Technol X:147–241
Dias JS (2011) Biodiversity and vegetable breeding in the light of developments in intellectual property rights, ecosystems biodiversity. Grillo O (ed) ISBN: 978-953-307-417-7, InTech. http://www.intechopen.com/books/ecosystems-biodiversity/biodiversity-and-vegetable-breeding-in-thelight-of-developments-in-intellectual-property-rights
Fitzgerald P (2011) Agrifood Awareness Australia Limited- (2011) http://www.horticulture.com.au/librarymanager/libs/174/Paula%20Fitzgerald%20 %20Glocal%20Status%20of%20 gene%20tachnology%20in%20horticulture.pdf. Accessed 28 Nov 2013
Food and Agriculture Organization (2012) The State of Food Insecurity in the World 2012. http://www.fao.org/docrep/016/i3027e/i3027e.pdf. Accessed 9 June 2013
Food Security Network (2012) http://www.foodsecuritynews.com/What-is-food-security.htm. Accessed 9 May 2013
Fry WE, Godwin SB (1997) Re-emergence of potato and tomato late blight in the United States. Plant Dis 81(12):1349–1357
João SD (2010) Impact of improved vegetable cultivars in overcoming food insecurity. Euphytica 176:125–136
Kingsbury N (2009) Hybrid: the history and science of plant breeding. The University of Chicago Press, Chicago
Lippman ZB, Zamir D (2007) Heterosis: revisiting the magic. Trends Genet 23:60–66
Milbank J (2007) Globally competitive horticultural production and marketing utilizing renewable energy. Nuffield Australia Project No SCVCT0601
Riboua SB, Douama F, Hamanta O, Frohlich MW, Negrutiua I (2013) Plant science and agricultural productivity: why are we hitting the yield ceiling? Plant Sci 210:159–176
Rodriguez-Enriquez J, Dickinson HG, Grant-Downton RT (2011) MicroRNA misregulation: an overlooked factor generating somaclonal variation? Trends Plant Sci 16:242–248
Shu QY (2009) Induced plant mutations in the genomics era. Food and Agriculture Organization of the United Nations, Rome, pp 425–427
Shu QY, Lagoda, PJL (2007) Mutation techniques for gene discovery and crop improvement. Mol Plant Breed 5:193–195
Sleper DA, Poehlman JM (2006) Breed Field Crops Crop Sci 47:900–901
Sparrow DHB (1979) Special techniques in plant breeding. In: Knight R (ed) Plant breeding. Academy Press, Brisbane, p 163
Spore (2007) Wageningen, The Netherlands: The Technical Centre for Agricultural and Rural Co-operation, CTA, ACP-EU 1-2. No. 128
Wilde DH, Chen Y, Jiang P, Bhattacharya A (2012) Emir. J. Food Agric 24(1):31–41

Transcriptome Analysis of *Musa* and its Applications in Banana Improvement

S Backiyarani, S Uma, M. S. Saraswathi and A Chandrasekar

1 Introduction

Banana and plantains are of great socio economic importance being a staple food for more than 400 million people with a global production of 129.0 million t. Bananas are considered to be a high energy food and also rich in minerals and vitamins. They are one of the earliest domesticated plant species (Denham et al. 2003) originated in South East Asia and Western Pacific and are now widely distributed throughout the tropical and subtropical regions. Although banana and plantains are best known as a food-fruit crop, almost every part of the plant can be used in one way or another, starting from banana sap to seed for the benefit of mankind.

The vast majority of cultivated bananas have evolved either from natural inter/intraspecific hybridization between two wild species, *M. acuminata* and *M. balbisiana*. Cultivated bananas are sterile triploids or diploids, with parthenocarpic fruit development and propagated through vegetative means. Asexual driven evolution has resulted in a narrow genetic base which is currently threatened by major pests and diseases. They include diseases like *Mycosphaerella* spp., Fusarium wilt, bacterial wilt, Moko, viral diseases etc. The pests include pseudostem borers, corm weevils, aphids, nematodes etc. Although management of selected pests and diseases is possible through chemical means, it often leads to environmental pollution and enhanced cost of cultivation. The ultimate solution for these threats is the development of resistant varieties through introgression of resistant genes either through conventional breeding or genetic engineering approaches. Understanding the host-pathogen interaction is essential for developing pest and disease resistance through the identification of signal transduction pathways in bananas. Isolation of trait specific genes and markers and their application in marker assisted selection (MAS) has accelerated both conventional breeding and genetic engineering meth-

S. Uma (✉) · S. Backiyarani · M. S. Saraswathi · A. Chandrasekar
Crop improvement, National Research Centre for Banana, (ICAR), Thogamalai Road, Thayanur Post, Tiruchirapalli, TN 620102, India
e-mail: umabinit@yahoo.co.in

D. Nandwani (ed.), *Sustainable Horticultural Systems*, Sustainable Development and Biodiversity 2, DOI 10.1007/978-3-319-06904-3_12

ods for crop improvement. In the present day context, identification of trait specific gene(s) has been made easier with the available data on genome sequencing. In banana, team led by French Agronomic Research Centre, CIRAD has successfully completed the whole genome sequencing of double haploid Pahang (W), of *M. acuminata* subsp. *malaccensis* (A genome) (D'Hont et al. 2012). It has genome coverage of 472.2 Mb which represented 70 % of the DH-Pahang genome size. Similarly, whole genome sequencing of wild diploid *M. balbisiana* (BB) type 'Pisang Klutuk Wulung (PKW)' has been completed (Davey et al. 2013) with a genome coverage of 341.4 Mb representing 78 % of the expected B genome size. The sequence information on the ancestral genomes (A and B) of present day bananas will provide an idea about the genome structure, organization, allelic diversity and regulatory elements of the *Musa* species. Though genome sequencing provides the vital and voluminous information, it fails to give a clear picture about the expression of genes in a particular tissue for a particular physiological condition. To understand this relationship between the genome and the phenotype, it is essential to study the genome products like messenger RNAs. Transcriptome studies has an added advantage over whole genome sequencing with respect to the information on the mRNA variants which are the resultant products of alternative splicing, RNA editing or transcription initiation and termination sites. Thus understanding the transcriptome is essential for interpreting the functional elements of the genome, revealing the molecular constituents of tissues and also for understanding the resistance mechanism for a specific stress. Transcriptome profiling has been made easy with the availability of user friendly bioinformatic tools to understand the molecular mechanism of plants under various physiological conditions. *Musa* improvement program through conventional approach is a slow process owing to its complex genetic structure. Moreover bananas exhibit some unique characteristics such as different ploidy levels, parthenocarpy, sterility, vegetative propagation and biparental cytoplasmic inheritance (Roux et al. 2008). Thus, transcriptome analysis and gene expression studies of *Musa* will be a powerful model for understanding the above mentioned fundamental aspects for which the other two model species (*Arabidopsis,* Rice) use differently.

2 Global Approaches to Gene Expression Analysis

Various technologies have been developed to deduce and quantify the transcriptome, including hybridization or sequence-based approaches. Over the years, several techniques, such as suppression subtractive hybridization (SSH), representational difference analysis (RDA), expressed sequence tags (EST), cDNA fragment fingerprinting (cDNA AFLP), serial analysis of gene expression (SAGE), DNA microarrays etc., which allow high throughput identification and/or quantification of transcriptome components have been developed. The merits and demerits of these techniques are mentioned in Table 1. The aforesaid techniques have been successfully used to identify many differently expressed genes in various tissues against various stresses/developmental stages (Table 2). The details of those techniques applied exclusively in banana are enumerated.

Table 1 Principle, merits and demerits of banana transcriptome techniques

Techniques	Principle	Merits	Demerits	Reference
SSH	PCR-based amplification of only cDNA fragments that differ between a control (driver) and experimental transcriptome	To compare two mRNA populations and obtain cDNAs representing genes that are either over-expressed in one population as compared to other	RNA is required in High quantities	Diatchenko et al.1996 Munir et al. 2004
RDA	To find sequence differences in two genomic or cDNA samples	It simplifies analysis of the subtracted library	Experiment is expensive	Lisitsyn et al. 1993
cDNA AFLP	To distinguish closely related individuals at the sub-species level and can also map genes	Identification of up and downregulated genes in contrast to subtractive hybridization	DNA sequencing and assembly of complex, large, polyploid, and/or repetitive genomes is still technically difficult	Althoff et al. 2007
SAGE	Short sequence tags are sufficient to identify a gene transcript, provided the tags represent a known location within the gene	Data accurate, quantitative and cumulative	More sequencing errors due to the misidentification of the region in the database sequence	Velculescu et al. 1995
EST	The EST sequences are generated by randomly picking the clones from a cDNA library and performing a single sequencing reaction to produce 300–500 bp of sequence per clone	To identify gene transcripts, which are instrumental in gene discovery and gene sequence determination	"EST" to describe genes for which little or no further information exists besides the tag	Bortoluzzi and Danieli 1999
DNA/cDNA microarray	Hybridization between two DNA strands, the property of complementary nucleic acid sequences to specifically pair with each other by forming hydrogen bonds between complementary nucleotide base pairs	Reduces the bias towards gene sets that are already known	Experimental replication is expensive	Maskos and Southern 1992

Table 2 Transcriptome techniques exploited in banana

Variety/cultivar	Techniques	Tissues	Stage/stress	No. of expressed genes	Reference	Laboratory
M. acuminata, subgroup Cavendish 'Grand Nain'	cDNA AFLP	Fruit	Crown rot disease	10	Lassois et al. 2011	University of Liege, Gembloux, Belgium
M. acuminata ssp. burmannicoides 'Calcutta-4, cv. Rose, FHIA 17, Williams'	cDNA AFLP	Root	*Fusarium*	76	Munro 2008	University of Pretoria, Pretoria, South Africa
M. acuminata ssp. burmannicoides 'Calcutta-4'	SSH	Root	*Fusarium*	83	Swarupa et al. 2013	IIHR, Bangalore, India
M. acuminata cv. Manoranjitham (AAA)	SSH	Leaf	*Eumusae* leaf spot	805	Uma et al. 2011	NRCB, Trichy, India
M. acuminata cv. Karthobiumtham (ABB)	SSH	Root	Nematode	256	Backiyarani et al. 2014	NRCB, Trichy, India
M. acuminata cv. Tuu Gia	SAGE	Leaf		5292	Coemans et al. 2005	Laboratory of Tropical Crop Improvement, Katholieke Universiteit Leuven, Belgium
M. acuminata ssp. burmannicoides 'Calcutta-4'	EST Sequencing	Hot and cold stress exposed leaf tissues	Fruit ripening	220	Santos et al. 2005	EMBRAPA, Brazil
M. acuminata 'Cavendish Williams'	Combination of SSH and microarray	Fruit	Fusarium	79	Van den berg et al. 2007	FABI, University of Pretoria, South Africa
M. acuminata, cv. Grand Nain	Combination of SSH and microarray	Fruit	Early ripening of banana	84	Manrique-Trujillo et al. 2007	Centro de Investigación y de EstudiosAvanzados del IPN, México
M. acuminata AAA group	Combination of SSH and microarray	Fruit	Early ripening of banana	265	Xu et al. 2007	Institute of Tropical Bioscience and Biotechnology, Chinese Academy of Tropical Agricultural Sciences, China

Table 2 (continued)

Variety/cultivar	Techniques	Tissues	Stage/stress	No. of expressed genes	Reference	Laboratory
M. acuminata L. AAA group	Combination of SSH and microarray	Fruit	Ethylene biosynthesis initiation in fruits.	22	Jin et al. 2009	Institute of Tropical Bioscience and Biotechnology, Chinese Academy of Tropical Agricultural Sciences, China
M. acuminata, Dwarf Cavandish (AAA)	Combination of SSH and microarray	Fruit	Ethylene biosynthesis initiation in fruits	37	Kesari et al. 2007	NBRI, Lucknow, India

2.1 cDNA-amplified Fragment Length Polymorphism Analysis (cDNA-AFLP)

cDNA-AFLP is an improvement over traditional differential display techniques. It is a PCR-based method, in which cDNA is digested with two restriction enzymes (four-cutter and a six-cutter) and adapters are ligated to the ends of the fragments. In the first step, only those fragments that are digested by both restriction enzymes are thus have different adapters at the end. In the following amplification, the complex starting mixture of cDNA is fractionated into smaller subsets by selective PCR amplification using primers for the adapters with one or more extra nucleotides. By increasing stringency of PCR amplification (with additional nucleotides to the primers), the sensitivity of the analysis can be increased which allows the detection of genes with low expression levels. Differences in intensity of the bands provide a good measure of the relative difference in the levels of gene expression.

Munro (2008) examined the *Fusarium oxysporum* f. sp. *cubense* (FOC) resistant (cvs. Rose and Calcutta 4 and tolerant hybrid FHIA 17) and susceptible (Cavendish cv. Williams) to elucidate the genetic factors involved in banana defense response against FOC. Transcriptome profile using cDNA-AFLP was carried out on tissues collected at 6 and 72 h post inoculation of these cultivars with FOC 'subtropical' race 4 (VCG 0120). Seventy-six differentially expressed transcript derived fragments (TDFs) were isolated, sequenced and annotation of these sequences resulted in identification of defense-related genes such as S-adenosylmethionine synthase (SAMS) and isoflavone reductase, which are potentially involved in the production of cell wall strengthening compounds like lignin etc.

Crown rot disease, caused by *Colletotrichum musae*, is one of the main postharvest diseases of banana that has a negative impact on the market value of fruits and affects export. Lassois et al. (2008) tried to identify genes influencing the susceptibility to crown rot disease, at distinct physiological growth stages of 'Grande Naine' fruits. The cDNA-AFLP fingerprinting of this study resulted in identification of genes involved in signal transduction, pathogenesis- related proteins, cellulose synthase gene and dopamine-β-monooxygenase. Delgado et al. (2003) also evidenced that reduction in cellulose synthesis leads to activation of defense responsive genes through various signalling pathways. Identification of differentially expressed genes like dopamine-monooxygenase, which play a key role in catecholamine biosynthesis pathway, suggested the possible involvement of these genes in resistance mechanism (Lassois et al. 2011).

2.2 EST Sequencing

Expressed sequence tags (ESTs) is a quantitative method to measure specific transcripts within a cDNA library (Rudd 2003) and is achieved by single-pass sequencing of a large number of cDNAs traditionally yielding sequence tags of around 300–500 bp in length. In a study to understand the function and regulation of the genes involved in temperature stress, Santos et al. (2005), developed and characterized two full length enriched cDNA libraries from the RNA isolated from cold (5–25 °C) and hot (25–45 °C) stress exposed leaves of *M. acuminata* ssp. *burmannicoides* var. Calcutta 4. The global analysis of these two libraries indicated that only 10 % of the cDNAs had common expression, while 42 and 48 % of cDNAs were unique to cold and hot libraries respectively. From cold EST library, two genes related to the ABA response to cold acclimation and one low -temperature responsive genes were identified. The data is in public domain of EMBRAPA, Brazil (http://www.cenargen.embrapa.br/index.html) as a part of the DATA_*Musa* database (a genome database for *Musa* spp consisting DNA sequences and sequence information).

2.3 Serial Analysis of Gene Expression (SAGE)

SAGE is a method that generates sequences from cDNA fragments for the discovery of new genes and the quantification of their expression levels. SAGE is a strategy for extensive tag profiling of cDNA over EST sequencing (Velculescu et al. 1995) which is based on counting sequence tags of 10–15 bp from cDNA libraries. Briefly, a short cDNA fragment or tag (13 bp) is extracted from a defined position in each transcript by a series of linker ligations and restriction digestions. Tags are then amplified once by PCR, concatenated, cloned in a vector and sequenced. These tags, in most cases, contain sufficient information for unambiguous annotation to EST sequences. The frequency of each tag directly reflects the abundance of the corresponding mRNA (Velculescu et al. 1995). But annotation of the tags depends largely on the availability of cDNA libraries or EST collections.

For the first time, *M. acuminata* gene expression pattern in youngest unfurled leaf tissue of *M. acuminata* cv. "Tuu Gia" (accession No. ITC.610) was analysed by Coemans et al. (2005) through SAGE. A total of 10,196 tags were generated from 5292 expressed genes and abundantly expressed transcripts were annotated by homology to cDNA or EST sequences. Analysis of the transcript profiles showed that the majority of the abundant transcripts are involved in energy production, mainly photosynthesis. However, the most abundant tag was derived from a type 3 metallothionein transcript, which accounted for nearly 3% of total transcripts analysed. The unknown tags were also identified by applying 3' rapid amplification of cDNA ends (3' RACE) which resulted in the identification of a novel NADPH protochlorophyllide oxidoreductase, the key enzyme in chlorophyll biosynthesis. Limited availability of banana genomic DNA and cDNA/EST sequences at that time restricted the use of classical SAGE yielding short tags.

2.4 Suppression Subtractive Hybridisation (SSH)

The principle of subtractive hybridization is commonly been used to analyze the differences in gene expression between two samples. The selection of differentially expressed sequences is performed by allowing the samples to cross-hybridize, followed by isolation of hybridization products unique for one sample (termed the tester) from sequences that are shared between the samples or unique for the other sample (the driver). Suppression subtractive hybridization was developed for the enrichment of differentially expressed genes of both high and low abundance (Diatchenko et al. 1996).

Swarupa et al. (2013) demonstrated the FOC resistance mechanism involved in *M. acuminata* spp. *burmannicoides* var. Calcutta 4, using SSH approach. From this library, a total 83 unique genes were obtained and homology comparison revealed that nearly 17 % of genes were involved in stress and defense while 15 % in signal transduction. Based on RT-PCR studies in contrasting cultivars like Calcutta 4 and Kadali for FOC, many genes which are involved in recognition of pathogens like, transmembrane protein, receptor like protein kinase, mannose binding protein etc. were confirmed based on their expression pattern in the resistant cultivar. Ravishankar et al. (2011) has identified some differentially expressed genes (lipoxygenase, rubiscoactivase, glycinedehydrogenease, catalase and ethylene responsive factor) from drought stressed leaf samples of Bee Hee Kela (BB) through SSH approach.

Similarly, Uma et al. (2011) identified a total of 498 unigenes, which consisted of 78 contigs and 420 singletons from the SSH library obtained from *M. eumusae* challenged resistant cultivar (Manoranjitham AAA). Top BLAST hits of BLAST2GO analysis resulted in a total of 89 hits, of which 13 defense related genes were chosen for gene expression studies. It revealed that cytochrome oxidase, lipoxygenase, metallothionein, flavin containing monoxygenase were upregulated only in the resistant cultivar within 48 h post inoculation (HPI) and these genes could be involved in *M. eumusae* leaf spot resistance mechanism. Backiyarani et al. (2014) also identified 256 unique genes from the root lesion nematode (*Pratylenchus coffeae*) challenged resistant cultivar (Karthobiumtham ABB) through SSH approach. Of these, 26.8 % were unigenes involved in defense and/or signal transduction including resistant gene homologues, disease resistance response proteins and protein kinase signaling. 12.8 % of the unigenes hit with unknown protein classes. These findings imply that invasion of nematode could trigger multiple signaling pathways. Like in many other methods, SSH also relies on restriction enzymes leading to the loss of transcripts lacking restriction sites. Combining SSH technique with high throughput screening of the harvested clones through the use of cDNA microarrays could greatly reduce the tedious work of northern blot analysis and also the likelihood of false-positive clones enriched via SSH (Yang et al. 1999).

2.5 Combination of SSH and Microarray

Combination of the suppressive subtractive hybridization and cDNA microarray techniques has greatly improved the efficiency of fishing out of the differentially expressed clones (Yang et al. 1999; Geschwind et al. 2001). Two possible combinations are (i) using the subtracted cDNA clones as probes printed on chips and (ii) using the labeled and enriched amplicons. In the second method, instead of using the subtracted clones as probes spotted on chips, enriched amplicons were labeled and used as targets to hybridize the pre-made microarray chips for microarray analysis.

This combination technology, allowed not only the identification of genes without extensive sequencing works in the subtracted amplicons regardless of many redundant clones, but also to easily evaluate the subtraction efficiency and specificity. Moreover, the SSH/microarray approach made it possible to conduct a transcriptome-wide identification of differentially expressed genes particularly for those with low expression. It has been reported that the absolute expression level is not a crucial determinant for identifying the genes, while the relative difference in expression levels does impact on whether or not a gene is recovered by subtractive hybridization.

In banana, the differentially expressed genes during fungal and fruit ripening stages were identified by this SSH/microarray approach. Van den berg et al. (2007) constructed banana SSH library with 736 differentially expressed genes in FOC tolerant Cavendish selection 'GCTCV-218' at 6 HPI of FOC race 4 (VCG 0120). Subsequently, this library was screened using glass slide microarray to identify and discard housekeeping and rRNA genes that had escaped subtraction and to select defense response-associated genes. As a result, out of 736, 79 clones were shortlisted and sequenced. Annotation of these sequences revealed that many showed homology to defense-associated genes, including cell wall-strengthening genes. These genes were confirmed through quantitative RT-PCR based on up-regulation and differential expression throughout a time-course, following FOC infection in the tolerant GCTCV-218 compared with susceptible cv. Williams. It was confirmed that genes coding for PR-1, PAE, xylanase inhibitor, peroxidase, catalase 2, metallothionein, response regulator 6 and trypsin inhibitors were differentially expressed in the tolerant cultivar GCTCV-218. This study also proved that tolerance of GCTCV-218 was linked to significantly increased induction of cell wall-associated phenolic compounds.

A better understanding of the mechanism of postharvest ripening is necessary to manipulate postharvest management in banana. Xu et al. (2007) tried to identify the differentially expressed genes during the early stages of fruit ripening through SSH approach. SSH was performed along with cDNA microarray analysis on the day of harvest as the "driver" and cDNA of the fruit, 2 days afters harvest (DAH) as the "tester." A total of 265 readable differentially expressed genes were obtained through forward subtraction from tester and their expression levels were estimated by overlaying images of cDNA microarrays screened by mRNA from 0 to 2 DAH. Their expression level was confirmed through RT-PCR and the results were found to be in accordance with the microarray data. Upregulation of endoglucanase and cellulose synthase genes were generally associated with cellulose biosynthesis of primary and secondary cell walls suggesting the importance of these proteins for the biosynthesis of highly crystalline cellulose. The over expression of alcohol dehydrogenases (Adh) at both early and late stages of fruit ripening suggested that aroma production begins at the beginning of postharvest and continues up to the late phase of ripening. In order to gain an understanding of the genetic components responsible for metabolic shifts associated with banana ripening, Manrique-Trujillo et al. (2007) identified differentially expressed transcripts in a late stage of fruit ripening through SSH approach. The differentially expressed SSH clones were confirmed through hybridization with DNA arrays and genes involved in the processes associated with fruit ripening, such as stress, detoxification, cytoskeleton and biosynthesis of volatile compounds (orcinol O-methyltransferase, putative alcohol dehydrogenase, ubiquitin-protein ligase, chorismatemutase etc.) were identified.

Davey et al. (2009) tried to profile the banana leaf transcriptome to drought stress, pooled RNA of control and drought stressed leaves of the *Musa* cultivar 'Cachaco' were hybridized to Affymetrix Rice Genome Array. This resulted in the identification of 2910 transcripts displaying >2 fold difference in expression levels in response to drought. These drought responsive transcripts included many functional classes associated with plant biotic and abiotic stress responses, as well as regulatory genes like ERF, DREB, MYB, bZIP and bHLH transcription factor families. The outcome of this study proved that cross species (heterologous) microarray studies using gDNA-based probe selection as a highly promising strategy to study complex biological responses in a non-model species.

2.6 Next Generation Sequencing

Recent technologies of next generation sequencing (NGS) provide a comprehensive snapshot of the transcriptome. Transcriptome analysis through NGS is called as RNA sequencing or RNA-seq, which enables simultaneous sequencing of millions of molecules without involving any cloning. Thus RNA-seq has become an attractive analytical tool in trancriptomics, due to minimum running cost and the possibility to uncover novel transcriptional-related events (Wang et al. 2011).

- NGS transcriptome analysis is more accurate and highly sensitive over other techniques.
- NGS technique offers true quantification in contrast to microarrays
- Simultaneous data collection for genes with differential expression levels is possible only in transcriptome analysis
- This technique has the ability to distinguish isoforms of the same transcript and to detect new splicing variants.
- Very effective for the discovery of novel splicing sites
- New transcripts could also be detected from NGS transcriptome techniques which are not possible through microarrays.
- Transcriptome sequencing does not necessarily require prior knowledge of the genome sequence whereas designing of microarrays needs a reference sequence. Thus this method is particularly attractive for non-model species.
- NGS transcriptome enables SNP discovery whereas it is not possible through other transcriptome techniques

RNA–Seq techniques involve three major steps namely, library construction, DNA sequencing and sequence data analysis. Library construction includes isolation of high quality RNA samples, fragmentation of RNAs, and reverse transcription of RNA into cDNA followed by second strand synthesis and end modification by ligation of oligonucleotide adapters. Optionally this cDNA library may be amplified through PCR to selectively enrich the adapter molecules ligated to both ends. Then their quality and quantity can be validated by capillary electrophoresis, spectrophotometry and qPCR prior to sequencing.

NGS platforms are ideally suited for the sequencing of RNA-Seq libraries. The term NGS is used to collectively describe technologies other than Sanger sequencing that have the potential to process millions of sequence reads in parallel, rather than 96 at a time. Also, next generation sequence reads are produced from fragment libraries that have not been subjected to the conventional vector based cloning and *E. coli* based amplification stages used in capillary sequencing. The length of next generation sequenced reads is 35–250 bp (depending on the platform) (Table 3) than capillary sequenced reads which is about 650–800 bp. Of many NGS platforms available, Roche 454, ABI SOLid, Illumina (Hiseq, Miseq) are the popular platforms and Ion Proton, PacBio RS, Oxford nanopore are some of the newer platforms.

Million or billion reads obtained through this platforms will be filtered for obtaining high quality reads for further downstream analysis (Fig. 1). Initially, the reads are mapped to the reference genome or transcriptome and the reads that align to the same locus are assembled into transcripts. The expression level of that particular transcript is estimated by counting the number of reads that aligned to this transcript. Mortazavi et al. (2008) proposed the quantification of transcript levels is an accurate method, ie. reads per kilobase of exon model per million mapped measures (RPKM). $RPKM = 10^9 x C/N.L$ where C is the number of mappable reads that fall on to the gene's exon, N is the total number of mappable reads and L is the total length of exons.

Table 3 Comparison of various NGS platforms

Platform	Chemistry	Read length (bp)	Run time	Gb/Run	Primary error	Base error rates	Advantage	Disad-vantage
454 GS Junior (Roche)	Pyro-sequencing	500	8 h	0.04	Indel	1.00 %	Long read length	High error rate in homo-polymer
454 GS FLX + (Roche)	Pyro-sequencing	700	23 h	0.7	Indel	0.5 %	Long read length	High error rate in homo-polymer
Hi Seq (Illumina)	Reversible terminator	2*100	2 days (rapid mode)	120 (rapid mode)	Substi-tution	~1–2 % over 100 bp	High throughput/ low cost	Short reads long run time
MiSeq (Illumina)	Reversible terminator	2*150	27 h	1 Gb	Substi-tution	~1–2 % over 100 bp	High throughput/ low cost	Short reads
SOLiD (Life)	Ligation	85	8 days	150	A-T bias	0.06 %	Low error rate	Short reads long run time
Ion Proton (Life)	Proton detection	200	2 h	100	Indel	1.2 % over 150 bp	Short Run times	Higher error rate
PacBio RS	Real-time sequencing	3000 (up to 15,000)	20 min	3	CG deletion	13.00 %	Long read length without PCR	High error rate

To enable comparison of two or more different transcriptome libraries, the values are normalized by taking into account the fact that the number of reads per transcript depends on the transcript length and the total number of reads obtained for a particular library (Mortazavi et al. 2008). Apart from quantification of transcripts, RNA-Seq also permits isoform identification, detection of new splicing sites. Specific softwares are available as free or open source programs for each analysis (Table 4).

In a nutshell, the NGS transcriptomics aims at the detection of transcripts expressed under specific physiological conditions, determination of the transcriptional structure of genes, splicing patterns and other post-transcriptional modifications apart from quantification of varying expression levels of each transcript.

2.6.1 Transcriptome Profiling in Banana Through NGS

Understanding the complexity of the disease resistance will pave way for the development of biotic stress resistant bananas. Many studies have been carried out to understand the molecular mechanism of resistance/tolerance through RNAseq or transcriptome analysis.

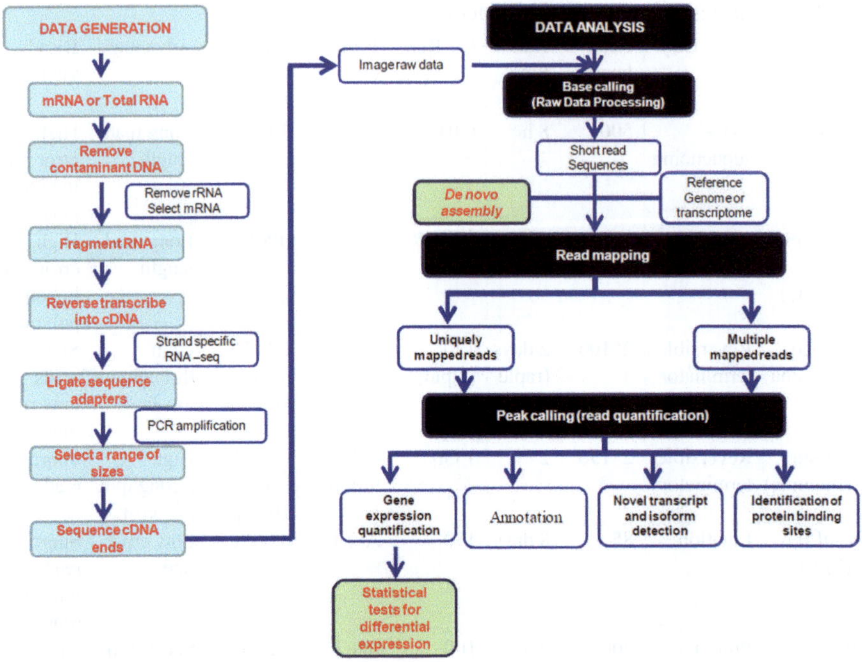

Fig. 1 Methodology involved in transcriptomics using NGS

Fusarium wilt caused by *FOC* is affecting bananas in tropical and subtropical countries (Ploetz 2006). Li et al. (2012) identified defense related transcripts which are involved in *FOC* TR-4 resistance by comparing the root transcriptome profiles of 'Nongke No 1' (resistant mutant of Cavendish) and susceptible variety 'Brazilian'. Study could generate 2691 differentially expressed unigenes including defense and signal transduction in the resistant mutant. Most of the defense related genes and pathways identified in this study differed from those in model plants such as rice and Arabidopsis, suggesting that the mechanism of resistance in banana could be highly variable. Based on this profiling study, they demonstrated that in mutant variety, jasmonic acid (JA) and ethylene (ET) hormone signaling pathways are involved in *FOC* resistance but not salicylic acid (SA) pathway.

Similarly Wang et al. (2012) also indicated that JA biosynthesis is regulated in *FOC* TR-4 resistant cultivar. They performed a combination of DGE (Digital Gene Expression) and RNA seq transcriptome analysis in *M. acuminata* AAA group 'Brazilian' which was challenged with GFP tagged *FOC* TR-4. Some of the genes like peroxidases, 4-coumarate, cinnamate 4- hydroxylases and phenylalanine ammonia lyases involved in phenylpropanoid biosynthesis were reported to be upregulated. Costa et al. (2013) tried to identify the differentially expressed genes in contrasting cultivars BRS Platina (R) and Prata-Ana (S) under Fusarium challenged condition by profiling the global transcriptome. This result is worth mentioning as it enabled

Table 4 Bioinformatics tools for transcriptome analysis

Sl. No	Tool name	Description	Website
	Quality control and pre-processing data		
1	FastQC	Quality control tool for high-throughput sequence data	http://www.bioinformatics.babraham.ac.uk/projects/fastqc/
2	FASTX	Toolkit is a set of command line tools to manipulate reads in files FASTA or FASTQ format	http://hannonlab.cshl.edu/fastx_toolkit/
3	RNA-SeQC	Tool with application in experimental design, process optimization and quality control before computational analysis	http://www.broadinstitute.org/cancer/cga/rna-seqc
4	PRINSEQ	Generates statistics of sequence data for sequence length, GC content, quality scores, replicates, complexity, tag sequences, poly-A/T tails, odds ratios	http://edwards.sdsu.edu/cgi-bin/prinseq/prinseq.cgi
5	FastQ Screen	Screens FASTQ format sequences to a set of databases to confirm that the sequences contain what is expected (such as species content, adapters, vectors, etc)	http://www.bioinformatics.babraham.ac.uk/projects/fastq_screen/
Alignment tools			
6	TopHat	Fast splice junction mapper for RNA-Seq reads	http://tophat.cbcb.umd.edu/
7	Bowtie	Fast short aligner using an algorithm based on the Burrows-Wheeler transform and the FM-index	http://bowtie-bio.sourceforge.net/index.shtml
8	BWA	Short aligner using an algorithm based on the Burrows-Wheeler transform	http://bio-bwa.sourceforge.net/
9	MAQ	First aligns reads to reference sequences and after performs a consensus stage	http://maq.sourceforge.net/maq-man.shtml
Transcriptome assemblers (Genome-Guided assemblers)			
10	Cufflinks	Assembles transcripts, estimates their abundances, and tests for differential expression and regulation in RNA-Seq samples	http://cufflinks.cbcb.umd.edu/
11	RNAeXpress	To extract and annotate biologically important transcripts from next generation RNA sequencing data	http://www.rnaexpress.org/
Transcriptome assemblers (Genome-Independent assemblers)			
12	SOAPdenovo	De novo Assembler	http://soap.genomics.org.cn/soapdenovo.html
13	Velvet	Sequence assembler for very short reads	http://www.ebi.ac.uk/~zerbino/velvet/
14	Trinity	Novel method for the efficient and robust de novo reconstruction of transcriptomes from RNA-seq data	http://trinityrnaseq.sourceforge.net/

Table 4 (continued)

Sl. No	Quality control and pre-processing data		
	Tool name	Description	Website
Quantitative analysis and differential expression			
15	Cuffdiff	Measure global de novo transcript isoform expression. It performs assembly of transcripts, estimation of abundances and determines differential expression (Cuffdiff) and regulation in RNA-Seq samples	http://cufflinks.cbcb.umd.edu/
16	DESeq	Bioconductor package to perform differential gene expression analysis based on negative binomial distribution	http://bioconductor.org/packages/release/bioc/html/Deseq.html
17	EdgeR	R package for analysis of differential expression of data from DNA sequencing methods, like RNA-Seq, SAGE or ChIP-Seq data	http://www.bioconductor.org/packages/release/bioc/html/edgeR.html
Workbench (analysis pipeline/integrated solutions)			
18	Avadis NGS	The data mining and visualization platform at the core of all bioinformatics products developed by Strand Scientific Intelligence	http://www.avadis-ngs.com/
19	CLC Genomics Workbench	CLC bio deliver genomics sequencing analytics solution	http://www.clcbio.com/
20	DNASTAR	DNASTAR's Lasergene Genomics Suite includes software that allows users to quickly and easily align next-gen RNA reads against a reference genome, and then perform in-depth analysis of the aligned data	http://www.dnastar.com/t-sub-nextgen-genome-solutions-rnaseq.aspx
21	NextGENe	Perfect analytical partner for the analysis of desktop sequencing data produced by the ION PGM™, Roche Junior, IlluminaMiSeq as well as high throughput systems as the Ion Torrent Proton, Roche FLX, Applied BioSystemsSOLiD™ and Illumina® GA, and HiSeq systems	http://www.softgenetics.com/NextGENe.html
Open source solutions			
22	ArrayExpressHTS	BioConductor package that allows preprocessing, quality assessment and estimation of expression of RNA-Seq datasets	http://bioconductor.org/packages/2.11/bioc/html/ArrayExpressHTS.html
23	easyRNASeq	Count summarization and normalization for RNA-Seq data	http://www.bioconductor.org/packages/2.11/bioc/html/easyRNASEq.html
24	Galaxy:	Galaxy is a general purpose workbench platform for computational biology	http://galaxyproject.org/

Table 4 (continued)

Sl. No	Quality control and pre-processing data		
	Tool name	Description	Website
Functional, network & pathway analysis tools			
25	Blast2GO	Tool for functional annotation of (novel) sequences and the analysis of annotated data	http://www.blast2go.com/b2ghome
Visualization tools			
26	GBrowse	Combination of database and interactive web pages for manipulating and displaying annotations on genomes	http://gmod.org/wiki/GBrowse
27	Artemis	Genome Browser and Annotation Tool	http://www.sanger.ac.uk/resources/software/artemis/
28	Integrated Genome Browser	An application intended for visualization and exploration of genomes and corresponding annotations from multiple data sources	http://www.affymetrix.com/estore/partners_programs/programs/developer/tools/igbsource_terms.affx
RNA-Seq databases			
29	Queryable-rna-seq-database	System is designed to simplify the process of RNA-seq analysis by providing the ability to upload the result data from RNA-Seq analysis into a database, store it, and query it in many different ways	https://github.com/fatPerlHacker/queryable-rna-seq-database
30	RNA-Seq Atlas	A reference database for gene expression profiling in normal tissue by next generation sequencing	http://medicalgenomics.org/rna_seq_atlas
31	SRA	The Sequence Read Archive (SRA) stores raw sequencing data from the next generation sequencing platforms	http://www.ncbi.nlm.nih.gov/sra
SSR identification tools			
32	MISA	MIcroSAtellite identification tool	http://pgrc.ipk-gatersleben.de/misa/misa.html
SNP identification tools			
33	SNPsFinder	Web-based program developed to facilitate the SNPs discovery process	http://snpsfinder.lanl.gov/
Primer designing			
34	BatchPrimer3	Comprehensive web primer design program using Primer3 core program as a major primer design engine to design different types of PCR primers and sequencing primers in a high-through manner	http://probes.pw.usda.gov/batchprimer3/

the identification of unique genes, effectors of the immune response, in resistant cultivar and many of which were validated through RT-qPCR.

Genus *Mycosphaerella* is the most important foliar fungal pathogen on banana. Three species of this genus, such as *M. fijiensis, M. musicola and M. eumusae* cause black leaf streak disease (BLSD), Sigatoka leaf spot (yellow Sigatoka) and *eumusae* leaf spot respectively. Passos et al. (2012) performed Roche 454 pyrosequencing of expressed genes in contrasting genotypes Calcutta 4 (R) and Cavendish cv. Grand Naine (S) for BLSD resistance. Many unigenes which are potentially involved in signal transduction like receptor protein kinase containing LRR repeats, serine/threonine protein kinase, mitogen activated protein kinase kinase, WRKY super family transcription factors, which are typically associated with plant immunity mechanisms were identified. More accumulation transcripts of OXOs (Oxalate oxidase) only in incompatible interactions suggested their possible involvement in ROS (Reactive Oxygen Species) and associated HR (Hypersensitive Response) components. OXO's catalyse the conversion of ROS into hydrogen peroxide (H_2O_2), an important component of hypersensitive response HR in plants. Similarly upregulation and early expression of Glutathione S transferase, metallathionein families in incompatible interactions revealed their involvement in cell protection like ROSscavenging during HR re-sponses.

M. balbisiana, is one of the ancestors of the present day triploid banana cultivars, harbouring many pest and disease resistant and drought tolerant genes (Vanhove et al. 2012; Liu et al. 2010; Uma et al. 2013). Whole transcriptome profiling of *M. balbisiana* has been completed by National Research Centre for Banana (NRCB), India, using Ion-torrent platform. The quality reads obtained from these sequencing were assembled with AA reference genome (*M. acuminata* DH Pahang). The annotated results revealed that approximately 3301 unigenes were found to be homologous to known defense-related genes in other plants. Plant CYC database analysis also stated that these unigenes were significantly enriched in various known resistance related metabolic or signaling pathways. Overlaying of these transcripts with the recently sequenced gene models of B genome (*M. balbisana* type Pisang Klutuk Wulung) is in progress to identify the unique candidate genes, detection of SNPs and SSRs in resistant/defense related genes.

Under the Network project of Transgenic crops—Functional Genomics component, of ICAR, India, NRCB has carried out banana transcriptomic studies for various biotic (*Eumusae* leaf spot and root lesion nematode) and abiotic (water defecit) stresses. The experiment for each stress was carried out by constructing four cDNA libraries in both resistant and susceptible cultivars with challenged and unchallenged conditions. These libraries were subjected to next generation sequencing using Illumina platform and quality reads were assembled with AA reference genome. This project was initiated with an aim to have a near-complete snapshot of transcriptome, including the rare transcripts that have regulatory roles in various biotic and abiotic stresses in compatible and incompatible genotypes. Besides, the DGE analysis of these transcriptome data of the contrasting cultivars under challenged and unchallenged conditions was also carried out to identify the differentially expressed genes in resistant cultivars.

By comparing the expression profiling of *M. eumusae* challenged and unchallenged resistant and susceptible cultivars, 46 unique genes which are involved in resistance related pathways were detected only in resistant genotype (cv. Manoranjitham). Similarly on KEGG analysis, it was observed that among all defense related transcripts, 953 were involved in PTI (PTI is a branch of plant immunity which involves interactions between host pattern recognition receptor-like kinases (PRRs) and pathogen associated molecular patterns (PAMPs)), 816 transcripts in plant hormones biosynthesis and signaling, 88 transcripts in pathogenesis—related and 346 in cell-wall modification. 54 transcripts in effectors triggered immunity (ETI) etc.

Similarly, 16 genes which are expressed only in root lesion nematode challenged resistant cultivar (Karthobiumtham-ABB) were considered as unique genes and in-depth study on these putative genes is in progress to confirm their role in on resistance mechanism. The digital gene expression analysis also revealed that the genes involved in phenylpropanoid pathway, flavanoid and lignin biosynthesis pathway are over expressed in resistant genotypes under challenged condition compared to unchallenged resistant genotype and challenged susceptible genotype (cv. Nendran). Portal et al. (2011) studied the compatible interaction of *M. fijiensis* with banana through the construction of suppression subtractive hybridization. Upon infection, it was found that, jasmonic acid and ethylene signaling transduction dependent defense mechanisms, such as production of membrane damaging PR proteins and GDSL like-lipase were induced only in later stages. However, activation of the JA and ET pathways have not prevented BLSD development in the compatible interaction of cultivar 'Grand Naine', probably because the defense mechanisms induced were repressed again at very late stages of infection. At this stage, plant defense mechanisms appear to be actively suppressed, while the fungal UGPase encoding gene with a possible role in membrane protection as an osmoprotectant is induced.

3 Databases and Tools for *Musa* Transcriptome Analysis

Till date (2013), 55854 *Musa* EST sequences have been submitted in the Expressed Sequence Tags database (dbEST) (http://www.ncbi.nlm.nih.gov/dbEST). The dbEST contains sequence data and other information on single-pass cDNA sequence or Expressed Sequence Tags. The number of ESTs with respect to each *Musa* genomic group submitted in the dbEST is given in Table 5.

The *Musa* transcriptome database maintained by Global *Musa* Genomics Consortium has 91,041 banana cDNA sequences till date in public domain. To support post-genomic efforts, Droc et al. (2013) improved existing systems (eg. web front end, query builder), by integrating available *Musa* data into generic systems (e.g. markers and genetic maps, synteny blocks). They have made inter operable with the banana hub (http://banana-genome.cirad.fr/) and other existing systems containing *Musa* data (e.g. transcriptomics, rice reference genome, workflow manager) and generated new results from sequence analyses (e.g. SNP and polymorphism analysis). Thus study of the banana genome has been enhanced by a number of tools and resources that allow harnessing its sequence information.

Table 5 Number of *Musa* ESTs available in the NCBI public domain

Musa species, hybrids	No. of ESTs
M. acuminata (AA group)	29,525
Musa x paradisiaca	11,154
Musa (ABB Group)	11,070
M. acuminata (AAA Group)	7651
M. balbisiana (BB group)	5289
M. acuminata subsp. burmannicoides (AA group)	2289
Musa (AAB Group)	24
Musa hybrids and cultivars	6
Total	55,854

The Banana Genome Hub is supported by the South Green Bioinformatics Platform (http://southgreen.cirad.fr/), which provides access to original bioinformatics methods and tools to manage genetic and genomic resources of tropical plants. The comparative genomics is critical for the generation of reliable gene function annotations. The Banana Genome Hub relies on a robust comparative genomics database called GreenPhylDB (http://greenphyl.cirad.fr), which is now added with the *Musa* protein coding genes. This database comprises complete proteome sequences from the major plant phyla. The clustering of these proteomes was performed to define a consistent and extensive set of homeomorphic plant families. Based on this, lists of gene families such as plant or species specific families and several tools are provided to facilitate comparative genomics within plant genomes. Comparison involves two main steps; gene family clustering and phylogenomic analysis of the generated families. Once a group of sequences (cluster) is validated, phylogenetic analyses are performed to predict homolog relationships such as orthologs and ultraparalogs. GreenphylDB is linked to Uniprot DB to access *M. acuminata* proteome.

TropGeneDB (http://tropgenedb.cirad.fr) is a database that manages genomic, genetic and phenotypic information about tropical crops. Banana markers, maps, germplasm and genotypes data module are included in this DB. Genotyping 546 accessions of *Musa* germplasm with 22 SSR markers data are included in this DB.

Expressed Sequence Tag Treatment and Investigation Kit (Esttik) (http://esttik.cirad.fr) is a tool for editing and assembling of raw data of cDNA to form unigenes. The pipeline executes a series of programs to assess quality and nucleotides from chromatograms, then edits and assembles the input DNA sequence information into a non-redundant data set. This unigene is then searched for SSRs and SNPs. It is used as input for annotation against public databases including an extraction of Gene Ontology terms (Argout et al. 2007).

HarvEST: *Musa* displays EST sequences from 13 cDNA libraries from *M. acuminata*, *M. balbisiana* and related species. All sequences were derived from trace files received from various sources. Although HarvEST contains BLASTX hits from UniProt it needs updating (Close et al. 2007).

GNPAnnot (http://www.gnpannot.org/) is a project on green genomics which intends to develop a community system of structural and functional annotation supported by comparative genomics and dedicated to plant, insect and fungal genomes allowing both automatic predictions and manual curations of genomic objects. Community annotation system for *M. acuminata* is also provided in this website.

4 Application of Transcriptome Analysis in Banana Improvement

Gene-derived SSRs can be associated with functional genetic variation, as they are located in transcribed regions potentially influencing the gene function, transcription or translation (Tranbarger et al. 2012). Consequently, these EST SSRs offer potential tool for marker assisted breeding, with markers either originating from a gene for a desirable phenotypic trait, or co-localizing with a particular quantitative trait locus. With such markers isolated from coding regions, conservation is also potentially greater, increasing transferability to related species (Gupta et al. 2003). Amorim et al. (2012) and Wang et al. (2007) used EST-SSR markers for banana germplasm accessions and revealed that these markers are highly informative SSR's which can be used to accelerate the banana breeding programs, improve the studies on genetic diversity and leading to the development of saturated genetic linkage maps.

Passos et al. (2013) tried to develop defense related gene derived SSR loci, selected on the basis of BLAST similarities and KOG (eukaryotic orthologous groups of proteins). A total of 4068 genic-SSR loci were identified in Calcutta 4 and 4095 in Cavendish cv. Grande Naine. A subset of 95 potential defense-related, gene-derived SSRs loci was validated for specific amplification and polymorphism across *M. acuminata* accessions. This result evidenced that only 14.7% of markers showed polymorphism among 20 *M. acuminata* accessions with alleles per polymorphic locus ranging from 3 to 8. In spite of limited diversity observed while using gene derived markers, the polymorphic markers were applicable for association with trait loci and down stream marker assisted selection, as well as evolutionary analysis, parentage assessment and general genotyping applications in breeding programs.

Similarly at NRCB, SSR database containing 5362 SSRs with primer details, product size, annealing temperature etc. has been developed using the *Musa* ESTs available at NCBI. This assisted in the identification of two putative candidate markers which could differentiate the nematode resistant and susceptible banana cultivars.

Based on transferability studies Backiyarani et al. (2012) reported that SSR markers developed from *Musa* ESTs could be used across different genera of the family Zingiberaceae. These transferable *Musa* EST-SSR markers could also be used in cardamom improvement program as the availability of cardamom specific primers are very meager. Fatokun et al. (1992) reported that different plant species often share orthologous genes for very similar functions. This suggested that *Musa* EST-SSR markers have potential to develop conserved orthologous set (COS) markers, which facilitate comparative genomic studies between species and members of the order Zingiberales.

Table 6 Number of SSRs and SNPs developed from various cDNA libraries of *Musa* cultivars

Libraries	*M. eumusae*		Nematode (*P. coffeae*)		Water deficit stress	
	SSR	SNP	SSR	SNP	SSR	SNP
Unchallenged resistant	4825	83	11,650	432	9692	451
Challenged resistant	9153	355	3408	287	8128	351
Unchallenged susceptible	5008	79	8250	143	8478	92
Challenged susceptible	9239	271	7512	305	3775	334
Total	28,225	788	30,820	1167	30,073	1228

The transcriptome data on 12 cDNA libraries available at NRCB were subjected to *in silico* analysis to predict the SSR and SNP markers and resulted in the identification of a maximum of 11,650 and 355 SSRs and SNPs respectively (Table 6). *Musa* SNP database is being developed with all associated information.

Wan and Pentecost (2013) suggested that identification of multiple components involved in the chitin-mediated defense pathway (chitin degradation and signal generation; signal perception and amplification and downstream signaling components) is an important strategy to develop resistance to diverse biotic stresses. Hence, efforts were made to identify the chitinase isoforms from the available transcriptome data. DGE analysis of *M. eumusae* challenged resistant cultivar and *P. coffeae* challenged resistant cultivar indicated that the class II chitinase belonging to PR3 protein family was found to be commonly over expressed in both nematode and *M. eumusae* resistant genotypes. This information could be used for selection of genes to develop multiple resistances in banana through transgenic approach.

Shekhawat et al. (2011a) identified a novel SK3 type dehydrin (*Musa*DHN-1) from banana leaf cDNA library (Gene bank accession No. ES 436956) and found that this gene is significantly upregulated in abiotic stresses like cold, drought and high salinity in a robust and hardy cultivar 'Karibale' (Monthan, ABB). This gene is being used to improve drought and salt stress tolerance in cv. Rasthali through genetic transformation. Shekhawat et al. (2011b) also identified a banana EST sequence corresponding to a WRKY gene based on comparative analysis of stressed and non-stressed tissue derived EST data sets obtained from cv. Cachaco, a drought tolerant variety. Shekhawat et al. (2013) developed a transgenic Rasthali by incorporating this novel WRKY gene through transgenic approach and displayed enhanced tolerance towards oxidative and salt stress. Although *Musa*WRKY71 is induced in response to elicitor molecules of biotic stress response pathways like ethylene, salicylic acid, *Musa*WRKY71 over expressing plants were found to be equally susceptible to the infection of *Fusarium oxysporum* f. sp. *cubense* as the untransformed control plants. These results indicated that *Musa*WRKY71 is an important constituent in the transcriptional reprogramming involved in diverse stress responses in banana.

Conclusion

Transcriptome analysis of various tissues exposed to different stresses enables the identification of differentially expressed genes. Among various methods available, genome wide expression profiling would allow the measurement of the activity of majority of genes and to understand the molecular mechanism involved. In 1990s EST sequencing was the best approach for rapid identification of expressed genes, or at least fragments of those genes. Although at that time EST sequencing was considered a very high-throughput technique, both costs and technical limitations prevented the development of a complete transcript catalog. As a consequence, much of our knowledge on the protein-coding region of the transcriptome relied on different computational gene prediction methods. Other technologies like tag based methods (SAGE and MPSS) are the best methods over ESTs as these approaches have the capability to quantify the level of gene expression, but fail to distinguish the isoforms. In addition, most of them are based on traditional Sanger sequencing technology, which is expensive to apply on a large scale. A hybridization-based approach like microarray is one of the inexpensive ways to detect and quantify the transcripts on a large scale. Though this technique has some advantages like high throughput and ability to quantify distinct spliced isoforms, it has some disadvantages too. Because of differences in hybridization strength, cross-hybridization and other experimental variables, microarrays provide a noisy output signal. They can measure only the genes for which the sequence and the precise exon-intron boundaries are known, thus failing to identify novel genes or novel splicing events. Recently, RNA-seq technologies provide unprecedented opportunities for characterizing the set of RNA transcripts produced in tissues. Many platforms are available to sequence the RNAs, the choice of platforms varies with the objectives (de novo sequencing and resequencing) and budget availability. Unlike hybridization-based methods, it is not limited to the detection of known transcripts, but can also measure a broader range of expression levels. Among its other advantages, RNA-seq data has relatively low background noise, achieves base-pair resolution, allows precise identification of exon and intron boundaries, detecta single nucleotide polymorphisms (SNPs) and other variants within transcripts.

In banana, transcriptome analysis has been carried out to understand the molecular mechanisms involved during fruit ripening and various biotic and abiotic stress conditions using different approaches like SSH, cDNA-AFLP, ESTs and microarray. After the development of next generation sequencing, many have been carried out to understand the *Musa*-pathogen interactions and to identify the unique genes responsible for inducing resistance. The generation of enormous amount of sequence data from these transcriptome analyses has resulted in development of genetic markers like EST-SSRs, SNPs etc. These have led to the development of trait specific markers being used in conventional breeding for the selection of parents and progenies economising the time and resources needed for field evaluation. Besides these markers are valuable resources for comparative mapping in evolutionary studies.

It is evidenced from previous studies that there are differences in expression levels across homeologous chromosomes suggesting independent contribution of the two genomes to banana metabolism. This emphasises the need of both A and B genome as reference sequences for the transcriptome mapping of any banana and plantain cultivar irrespective of its genomic group (AAA, AA, AAB, ABB, BB). The fact that nearly 20 % of the transcripts do not hit with any known functional genes emphasise the need to study the function of uncharacterized genes either through RNAi and/or knockout gene approaches. This may lead to the identification of novel genes for important traits like parthenocarpy, sterility etc. The outcome of the research will not only be useful for improving banana but also paves way to develop seedlessness in other fruits crops.

References

Althoff DM, Gitzendanner MA, Segraves KA (2007) The utility of amplified fragment length polymorphisms in phylogenetics: acomparison of homology within and between genomes. Syst Biol 56:477–484

Amorim EP, Silva PH, Ferreira CF, Amorim VBO, Santos VJ, Vilarinhos AD, Santos CMR, Souza Júnior MT, Miller RNG (2012) New microsatellite markers for bananas (*Musa* spp). Genet Mol Res 11:1093–1098

Anon (1994) FAO Trade yearbook 1993. FAO, Rome

Argout X, Ruiz M, Fouet O, Lanaud C, Wincker P, Da Silva C, Courtois B (2007) ESTtik: a semi-automatic cDNA sequence analysis and annotation pipeline including SSR and SNP search tools. 15th International Cocoa Research Conference 1:501–506

Backiyarani S, Uma S, Varatharj P, Saraswathi MS (2012) Mining of ESTSSR markers of *Musa* and their transferability studies among the members of order the Zingiberales. Appl Biochem Biotechnol 169(1):228–238

Backiyarani S, Uma S, Arunkumar G, Saraswathi MS, Sundararaju P (2014) Differentially expressed genes in incompatible interactions of *Pratylenchus coffeae* with *Musa* using suppression subtractive hybridization. Physiol Mol Plant Pathol 86:11–18

Banana Market Review and Banana Statistics 2012–2013 Market (FAO statistics), Rome, 2014

Bortoluzzi S, Danieli GA (1999) Towards an in silico analysis of transcription patterns. Trends Genet. 15:118–119

Cano-Delgado A, Penfield S, Smith C, Catley M, Bevan M (2003) Reduced cellulose synthesis invokes lignification and defense responses in *Arabidopsis thaliana*. Plant J 34:351–362

Close TJ, Wanamaker S, Roose ML, et Lyon M (2007) HarvEST. Methods Mol Biol 406:161–177

Coemans B, Matsumura H, Terauchi R, Remy S, Swennen R, Sagi L (2005) SuperSAGE combined with PCR walking allows global gene expression profiling of banana (*Musa acuminata*), a non-model organism. Theor Appl Genet 111:1118–1126

Costa JL, Herai RH, Haddad F, Nebó JF, Nishimura DS, Ferreira CF, Amorim EP, e Figueira AVO (2013) Analysis of differentially expressed genes in banana cultivars infected by *Fusariumoxy sporum* f. sp. *cubense*. In: IV Simposio Brasileiro De Genetica Molecular De Plantas, p 33

Davey MW, Graham NS, Vanholme B, Swennen R, May ST, Keulemans J (2009) Heterologous oligonucleotide microarrays for transcriptomics in a non-model species; a proof-of-concept study of drought stress in *Musa*. BMC Genomics 10(436):1–19

Davey MW, Gudimella R, Harikrishna JA, Sin LW, Khalid N, Keulemans J (2013) A draft *Musa balbisiana* genome sequence for molecular genetics in polyploid, inter- and intra-specific *Musa* hybrids. BMC Genomics 14:683

Denham TP, Haberle SG, Lentfer C et al (2003) Origins of agriculture at Kuk Swamp in the Highlands of New Guinea. Science 301:189–193

D'Hont A, Denoeud F, Aury J, Baurens F, Carreel F, Garsmeur O et al (2012) The banana (*Musa acuminata*) genome and the evolution of monocotyledonous plants. Nature 488:213–217

Diatchenko L, Lau YF, Campbell AP, Chenchik A, Moqadam F, Huang B, Lukyanov S, Lukyanov K, Gurskaya N, Sverdlov ED, Siebert PD (1996) Suppression subtractive hybridization: a method for generating differentially regulated or tissue-specific cDNA probes and libraries. Proc Natl Acad Sci 93:6025–6030

Droc G, Lariviere D, Guignon V, Yahiaoui N, This D, Garsmeur O, Dereeper A, Hamelin C, Argout X, Dufayard JF, Lengelle J, Baurens FC, Cenci A, Pitollat B, D'Hont A, Ruiz M, Rouard M, Bocs S (2013) The banana genome hub. Database (Oxford) 2013:1–14

Fatokun CA, Menacio DI, Danesh D, Young ND (1992) Evidence for orthologous seed weight genes in cowpea and mungbean, based upon RFLP mapping. Genetics 132:841–846

Geschwind DH, Ou J et al (2001) A genetic analysis of neural progenitor differentiation. Neuron 29(2):325–339

Gupta PK, Rustgi S, Sharma S, Singh R, Kumar N, Balyan HS (2003) Transferable EST-SSR markers for the study of polymorphism and genetic diversity in bread wheat. Mol Genet Genomics 270:315–323

Jin ZQ, Xu BY, Liu JH, Su W, Zhang JB, Yang XL, Jia CH, Lia MY (2009) Identification of genes differentially expressed at the onset of the ethylene climacteric in banana. Postharvest Biol Technol 52:307–309

Kesari R, Kumar Trivedi P, Nath P (2007) Ethylene induced ripening in banana evokes expression of defense and stress related genes in fruit tissue. Postharvest Biol Technol 46:136–143

Lassois L, de Lapeyre deBL, Jijakli MH (2008) Biological control of crown rot of bananas with *Pichia anomala* strain K and *Candida oleophila* strain O. Biol Control 45:410–418

Lassois L, Frettinger P, de Lapeyre deBL, Lepoivre P, Jijakli H (2011) Identification of genes involved in the response of banana to crown rot disease. Mol Plant Microbe Interact 24(1):143–153

Li CC, Deng G, Yang J, Jin AVY, Kuang R, Zuo C, Zhi-cheng L (2012) Transcriptome profiling of resistant and susceptible Cavendish banana roots following inoculation with *Fusarium oxysporum* f. sp. *cubense* tropical race 4. BMC Genomics 13:374

Lisitsyn N, Lisitsyn N, Wigler M (1993). Cloning the differences between two complex genomes. Science 259:946–951

Liu HY, Dai JR, Feng DR, Liu B, Wang HB, Wang JF (2010) Characterization of a novel plantain Asr gene, MpAsr, that is regulated in response to infection of *Fusarium oxysporum* f. sp. *cubense* and abiotic stresses. J Integr Plant Biol 52:315–323

Manrique- Trujillo SM, Ramírez LAC, Ibarra Laclette E, Gómez-Lim MA (2007) Identification of genes differentially expressed during ripening of banana. J Plant Physiol 164:1037–1050

Maskos U, Southern EM (1992) Oligonucleotide hybridizations on glass supports: a novel linker for oligonucleotide synthesis and hybridization properties of oligonucleotides synthesised in situ. Nucl Acids Res. 20(7):1679–84

Mortazavi A, Williams BA, McCue K, Schaeffer L, Wold B (2008) Mapping and quantifying mammalian transcriptomes by RNA-SEq. Nat Methods 5:621–628

Munro CL (2008) Identification of defense-related genes in banana against *Fusarium* Wilt. MSc dissertation, University of Pretoria, Pretoria, p 137

Passos MA, de Oliveira CV, Emediato FL, de Camargo TC, Souza MT Jr, Matsumoto T, RennóAzevedo VC, Ferreira CF, Amorim EP, de Alencar Figueiredo LF, Martins NF, de Jesus B, Cavalcante M, Baurens FC, da Silva OB Jr, Pappas GJ Jr, Pignolet L, Abadie C, Ciampi AY, Piffanelli P, Miller RN (2012) Development of expressed sequence tag and expressed sequence tag-simple sequence repeat marker resources for *Musa acuminata*. AoB Plants 2012:1–63

Passos MA, de Cruz VO, Emediato FL, de Teixeira CC, Azevedo VC, Brasileiro AC, Amorim EP, Ferreira CF, Martins NF, Togawa RC, Júnior GJ, da Silva OB Jr, Miller RN (2013) Analysis of the leaf transcriptome of *Musa* acuminata during interaction with *Mycosphaerella musicola*: gene assembly, annotation and marker development. BMC Genomics 14(78):1–17

Ploetz RC (2006) *Fusarium* wilt of banana is caused by several pathogens referred to as *Fusarium* oxysporum f. sp. *cubense*. J Phytopathol 96:653–656

Portal O, Izquierdo Y, De Vleesschauwer D, Sánchez-Rodríguez A, Mendoza-Rodríguez M, Acosta-Suárez M, Ocaña B, Jiménez E, Höfte M (2011) Analysis of expressed sequence tags derived from a compatible *Mycosphaerella fijiensis* banana interaction. Plant Cell Rep 30:913–928

Ravishankar KV, Rekha A, Swarupa V, Savitha G (2011). Gene expression analysis in roots of *Musa* acuminata ssp. burmanniccoides 'Calcutta-4', a banana genotype tolerant to *Fusarium* wilt. Acta Horticult (ISHS) 897:363–370

Roux N, Baurens F, Doležel J, Hřibová E, Harrison PH, Town C, Sasaki T, Matsumoto T, Aert R, Remy S, Souza M, Lagoda P (2008) Genomics of banana and plantain (*Musa* spp.), major staple crops in the tropics. Plant Genet Genomics Crops Models 1:83–111

Rudd S (2003) Expressed sequence tags: alternative or complement to whole genome sequences. Trends Plant Sci 8:321–329

Santos CMR, Martins NF, Horberg HM, de Almeida ERP, Coelho MCF, Togawa RC, da Silva FR, Caetano AR, Miller RNG, Souza MT (2005) Analysis of expressed sequence tags from *Musa acuminata* ssp. *burmanniccoides*, var. Calcutta 4 (AA) leaves submitted to temperature stresses. Theor Appl Genet 110:1517–1522

Shekhawat UKS, Ganapathi TR (2013) *Musa* WRKY71 overexpression in banana plants leads to altered abiotic and biotic stress responses. PLoS ONE 8(10):e75506

Shekhawat UK, Srinivas L, Ganapathi TR (2011a) *Musa* DHN-1, a novel multiple stress-inducible SK(3)-type dehydrin gene, contributes affirmatively to drought- and salt-stress tolerance in banana. Planta 234(5):915–932

Shekhawat UKS, Ganapathi TR, Srinivas L (2011b) Cloning and characterization of a novel stress-responsive WRKY transcription factor gene (*Musa* WRKY71) from *Musa* spp. cv. Karibale-Monthan (ABB group) using transformed banana cells. Mol Biol Rep 38:4023–4035

Swarupaa V, Ravishankar KV, Rekha A (2013) Characterization of tolerance to *Fusarium* xysporumf. sp., cubense infection in banana using suppression subtractive hybridizationand gene expression analysis. Physiol Mol Plant Pathol 83:1–7

Tranbarger TJ, Kluabmongkol W, Sangsrakru D, Morcillo F, Tregear JW, Tragoonrung S, Billotte N (2012) SSR markers in transcripts of genes linked to posttranscriptional and transcriptional regulatory functions during vegetative and reproductive development of *Elaeis guineensis*. BMC Plant Biol 12:1–12

Uma S, Sudhakar B, Thangavelu R, Saraswathi MS, Backiyarani S, Saravanakumar AS, Vasanthi R (2011) EST based probing and isolation of functionally important full length transcripts against sigatoka (*Mycosphaerella musae*) leaf spot and confirmation by semi quantitative real time PCR in *Musa accuminata* cv. Manoranjitham (AAA). In: XVIII International Conference on Plant & Animal Genomes held at Town & Country convention Center, San Diego, CA, 13-18 Jan 2011

Uma S, Backiyarani S, Saraswathi MS, Mustaffa MM (2013) Banana genome to enrich fruit basket. Indian Horticult 58(2):3–6

Van denBN, Hein I, Birch PRJ, Berger DK, Wingfield MJ, Viljoen A (2007) Tolerance in banana to *Fusarium* wilt is associated with early up-regulation of cell wall-strengthening genes in the roots. Mol Plant Pathol 8(3):333–341

Vanhove AC, Vermaelen W, Panis B, Swennen R, Carpentier SC (2012) Screening the banana biodiversity for drought tolerance: can in vitro growth model and proteomics be used as a tool to discover tolerant varieties and understand homeostasis. Front Plant Sci 3:176

Velculescu VE, Zhang L, Vogelstein B, Kinzler KW (1995) Serial analysis of gene expression. Science 270:484–487

Wan J, Pentecost G (2013) Potential application of chitin signaling in engineering broad spectrum disease resistance to fungal and bacterial pathogens in plants. Adv Crop Sci Tech 1:e103

Wang N, Hu J, Ohsawa R Y O, Ohta M, Fujimura T (2007) Identification and characterization of microsatellite markers derived from expressed sequence tags (ESTs) of radish (Raphanus satvius L.). Mol. Ecol. Notes 7: 503–506

Wang L, Wang X, Wang X, Liang Y, Zhang X (2011) Observations on novel splice junctions from RNA sequencing data. Biochem Biophys Res Commun 409(2):299–303

Wang Z, Zhang J, Jia C, Liu J, Li Y, Yin X, Xu B, Jin Z (2012) *De Novo* characterization of the banana root transcriptome and analysis of gene expression under *Fusarium oxysporum* f. sp. *cubense* tropical race 4 infection. BMC Genomics 13:650

Xu BY, Su W, Liu JH, Wang JB, Jin ZQ (2007) Differentially expressed cDNAs at the early stage of banana ripening identified by suppression subtractive hybridization and cDNA microarray. Planta 226:529–539

Yang GP, Ross DT et al (1999) Combining SSH and cDNA microarrays for rapid identification of differentially expressed genes. Nucleic Acids Res 27(6):1517–1523

Commiphora wightii (Arnott.) Bhandari in the Indian Desert: Biology, Distribution and Threat Status

R. N. Kulloli and Suresh Kumar

1 Introduction

Commiphora wightii (Arnott) Bhandari, a highly valuable medicinal plant yields an oleo-gum resin important in Ayurvedic medicines (Chakravarty 1975). Mostly known as *Commiphora mukul* (Hook. ex Stocks) Engl. or *Balsamodendron mukul* Hook. ex Stocks (1849), the plant was renamed *Commiphora roxburghii* by Santapau in 1964 (Bhandari 1964) and corrected as *Commiphora wightii* (Arn.) Band. by Bhandari (1964). The plant is known as 'Indian bdellium' in English, as Mahisaksha, Guggulu, Amish, Palanksha and Pur in Sanskrit and as Guggul in most Indian languages (Anon 1950). Its anti-arthritic, hypocholesterolaemic and hypolipidaemic properties have been established (Satyvati 1990) and a commercial product 'guglip' has been marketed in India since 1988. Owing to the enormous demand for the drug, the plant is subjected to crude and destructive tapping procedures resulting in drastic shrinkage of its area of occurrence and abundance. It has been categorized as threatened but International Union for Conservation of Nature (IUCN) has put it under data deficient (DD) category. On one hand demand for its oleo-gum-resin is increasing while its increasingly sparser populations are unable to meet the requirements. There are varying claims about its regeneration through seeds or stem cuttings as well as biotechnological needs. Newer knowledge on its medicinal uses is also becoming available.

S. Kumar (✉) · R. N. Kulloli
Division of Integrated Land Use Management and Farming System,
Central Arid Zone Research Institute, Jodhpur, Rajasthan 342003, India
e-mail: skumar@cazri.res.in

D. Nandwani (ed.), *Sustainable Horticultural Systems*, Sustainable Development and Biodiversity 2, DOI 10.1007/978-3-319-06904-3_13

Fig. 1 Guggal. **a** Close up of seed with aril. **b** Different colored seeds. **c** Gum exudation using cut method by farmers. **d** Plant infested by termite and fungus. **e** Farmer with axe for tapping

2 Morphology

The plant with spinescent branching having glandular and pubescent young parts (Varier 1994). Leaves are trifoliate, alternate, rhomboidal or oval with serrate or dentate margins. Lateral leaves reduced. Flowers are very small, brown to pink unisexual, polygamous in fascicles. Female flowers have eight staminodes. Male flowers have eight stamens arranged in two series. Fruit drupe, red ovate, with two celled stone, rarely four valved (Fig. 1a) (Varier 1994). Stone covered with four strips of yellow aril. Seeds are brown in colour.

There are male, female and hermaphrodite plants of *C. wightii* in wild. The plant flowers twice during year winter (Sept–Oct) and summer (Mar–Apr). Seed germination is very less (1–3 %). Seeds collected during summer (Apr–May) have higher germination than those collected in winter (Nov) (Meena and Jadeja 2010). Black seeds (Fig. 1b) have more germination (45 %) over white seeds (5 %) (Anon 2005).

3 Distribution in Indian Sub-continent

In Indian-subcontinent *Commiphora* spp. occurs in India (Hooker 1872); Pakistan (Choudhary 1959) and Baluchistan (Hooker 1872). In India it occurs in the arid rocky tracts of Rajasthan, Gujarat, Madhya Pradesh, Karnataka and Kalat division of Andhra Pradesh (Soni 2010a) as well as Sindh and Baluchistan states of Pakistan (Gupta et al. 1996).

In India *Commiphora* species is represented by *C. myrrha, C. stocksiana, C. wightii, C. berryi, C. agallocha* but out of them *C. myrrha* is well documented for its use values (Hocking 1993). *Commiphora agallocha* is reported from Baragarh (Orissa) (Tewari et al. 2001) while *Commiphora berryi* in Tiruchirapalli, Coimbatore (Tamilnadu) (Selvamani et.al. 2009) and *Commiphora myrrha* in Anand (Gujarat) and Jabalpur (Madhya Pradesh) while *Commiphora stocksiana* in Anand (Gujarat). Siddiqui (2011) reported the presence of four species yielding guggal in India namely *C. wightii* (Arn.) Bhandari, *C. mukul* Engl. *C. agallocha* Engl. and *C. berryi* (Arn.) Engl. Of these *C. wightii* is now the correct and new name of *C. mukul*. On the other hand, Ramawat et al. (2008) reported that three species of *Commiphora* occuring in India namely *C. wightii* (Arn.) Bhandari, *C. stocksiana* Engl. and *C. berryi* (Arn.) Engl. Apparently spatial variability (heterogenesity) controls the distribution and abundance of *C.wightii* in Kuchcch Gujarat (Dixit and Subba Rao 2000).*Commiphora wightii* has comparatively a wider distribution in the country. In Gujarat *C.wightii* is mainly found in Kuchcch and some parts of Saurashtra regions (Sabnis and Rao 1983). It is found in the whole Kuchcch division apart from Kara hills of Khawada region (Atal et al.1975; Shah et al.1983).

In Rajasthan, it is reported from Sawai-Madhopur, Bundi, Kota, Jalore, Sirohi, Pali, Nagaur, Sikar, Churu, Bikaner, Ajmer, Alwar, Jaipur, Jaisalmer, Udaipur, Jodhpur, Jhunjhunun (Tewari et al. 2001). Kulloli et al. (2011a) gave geospatial distribution of guggal in Western Rajasthan. Reddy et al. (2012) has reported its associates, area of occurrence and occupancy based on satellite data of 2007 and field sampling up to 2009 and inferred its endangered status. It also occurs at Morena, Bhind, Shivpuri, Sheopur, Damoh districts in Madhya Pradesh (Thomas and Shrivastava 2010). Bhatnagar et al. (1973) also reported *C. berryi* from Gwalior (Madhya Pradesh). Hills and piedmonts are natural habitats of *C. wightii*. It is associated with tropical *Euphorbia* scrub which is major subtype of desert thorn forest of India (Champion and Seth 1968). Though so many workers have reported occurrence of *C. wightii* from a number of localities in India, its abundance, relationships with its associates and invasive as well as spatial distribution is yet inadequately known. Using GIS along with satellite data and modern cartographic tools would perhaps be the most desirable approach to know quantitative extent of guggal.

4 Worldwide Distribution

Genus *Commiphora* belongs to family Burseraceae which is moderately sized family of 17–18 the genera with about 540 species. There are over 200 species of *Commiphora* around Red Sea in East Africa, 20 species in Madagascar (Hanus et al. 2005) and 5 species in India (Pernet 1972), 10 new species of *Commiphora* were reported from Somalia (Thulin 2000). Vollesen 1989 have documented a full botanical account of more than 50 *Commiphora* species that occur in Eastern Africa, with the largest concentration of species in Northeast Africa, particularly in Ethiopia, Somalia and Kenya. The species is widely distributed in Arabia and Somalia (Dharmananda 2003). *C. guillaumini* is found in dry deciduous forest of western Madagascar. Occurrence of *Commiphora* spp., in South-West Africa has been confirmed (Vander Walt 1975). Lisowaski et al. (1972) recorded different species of *Commiphora* from Zaire. Petiole anatomy and branching pattern (Thomasson 1972) have formed the basis for identification of various species of *Commiphora* in Africa. *Commiphora* spp. is reported to occur in Arabia, Tropical and Southern Africa.

Its distribution further extends to Australia and Pacific Islands. Bisabol myrrha, *Commiphora erythraea var. glabrescence* is found in Somalia (Surburg and Panten 2006). Somalia is the major exporter of scented myrrh at present (Mabberley 2008).*Commiphora myrrha* is found in South Arabia and Somalia. Ruthless exploitation of this species has led to decline in its population continuously making this plant vulnerable which has been categorized as 'Data Deficient' in assemblage of IUCN in 2008 due to lack of research in establishment of its conservation status. Though, *C. wightii* is assigned to the DD (Data Deficient) category ver. 2.3 (1994) of the Red Data Book of IUCN, the Government of India has included it under RET (Rare, Endangered, Threatened) category (Samantaray et al. 2011).

5 Propagation

Propagation through seeds is more advantageous because seedlings have a deeper root system compared to the plants established through cuttings (Yadav et al.1999). Poor germination rates coupled with high mortality rate of seedlings limits propagation of *Commiphora* by this method. Fruit set and yield of fruits per plant are very low in natural conditions. Poor seed set, very poor seed germination (5 %) and harsh arid conditions are responsible for complete failure of plant establishment from seeds in nature. Natural regeneration through seeds has been rarely observed below the parent plants in the farm and in the forests (Reddy et al. 2012).

Commiphora wightii has been successfully propagated through stem cuttings of 25 cm length and 1 cm thickness. (Kumar et al. 2002). Indole-3-butyric acid (IBA) is still the most widely used auxin for rooting in stem cuttings and to increase the success percentage of cuttings (Al-Saqriand Alderson 1996). Effect of callusing on rooting is likely to have some effects on rooting which needs further confirmation (Kulloli et. al. 2011b). The cuttings are usually planted during late summer from healthy and disease free branches, when the plant is almost leafless. With the onset of monsoon, when foliation growth/flowering starts, the plant becomes physiologically active and the cuttings also show signs of sprouting in 25–50 days. After establishment in nursery beds the plants are transferred to polybags and then planted in the field during rainy season in the second–third week of June or early July (Yadav et al. 1999).

6 Tapping Methods and Tree Health

The traditional tapping methods and tools used for obtaining Guggal from *C. wightii* are unproductive and destructive (Plate-2). Tapping is discontinued during rainy season because yields are low and the resin is washed away from the stem (Krishnmurthy and Shiva 1997).Commercial tapping of trees for gum or resin affects its growth process (Karkkainen 1981) which also causes premature tree death and susceptible to insect, fungus infection (Jamal and Huntsinser 1993).

Excessive collection of resin and latex can be destructive as it may weaken the tree and causes carbohydrates to be spent on exudates that might otherwise have been allocated to growth and reproduction (Karkkainen 1981). Small trees would suffer more from tapping treatments than large trees, as the former generally have fewer stored carbohydrates available. The greater overall carbon budget of large sized trees yield more resins (Ruel et al. 1998). In juvenile trees, resin production originates from current photosynthesis, as the supply of stored carbohydrates is not sufficient (Lieutier et al. 1993). It highlights the reason for premature death of young trees due to excessive tapping. The resin is tapped during winter and each guggal tree yields about 700–900 g of resin (Siddiqui 2011). The ultra-structural details of secretion, seasons of production and methods of enhancing the yield have been studied by using different plant growth regulators. Of different plant growth regulators applied on the stem with lanolin paste, only kinetin increased the lumen size, while auxin and morphactin had adverse effect causing increase in the number of epithelial cells (Setia and Shah 1979), whereas ethephon enhanced 22 times oleo-gum production (Bhatt et al. 1989). Recently Siddiqui (2011) reported that guggal production can be enhanced by up to 22 times with an improved tapping technique using "Mitchie Golledge knife" coupled with ethephon (2-chloroethyl phosphoric acid), a plant growth regulator.

7 Economic Importance

C. wightii has many medicinal properties and exhibits interesting biological activities like anti-inflamatory, anti-bacterial, anti-microbial, anti-oxidant, anti-arthritic, anti-cancer, anti-ulcer, hypocholesterolmic and hypolipidaemic, hepatoprotective, muscle relaxing, mollucidal, anti-malerial, anti-mycobacterial, larvicidal

(Shen et al. 2007). Guggul contains resin, volatile oils and gum. Several pharmacologically active compounds have been identified in the plant including E and Z-guggulsterones (Singh et al. 1990). Guggul is often used in perfume due to its aroma and in toothpastes for its astringent action on the mucous membrane of mouth and throat.

8 Conservation Implication

Apart from the demographic and genetic constraints; the major threat to the species is overexploitation for its medicinally important resin (Haque et al. 2010). In view of so many clinical applications discussed above, designing conservation strategies for rare and endangered species requires a good knowledge about the levels and distribution of genetic diversity (Qiu et al. 2006). In the present study high levels of genetic structure was seen among populations of *C. wightii* and there seems to be little gene flow among them, the genetic variation shared among the population may largely represent common ancestry rather than recurrent gene flow. Conservation management in these species should aim not only the large but also the small populations as reduced levels of genetic diversity will affect the ability of the species to adapt to changes in habitat (Luijten et al. 2000). Vegetative reproduction, by stem cuttings, may postpone extinction but is basically an evolutionary dead end. Plans should be developed to encourage seedling recruitment in the small populations, specially those harboring low genetic variation (GH and HD). For this suitable habitats need to be identified for reintroduction in nature. Alternatively, this could be brought into cultivation on field margins in a participatory mode. Standardizing package of practices for its cultivation in a variety of agro-climates will therefore need a deeper study.

9 Gum Production

The traditional tapping methods and tools used for obtaining Guggal from *C. wightii* are unproductive and destructive. Commercial tapping of trees for gum affects its growth processes while over tapping for wood exudates leads to premature death of tree and susceptible to insect and fungus infections (Torquebiau 1984).

10 Tapping and Reproduction

Increase in tapping resin decreases the sexual reproduction in guggal. Large trees as well as non-tapped trees tended to produce not only slightly heavier seeds than small trees and non-tapped trees but also with higher germination percent (Toon et al. 2006). Tapping causes competition for carbohydrates between resin production, fruit and seed setting.

11 Guggal in Rajasthan

Modern tools of GIS may be used to provide, handle and analyze spatial data sets needed to understand and prioritize a range of conservation and management tasks (Best et al. 2007).These tasks of conservation activities can be carried out using satellite data at a range of scales (Kiage et al. 2007). The Landsat ETM + and ASTER GDEM Satellite Data were used in this study. *ERDAS Imagine* 9.1 and *Arc GIS* 9.2 software were used for digital image processing and spatial database handling respectively. Clipped Aster GDEM Data with District boundary and provided color range to get elevation information, Using GPS survey Collection points, requiring GIS layers, we created 3D elevation map. A digital elevation model of Rajasthan physiography was developed from ASTER (Advanced Spaceborne Thermal Emission and Reflection Radiometer) GDEM (Global Digital Elevation Model) Data of 30-m resolution. GIS layers of Area of occurrences for Guggal plant and elevation were created. This map along with topographic sheets of 1:50,000 were used for field traversing and ground truthing as per GPS location inferred from map.

Using locations inferred from DEM map, these sites (604) were visited. Those having absence of Guggal were recorded accordingly. Those sites having Guggal were sampled in 5–10 quadrats of 10 × 10 m placed beside each other. Presence, density, height and cover of Guggal and all associated species were recorded. Height and cover of canopy were used to infer plant health.

The maximum number of sites of presence of guggal occurred in Ajmer district (14 sites) followed by Barmer district (13 sites) (Table 1). It occurred at sites having altitude of 172–742 m, thus preferring hills and piedmonts at 94 % sites and pediment plains at 6 % sites.

12 Guggal Population Status

The Guggal density varied from 2–14 plants per ha in protected sites and 2–8 plants per ha in unprotected sites (Table 2). In protected sites its canopy cover and plant height were also more than that in disturbed sites. Though it was co-dominant at all the sites, its dominance improved in protection while it declined in open conditions. Similar trend was seen in its vigor. The densities of guggal and survey routes are shown in Figs. 2 and 3.

Density of *C. wightii* varied at different sites. It was higher at protected sites than at unprotected sites (Table 2). Dominance of guggal was high (RIV = 14.77) at protected sites compared to unprotected and degraded site (RIV = 9.97) (Table 5) indicating that guggal requires protection when reintroduced into natural habitats.

Table 1 Guggul collection sites

District	Total sites visited	Total sites of collection		Protected sites	Unprotected sites
		Hills	Plains		
Jaisalmer	34	5	7	2	10
Barmer	59	6	7	5	8
Jalore	10	7	–	6	1
Pali	29	5	2	5	2
Sirohi	18	2	1	2	1
Jodhpur	9	2	1	2	1
Nagaur	26	1	–	1	–
Ajmer	49	12	2	10	4
Rajsamand	58	5	1	4	2
Jhunjhunun	17	1	–	1	–
Sikar	47	1	–	1	–
Bikaner	59	–	–	–	–
Churu	64	–	–	–	–
Jhalawar	41	–	–	–	–
Kota	5	–	–	–	–
Bundi	8	–	–	–	–
Bhilwara	7	–	–	–	–
Chittorgarh	23	–	–	–	–
Ganganagar	17	–	–	–	–
Hanumangarh	24	–	–	–	–
Total	604	47	21	39	29

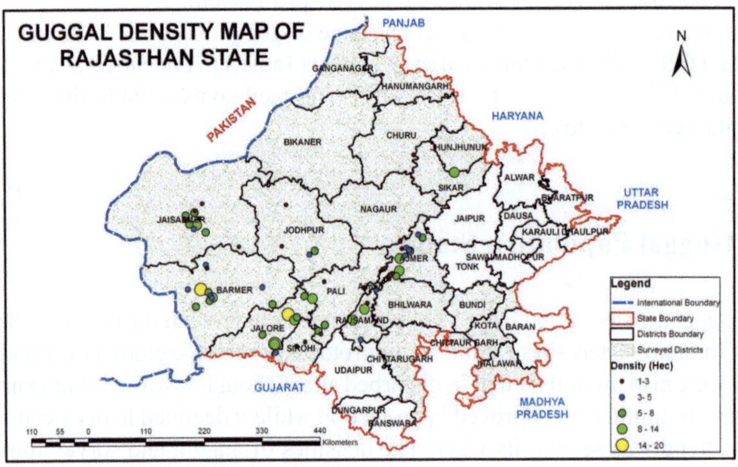

Fig. 2 Guggal density map of Rajasthan state

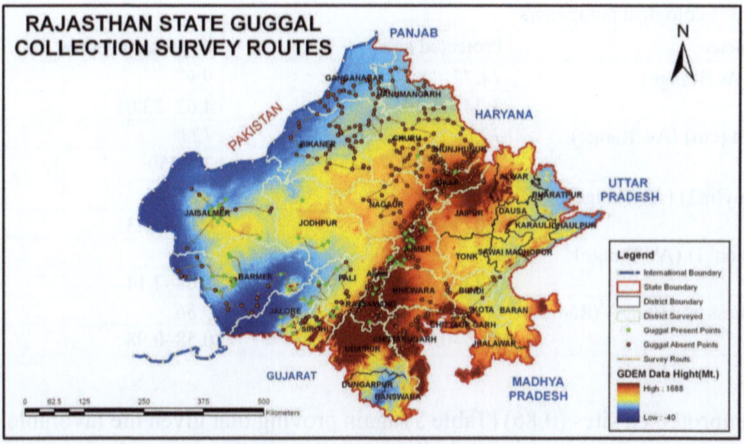

Fig. 3 Survey route map of guggal collection in Rajasthan state

Table 2 Density of *C. wightii* in different districts of Rajasthan

District	Protected	Unprotected
Jaisalmer	*8.5±2.121*	*4.3±2.351*
	7–10	2–8
Barmer	*9.6±4.560*	*3.57±1.618*
	4–16	1–6
Ajmer	*5.9±3.107*	*2.25±1.51*
	2–12	1–4
Pali	*5.2±5.215*	*3.5±3.53*
	2–14	1–6
Rajsamand	*6.5±4.123*	*3.5±2.121*
	2–16	2–5
Jodhpur	*5.5±0.707*	2[a]
	5–6	
Jalore	*12±5.656*	8[a]
	6–20	
Sirohi	*2.5±2.12*	2[a]
	1–4	
Jhunjhunun	9[a]	–
Nagaur	–	2[a]
Sikar	3[a]	–

[a] No of guggul plants found at only one site

The height and canopy cover of *C. wightii* was also more at protected sites than at unprotected sites (Table 3).

But, protected sites had low Shannon's diversity index (1.73) than unprotected sites (1.79). Higher diversity in moderately disturbed sites is reported by many workers (Brawn and Archer 1999). Evenness was more in protected sites (0.90)

Table 3 Ecological parameters

Parameter	Protected sites	Unprotected sites
RIV (Av./Range)	*14.77*	*9.97*
	3.24–35.10	4.62–23.05
Height (cm) (Av./Range)	*175*	*123*
	34.4–357.5	30–250
Canopy(m2) (AV/Range)	*3.54*	*1.87*
	0.077–10.46	0.014–5.35
Shannon_H (AV/Range)	*1.73*	*1.79*
	1.23–2.56	1.08–2.14
Evenness_e^H/S (AV/Range)	*0.90*	*0.86*
	0.65–0.99	0.58–0.98

than unprotected sites (0.86) (Table 3) again proving that given the favorable conditions such as protection, guggal facilitates optimum diversity and evenness.

13 Threats to *C. wightii* Population

1. **Destructive Tapping Method:** Tapping in villages is performed with the help of local knife, axe and use of guggul solution as an activator. Better yield and quick money made peoples to tap more trees using chemical based harsh activators and deeper incisions. The chemicals though helped in higher exudation of gum from the plant but the tapped plants begin to dry (Fig. 1c–e).
2. **Termite infestation:** *C. wightii* plant comparatively being soft woody and fleshy is susceptible to termite infestation. Termites eat away the plant from the base causing the plant to fall down and dry (Fig. 1d).
3. **Soil erosion:** In hilly region many guggal plants occurs standing on the peaks of hillocks and along its steep slope. Eroding soil base loosen the root grip of the plant and it is either displaced during monsoon. Many times guggal plants hanging from the top of the hillocks with the portion of its root still holding the plant is visible.
4. **Damage by shepherds:** In villages shepherds carry axe with them. Axe serves as weapon for their protection and also as an implement to cut fodder for the cattle. These shepherds also cause damage to *C. wightii* by cutting them.
5. **Fuel:** Losses of soil and vegetation have affected the sources of fuel in the village. Fuel is always in high demand in the villages. Collection of fuel wood in the revines is treacherous job for women and children. In the absence of sufficient fuel wood, chopping Guggal trees for fuel has become a routine practice.

Conclusion

Guggal is threatened plant of the Indian desert. Due to its overexploitation for oleo gum-resin in dhupbati, dry stem in fuel, fodder value etc. its populations are continuously becoming thinner and sparser. Increased emphasis is required on conservation efforts like *ex-situ* conservation, re-introduction in nature, habitat restoration, developed method for gum tapping, Local awareness. This plant can included in agro forestry systems due to its high medicinal importance.

Acknowledgments We are grateful to Director, CAZRI for providing necessary facilities for this work. The authors thank National Medicinal Plant Board, Govt. of India, New Delhi for financial support.

References

AL-Saqri F, Alderson PG (1996) Effects of IBA, cutting type and rooting media on rooting of *Rosa centifolia*. J Hort Sci 71:729–737

Anon (1950) The wealth of India: a dictionary of Indian raw material and industrial products, vol II. Council of Scientific and Industrial Research, Delhi, p 427

Atal CK, Gupta OP, Afaq SH (1975) *Commiphora mukul*: source of Guggul in Indian systems of medicine. Econ Bot 29:208–218

Best BD, Halpin PN, Fujioka E, Read AJ, Qian SS, Hazen LJ, Schick RS (2007) Geospatial web services within a scientific workflow: predicting marine mammal habitats in a dynamic environment. Ecol Inform 2:210–23

Bhandari MM (1964) Notes on Indian desert plants—Four new names and combinations. Bull Bot Surv India 6:327

Bhatnagar LS, Singh VK, Pandey G (1973) On the occurrence of Guggulu yielding *Commiphora* with reference to *Commiphora berryi* Engl. in Madhya Pradesh. J Res Indian Med 8:69–75

Bhatt JR, Nair MN, Mohan Ram HY (1989) Enhancement of oleo-gum resin production in *Commiphora wightii* by improving tapping technique. Curr Sci 58:346–349

Brawn JR, Archer S (1999) Shrub invasion of grassland: recruitment is continuous and not regulated by herbaceous biomass or density. Ecology 80:2385–2396

Chakravarty HL (1975) Herbal heritage of India. Bull Bot Soc Bengal 29:97–103

Champion HG, Seth SK (1968) A revised survey of the forest types of India. Government of India, Delhi, p 404

Choudhary II (1959) Scope of plant introduction in agriculture and forestry of West Pakistan. Pak J Forest 9:208–211

Dharmananda S (2003) Myrrh and Frankincense. http.itmonline.org/arts/myrrh.htm

Dixit AM, Subba Rao (2000) Observation on distribution and habitat characteristics of Guggal (*Commiphora wightii*) in the arid region of Kachchh, Gujarat (India). Trop Ecol 41(1):81–88

Gupta P, Shivanna KR, Mohan Ram HY (1996) Apomixis and polyembryony in the Guggal plant. Ann Bot 78:67–72

Hanus O, Rezankab T, Dembitsky VM, Moussaieff A (2005) Myrrh-*Commiphora* chemistry. Biomed Papers 149:3–28

Hocking D (1993) Trees for drylands. Oxford and IBH, New Delhi

Hooker JD (1872) The flora of British India, Part-I. Reeve, London, p 740

Jamal A, Huntsinger L (1993) Deterioration of a sustainable agro-silvo-pastoral system in the Sudan: the gum gardens of Kordofan. Agroforest Syst 23:23–28

Karkkainen M (1981) Increasing resin content in pine and spruce stem wood for higher byproduct yield. Commun Inst Forest Fenniae 96:181

Kiage LM, Liu KB, Walker ND, Lam N, Huh OK (2007) Recent land-cover/use change associated with land degradation in the Lake Baringo catchment, Kenya, East Africa: evidence from Landsat TM and ETM. Int J Remote Sens 28:4285–309

Krishnmurthy T, Shiva MP (1997) Salai-guggul from *Boswellia serrata*. Roxb. its exploitation and utilization. Indian Forest 103:466–474

Kulloli RN, Purohit Chandan Singh, Kumar S, Jindal SK, Deepika Acharya (2011a) Distribution of endangered plant *Commiphora wightii* (Arnott) Bhandari in the Indian desert. In: Abstracts and Souvenir of National workshop on "Conservation and Sustainable Utilization of Commiphora wightii (Guggal)" organized by Forest Department Govt. of Gujarat Gandhinagar Gujarat, 15–16 Sept 2011

Kulloli RN, Purohit Chandan Singh, Jindal SK, Kumar S (2011b). Effect of callusing on rooting and root traits in *Commiphora wightii* (Arnott) Bhandari in the Indian desert. In: Abstracts and Souvenir of National workshop on "Conservation and Sustainable Utilization of Commiphora wightii (Guggal)" organized by Forest Department Govt. of Gujarat Gandhinagar Gujarat, 15–16 Sept 2011

Kumar D, Jha BK, Chandra R (2002) Response of auxin and planting time on the regeneration of stem cuttings of *Commiphora wightii* (Indian Bdellium). J Trop Med Plants 3(ii):253–258

Lieutier F, Garcia J, Romary P, Yart A, Jacte Hl, Sauvard D (1993) Inter-tree variability in the induced defense reaction of Scots pine to single inoculations by *Ophiostoma brunneociliatum*, a bark-beetle-associated fungus. Forest Ecol Manage 59:257–270

Lisowaski S, Malaisse F, Symoena JJ (1972) Le genere *Commiphora* Jacq. (Burseraceae) in Zaire (ex-Congo-Kinshasa) [The genus *Commiphora* Jacq.(Burseraceae) in Zaire (former Congo-Kinshasa).]. Bol Soc Broteriana 46:81–92

Luijten SH, Dierick A, Gerard J, Oostermeijer B, Raijmann LEL, Den Nijs HCM (2000) Population size, genetic variation, and reproductive success in a rapidly declining, self-incompatible perennial (*Arnica montana*) in The Netherlands. Conserv Biol 14:1776–1787. doi:10.1046/j.1523-1739.2000.99345.x

Mabberley DJ (2008) Mabberley's plant-book: a portable dictionary of plants, their classification and uses. Cambridge University Press, Cambridge

Meena RL, Jadeja NL (2010) Department of Forest, Gujarat. Personal communication

Pernet R (1972) Phytochimie des *Burseraceaes*. Lioydia 35:280–287

Qiu YX, Li JH, Liu HL, Chen YY, Fu CX (2006) Population structure and genetic diversity of *Dysosma versipellis* (Berberidaceae), a rare endemic from China. Biochem Syst Ecol 34:745–752

Ramawat KG, Mathur M, Dass S, Suthar S (2008) Guggulusterone: a potent natural hypolipidemic agent from *Commiphora wightii*- problems, preservence, and prospects. In: Ramawat KG, Merillon JM (eds) Bioactive molecules and medicinal plants. Springer, Heidelberg, pp 101–121

Reddy CS, Meena SL, Krishna PH, Charan PD, Sharma KC (2012) Conservation threat assessment of *Commiphora wightii* (Arn.) Bhandari—an economically important species. Taiwania 57(3):288–293

Ruel JJ, Ayres MP, Lorio PL (1998) Loblolly pine responds to mechanical wounding with increased resin flow. Can J Forest Res 28:596–602

Sabnis SD, Rao KS (1983) Rare and endangered endemic of South Eastern Kutch. In: Jain SK, Rao KS (eds) Assessment of threatened plants of India. Proceeding of seminar held at Dehradun, 14–17 Sept, pp 71–77.

Samantaray S, Bishoyi A, Geetha KA and Satyabrata M (2011) Assessment of genetic diversity using RAPD and ISSR markers in Guggal (*Commiphora wightii*). epubs.icar.org.in/ejournal/index.php/JMAP/article/download/4390/1725

Satyvati GV (1990) Guggalipid: a promising hypolipidemic agent from gum Guggal (*Commiphora wightii*) Econ Med Plant Res 5 Plant Tradit Med, 47–82

Selvamani P, Sen DJ, Gupta JK (2009) Pharmacognostical standardization of *Commiphora berryi* (Arn) Engl. and phytochemistry studies on its crude extracts. Afr J Pharmacol III(2):37–46

Setia RC, Parthasarthy MV, Shah JJ (1977) Development histochemistry and ultrastructure of gum-resin ducts in *Commiphora mukul* Engl. Ann Bot 41:999–1004

Shah RR, Patel DB, Patel DH, Dalal KC (1983) Hormonal effect of germination of Guggal cuttings. Indian Drugs 20(II):435–437

Shen T, Wan W, Yuan H, Kong F, Guo H, Fan P, Lou H (2007) Secondary metabolites from *Commiphora opobalsamum* and their anti-proliferative effect on human prostrate cancer cells. Phytochemistry 68:1331–1337

Siddiqui ZM (2011) Guggal: an excellent herbal panacea. Asian J Pharma Health Sci 1:35–39

Singh V, Kaul S, Chander R, Kapoor NK (1990) Stimulation of low density lipoprotein receptor activity in liver membrane of Guggulusterone treated rats. Pharmacol Res 22:37–44

Soni V (2010a) Efficiency of in vitro tissue culture versus stem cuttings for propagation of *Commiphora wightii* in Rajasthan, India. Conserv Evid 7:91–93

Tewari DN, Kumar K, Tripathi A (2001) Guggal-Utthan. Center for sustainable Development and Poverty alleviation, Allahabad, p 117

Thomas M, Shrivastava A (2010) Madhya Pradesh. Personal communication

Thomasson M (1972) Remarquessur la vegetation des environs de Tulear: III. Modes de remification des vegetauxligneux. (Studies on the vegetation around Tulear: III. Branching in woody angiosperms). Bull Soc Bot France 119:207–213

Thulin M (2000) Ten new species of *Commiphora* (Burseraceae) from Somalia. Nord J Bot 20:395–411

Toon R, Ogbazghi W, Wessel M, Bongers F (2006) The effect of tapping for frankincense on sexual reproduction in *Boswelia papyrifera*. J Appl Ecol 43:1188–1195

Torquebiau E (1984) Man-made dipterocarp forest in Sumatra. Agroforest Syst 2:103–127

Van der Walt JJA (1975) The south west African species of *Commiphora*. Mitt Bot Staatssammlung Muenchea 12:195–266

Varier VPS (1994) Indian medicinal plants. http://www.vedamsbooks.com

Yadav BBL, Billore KV, Joseph JG, Chaturvedy DD (1999) Cultivation of GUGGULU. Central Council for Research in Ayurveda and Siddha, New Delhi, pp 1–87

Risk Assessments and Management Practices for the Major Invasive Plants Recorded in the Horticultural Ecosystem of the Western Pacific

Gadi V. P. Reddy

1 Introduction

Many introductions, for ornamental and other purposes, of herbs, shrubs, vines, and trees have had serious consequences for the environment in Micronesia. Non-native species introduced, accidentally or intentionally by humans, that then spontaneously expand their ranges are termed invasive species; they often menace the environment, agriculture, forestry, or other resources. Several recent studies have addressed the problem of invasive species, and these have been useful for structuring rigorous quantitative methods for predicting the invasiveness of alien species (Kolar and Lodge 2001). Our recent survey on Guam (Reddy 2011) identified the 20 most widespread invasive plants (see Table 1), but no work has been done on risk assessment for these major pest plants.

The weed risk assessment (WRA) system is an experiential device for identifying pest plants (Groves et al. 2001). It was adapted and developed for Australia and New Zealand by Pheloung et al. (1999). Daehler and Carino (2000) then developed a modified version (H-WRA) for identifying pest plants in Hawaii and on other Pacific islands. Daehler et al. (2004) developed an additional screening structure (HP-WRA).

Here, we describe the risk assessment for the 20 most widespread invasive plants identified on Guam and the best management practices available for them. The recommended mechanical control methods can be applied without regulation, but the chemical applications and biological-control agents must comply with legal requirements.

G. V. P. Reddy (✉)
Western Triangle Agricultural Research Center, Montana State University,
9546 Old Shelby Rd., Conrad, MT 59425, USA
e-mail: gadi.reddy@montana.edu

D. Nandwani (ed.), *Sustainable Horticultural Systems*, Sustainable Development and Biodiversity 2, DOI 10.1007/978-3-319-06904-3_14

Table 1 The 20 most prevalent invasive plants on Guam that our risk assessment was based on. We classified plants (on the basis of climate/distribution, domestication, weed status elsewhere, undesirable traits, plant type, reproduction, dispersal, and persistence attributes) into three categories: high risk, possible risk (requiring further evaluation), and low risk. Designations (based on the HP-WRA screening process, which is in turn based on published sources describing species biology and behavior in Hawaii and/or other parts of the world) are as follows: *H* likely to be invasive on Hawaii/Guam and on other Pacific islands, *G* documented to cause significant ecological or economic harm on Guam, as determined from published information on the species, *L* not currently recognized as invasive on Guam and unlikely to have major ecological or economic impacts on other Pacific islands

Plant	Family	Common name	Classification	Designation
Bidens alba	Asteraceae	Hairy beggar's-tick, Spanish needles	High risk	H
Panicum maximum	Poaceae	Guinea grass	High risk	H
Stachytarpheta jamaicensis	Verbenaceae	Blue porter weed, Jamaica vervain	High risk	H
Antigonon leptopus	Polygonaceae	Chain of love, Mexican creeper, coral vine	High risk	H
Paspalum paniculatum	Poaceae	Russell River grass, perennial grass	Possible risk	H
Miscanthus floridulus	Poaceae	Pacific island silver grass, Japanese silver grass	Possible risk	H
Euphorbia heterophylla	Poaceae	Wild poinsettia, Mexican fireplant, milkweed	Possible risk	H
Chromolaena odorata	Asteraceae	Siam weed	High risk	G
Mikania micrantha	Asteraceae	American rope, Chinese creeper, or mile-a-minute weed	High risk	H
Synedrella nodiflora	Asteraceae	Cinderella weed, node-weed, pig grass	Possible risk	H
Chamaesyce hirta	Euphorbiaceae	Garden spurge, red milk-weed, asthma plant	Possible risk	H
Mimosa pudica	Mimosaceae	Sensitive plant, sleeping grass	High risk	H
Leucaena leucocephala	Fabaceae	Leadtree	Possible risk	H
Pennisetum polystachion	Poaceae	Mission grass, foxtail grass	Possible risk	
Euphorbia cyathophora	Euphorbiaceae	Dwarf poinsettia, wild poinsettia	Possible risk	H
Sida rhombifolia	Malvaceae	Paddy's lucerne, common sida	Possible risk	H
Momordica charantia	Cucurbitaceae	Balsam pear, bitter gourd	Low risk	H
	Poaceae	Golden beardgrass, inifuk	Low risk	L
	Euphorbiaceae	Graceful spurge	Low risk	L
Chloris barbata	Poaceae	Swollen fingergrass	Low risk	L

2 Materials and Methods

Our risk assessment was based on climate/distribution, domestication, weed status elsewhere, undesirable traits, plant type, reproduction, dispersal, and persistence attributes (Daehler et al. 2004). We classified plants into three categories: low risk (score of <1), possible risk (requiring further evaluation, score of 1–6), and high risk (score of >6). Plants could be classified as high risk for many different reasons.

The literature was reviewed, and the best practices recommended in the published literature are described here.

3 Results and Discussion

Table 1 lists our risk classifications of the major invasive plant species recorded on Guam. The management practices recommended in the literature are summarized below for each species.

3.1 *Bidens alba* (L.) DC

Mechanical control by hand removal is effective, but the removed plants must be properly disposed of or burned because the stems of cut plants readily root when they touch the soil (Muniappan et al. 2002). Germination is prevented by mulches if they are thick enough and left undisturbed. The species is also susceptible to chipping and cultivation. It tolerates dry air and soil conditions (Swarbrick 1997), but because it prefers full sunlight, it can be controlled by competition from leafy crops, such as cassava and sweet potato. It can also be grazed by livestock (Motooka et al. 2003).

Bidens alba can easily be controlled by most broadleaf herbicides (2,4-D, dicamba, triclopyr, etc.), but it is a prolific seed producer, and the seeds germinate readily. Herbicide that does not possess soil activity will quickly control the plants present, but reestablishment from the seed can occur rapidly. Aminopyralid (a hormone-based herbicide manufactured by Dow AgroSciences) has been shown to control this species effectively. Applied at 490.4 g/ha, Milestone 0.9 kg/3.8 l aminopyralid) provides residual control for approximately 4 months after application (Hall et al. 1991).

Biological control has not been attempted.

3.2 *Panicum maximum* Jacquin

Manual control by digging out with a mattock or similar tool (Anonymous 2007) can be effective, particularly if isolated clumps are dug out before the plants spread.

The recommended chemical control is application of glyphosate 41 Plus/Roundup (EPA Reg Number 42750-61-72693) or fluazifop-p-butyl/Fusilade (EPA Reg Number 100-1070) to actively growing plants (Anonymous 2007).

Biological control has not been attempted.

3.3 Stachytarpheta jamaicensis (L.) Vahl

Small infestations can be removed manually before seeds form (PIER 2005).

For chemical control, apply 2,4-D amine at 2.2 l/ha or 22 ml/10 L to cover 100 m^2. This mixture can be used on seedlings and young plants and is most effective when sprayed in the summer. Control of the similar species *S. cayennensis* involves the use of translocated herbicides, including 2,4-D and glyphosate at standard rates, which may be effective for controlling *S. jamaicensis* as well (Land Protection Australia 2006a).

Biological control has not been attempted.

3.4 Antigonon leptopus Hooker & Arnott

The first step in preventative control of this vine is to limit planting and to remove existing plants. If possible, removal should occur before seeds are produced. Care must be exercised to prevent seed spread and dispersal during the removal process.

Mechanical control by cutting of vines removes only above-ground vegetation. New vines will sprout from seeds and from underground tubers. Only removal of the tubers will prevent further growth (MacDonald et al. 2008).

No herbicides are registered for use on *A. leptopus* (McConnell and Muniappan 1991), and research and data on its chemical control are limited. Spot treatment with glyphosate or triclopyr (EPA Reg Number 42750-127) is best (MacDonald et al. 2008).

Biological control has not been attempted.

3.5 Paspalum paniculatum L.

No management practices have been developed for *Paspalum paniculatum.*

3.6 Miscanthus floridulus (Labill.) Warb. ex K. Schum. & Laut.

On Fiji, old culms of *M. floridulus* are burned or cut.

Regrowth can be killed with glyphosate, followed by spot weeding some months later (Evans 1992).

Biological control has not been attempted.

3.7 *Euphorbia heterophylla* L.

Because seeds germinate at any time during the growing season, control is difficult, and the ability of seedlings to emerge from depth only adds to the aggressiveness of *E. heterophylla* (Parsons and Cuthbertson 2001). If not too numerous, the plants are best destroyed by hand pulling as soon as they are observed or by cutting so closely and frequently that no seed is allowed to mature.

The best approach to chemical control of most weeds among agricultural crops is a combination of soil-applied and postemergence herbicides. Pursuit (imazethapyr) and Strongarm (diclosulam) are the only two soil-applied herbicides labeled at this time that have good to excellent activity on *E. heterophylla*. Newer soil-applied herbicides presently under development, including Spartan (sulfentrazone) and Valor (flumioxazin), also provide good to excellent control and can be used as part of a total management system once they are labeled. Applications of paraquat + Storm (bentazon + acifluorfen) can be very effective but will not control plants that emerge later. Acceptable postemergence control *E. heterophylla* can be obtained with a timely application of either Ultra-Blazer (acifluorfen) or Cadre (imazapic) before the plants exceed 2 in. in height. Cadre applied after emergence also provides some residual control (Prostko 2009).

Current attempts at biological control in Papua New Guinea and Brazil may offer more effective means of control. In Brazil particularly, interest is centered on pathogenic fungi occurring naturally on this weed. Of these, *Alternaria* sp. and *Helminthosporium* sp. used as mycoherbicides have controlled *E. heterophylla* effectively in experiments (Parsons and Cuthbertson 2001). *Bipolaris euphorbiae* (anamorphic Ascomycota: Pleosporaceae) has been studied since the early 1980s as a potential biocontrol agent for *E. heterophylla*, which invades soybean fields in Brazil (Nechet et al. 2006).

3.8 *Chromolaena odorata* (L.) R.M. King & H. Robinson

Mechanical control of *C. odorata* is widely used in plantation crops, but it is labor intensive and requires repeated operations. In small-scale operations, hand tools such as picks, hand hoes, shovels, and mattocks are used, whereas in large-scale clearing situations, motorized brush cutters and tractor-drawn mowers are used (Erasmus 1988). Fire has been proven effective in controlling early establishment of the South African biotype of the species, but the biotype in other parts of the world is resistant to fire.

Herbicides such as triclopyr, glyphosate, 2,4-D amine, 2,4-D/ioxynil, and picloram/2,4-D have provided effective chemical control of *C. odorata* (Erasmus 1988). Both mechanical and chemical control methods are expensive and require frequent applications. Because classical biological control offers a long lasting, effective, and economical solution to this problem, it has been explored and implemented in several countries.

A natural enemy, *Pareuchaetus pseudoinsulata* Rego Barros (Lepidoptera: Arctiidae) was established on Guam in 1985, the Northern Mariana Islands in 1986–1987 (Seibert 1989), Yap in 1988 (Muniappan et al. 1988), Pohnpei in 1988–1990 (Esguerra et al. 1991), and Kosrae in 1992 (Esguerra 1998). It has effectively suppressed *C. odorata* that occurs in thickets but is not effective on isolated patches because of the insect-induced defense exhibited by the host plant. The eriophyid mite *Acalitus adoratus* Keifer (Acari: Eriophyidae) was fortuitously introduced into Micronesia from the neotropics via Southeast Asia, but does not effectively control this weed (Muniappan et al. 2009). The gall fly *Cecidochares connexa* (Macq.) (Diptera: Tephritidae) was subsequently released and established on Guam, Saipan, Rota, and Pohnpei (Muniappan et al. 2005; Cruz et al. 2006) and on Chuuk, Kosrae, and Yap (Muniappan et al. 2007). It causes galls in the terminal shoots, resulting in stunting of the plants and reduction of flower production. The fly has been very effective and has provided successful biological control of *C. odorata* in many areas (Zachariades et al. 2009).

3.9 *Mikania micrantha* Kunth ex H.B.K.

Hand weeding, uprooting, and digging are the main mechanical-control methods in practice, but *M. micrantha* is able to recover rapidly (Waterhouse and Norris 1987). Hand weeding before flowering and seed set gives temporary control, but quick regrowth from cut stumps frustrates this method. Uprooting during the initial stages of growth (before flowering and fruiting) is the most effective mechanical control method. The slash-and-burn technique is also practiced widely, but the weed stock may survive burning and produce young shoots in a couple of months. Mechanical control is very labor intensive and uneconomical (Sankaran et al. 2001). One advantage of this method is that it reduces the vegetative propagation of the species, but the cost of mechanical control is estimated to be 125–175 % higher than that of herbicidal control in Indonesia.

Both pre- and postemergence herbicides are generally used for control of *M. micrantha*. Control of this weed is difficult, because of the high output of viable seeds and because new plants can grow from even the tiniest stem fragments (Swarbrick 1997). Other than complete destruction of all the stems, herbicides provide the only suitable method of control. Oxyflourfen (0.06 kg/ha) + paraquat (0.24 kg/ha) is reported to be effective if applied before flowering or seed set. Glyphosate is widely used in many countries against *M. micrantha,* especially in forest plantations. The dosage used varies widely (0.5–4.5 kg/ha), depending on the intensity of infestation and number of applications required for effective control. This herbicide can also inhibit germination of seeds of the weed. Application of diuron at the rate of 1–2 kg/ha is reported to be as effective as glyphosate (Sankaran et al. 2001).

Actinote thalia pyrrha (Fabricius) and *Actinote anteas* (Doubleday and Hewitson) (Lepidoptera: Nymphalidae) have been used for the biological control of *M. micrantha* in various countries. Biological control using another natural insect enemy, *Liothrips mikaniae* (Priesner) (Thysanoptera: Phlaeothripidae) from Trinidad, was attempted in the Solomon Islands and Malaysia, but successful establishment was not achieved (Swarbrick 1997). *Liothrips mikaniae* does appear to be specific, however, and to have considerable potential as a biological-control agent. *Teleonemia* sp. (Hemiptera: Tingidae), several beetles, and the eriophyid mite *Acalitus* sp. (Acarina: Eriophyiidae) also warrant serious consideration. A number of other natural enemies of little known specificity also attack *M. micrantha* (Waterhouse and Norris 1987). Recent studies carried out by CABI Bioscience (UK) in collaboration with the Kerala Forest Research Institute (India) and institutions under the Indian Council of Agricultural Research have shown that a highly damaging microcyclic rust, *Puccinia spegazzinii* de Toni (Basidiomycetes: Uredinales), occurs naturally and causes damage to *M. micrantha* in the neotropics, has great potential as a biocontrol agent. The fungus was tested for host specificity against closely related members of the Asteraceae and a number of economically important plants and proved highly specific to *M. micrantha* (Sankaran et al. 2001). In Vanuatu, *Cuscuta campestris* (Convolvulaceae) is reported to suppress *M. micrantha*. In Sri Lanka and Assam (India), *Cuscuta chinensis* has been found to suppress *M. micrantha* and prevent its spread into tea plantations (Waterhouse and Norris 1987).

3.10 *Synedrella nodiflora (L.) Gaertn.*

Synedrella nodiflora can be effectively controlled by hand weeding every 4 weeks for the first 8–12 weeks. After this period, emerging weed seedlings cannot compete effectively with crops (Galinato et al. 1999).

The weed can be effectively controlled with 2,4-D at 0.5–0.8 kg/ha or MCPA at 0.4 kg/ha applied within 20–30 days after emergence (Galinato et al. 1999). Herbicides such as metolachlor and alachlor, at 1.25 kg/ha, significantly decreased density, plant growth, flower yield, and quality of various weeds including *S. nodiflora* (Sharadamma et al. 2002).

Biological control has not been attempted.

3.11 *Chamaesyce hirta (L.) Millsp.*

Because spurges, like *C. hirta,* flower when young and have little or no seed dormancy, weed populations can increase rapidly. Removal of plants when they are young reduces seed production. Spurges are time consuming to hand weed because they produce so many seedlings. Some species may forcefully discharge seeds (Neal and Derr 2005).

Spurges are well controlled by most preemergence nursery herbicides containing a dinitroaniline but less well controlled by oxadiazon or oxyfluorfen. Herbicides such as glyphosate (1.080 g/ha)+2.4-D (241.8 g/ha) at 14 days before sowing, paraquat+diuron (400+200 g/ha) on the sowing day, and glyphosate (960 g/ha) at 35 days after crop emergence gave good control of all the weeds tested, including *C. hirta* (Petter et al. 2007).

Biological control has not been attempted.

3.12 *Mimosa pudica* Linn.

Mimosa pudica can be dug out, but its thorns and woody roots often present difficulties, especially when the plant covers a large area and forms dense thickets (Anonymous 2006). Although the thorns make hand pulling unpleasant, the plant can be controlled by hoeing. Cultivation can also keep the plant under control.

Many herbicides have been effective at controlling *M. pudica,* including glyphosate, picloram (0.47 l/ha), triclopyr, and soil-applied tebuthiuron. Dicamba is sometimes effective (Parsons and Cuthbertson 1992). Thorough wetting of all leaf surfaces is essential. If plants are disturbed before spraying, the leaves will fold up, and the herbicide will be ineffective. All spraying must therefore be done with forward booms or ahead of operators with knapsack sprayers (Land Protection Australia 2006b).

Comparatively little is known about the natural enemies of *M. pudica,* and no work appears to be currently in progress (Waterhouse and Norris 1987). Some of the insects attacking *M. pudica* are *Microporus* sp. (Hemiptera: Cydnidae), *Chlamisus* sp. (Coleoptera: Cryptocephalinae), *Chalcodermus* sp. (Coleoptera: Curculionidae), *Lophocampa catenul*ate (Lepidoptera: Arctiidae), *Ptychamalia perlata* (Lepidoptera: Geometridae), *Platynota rostrana* (Lepidoptera: Tortricidae), *Hemiargus hanno* and *Thecla azia* (Lepidoptera: Lycaenidae), *Calepheles* sp. (Lepidoptera: Riodinidae) (Waterhouse and Norris 1987), but more work on these possible biological-control agents is needed. Sheep grazing is reported to control the dominance of *M. pudica* in pastures (Magda et al. 2006). The species is often found attacked by a fungus, *Ramularia mimosae*, which grows on the upper surface of the leaflets, forming irregular white spots (Waterhouse and Norris 1987).

3.13 *Leucaena leucocephala* (Lam.) De wit

Leucaena leucocephala is best controlled by mechanical removal. Small individual plants can be hand pulled, if care is taken to remove the roots. Cutting or burning of *L. leucocephala* are most useful at the beginning of the year. The first step in preventative control is to limit planting and to remove existing plants. If possible, they should be removed before seeds are produced. The public must be informed about this species' invasive tendency and discouraged from purchasing, propagating, or planting it. If it is used as forage, grazing should be managed to prevent flowering and seed formation. An approach that combines mechanical and chemical control methods with land-management practices is most effective. Grazing is effective if applied before trees grow out of cattle's reach (Hughes 1998).

Leucaena leucocephala is sensitive to foliar-applied triclopyr. Triclopyr ester applied to basal bark and stump bark is effective, whereas 2,4-D in combination with diesel fuel is effective for basal bark treatments (Motooka et al. 2003).

A bruchid beetle seed predator, *Acanthoscelides macrophthalmus,* has been deliberately introduced and released in South Africa as a biocontrol agent, and the same insect has been accidentally introduced into Australia. The accidental spread of a psyllid insect defoliator, *Heteropsylla cubana,* in the mid-1980s caused cyclical defoliation but did not kill the trees, and the insect in turn appears to have been brought under control by a number of generalist local (and in some cases introduced) predators and parasites (Jones et al. 1992; Hughes 1998). Goats can provide a large degree of control if allowed to graze continuously. The species is good forage in dry pastures, but its toxicity is something of a problem, and it grows out of reach of livestock and shades out the understory (Motooka et al. 2003).

3.14 *Pennisetum polystachion (L.) Schult.*

Control should aim at preventing seed formation and destroying the soil seed bank. Scattered plants can be hand pulled or dug out (PIER 2003). Seedlings, small plants, and small infestations can be dug out (grubbed); all the roots must be removed. This method is more feasible for species that do not form rhizomes. Follow-up control is essential, as regrowth may occur. Most *P. polystachion* seed germinates at the beginning of the wet season, and control undertaken before seeding is much more effective (Anonymous 2008).

Herbicides can be effective in controlling *P. polystachion,* if it is carefully chosen and selectively applied to minimize regrowth and off-target damage. The range of herbicide treatments available is limited, and the plants must be actively growing at the time of application. Seedlings are best sprayed with paraquat. Dense stands can be treated with glyphosate (Weber 2003). Herbicides such as ametryn, paraquat, glufosinate ammonium, imazapyr, and glyphosate isopropylamine have proven effective at preventing seed germination. Trials have shown that spraying with glyphosate and wetting agents during the plant's maximum growth phase is effective (Anonymous 2008).

No biological control agents have been identified.

3.15 *Euphorbia cyathophora J. A. Murray*

No definite control methods have been developed for *Euphorbia cyathophora,* the methods listed above for *E. heterophylla* are applicable for this species.

3.16 *Sida rhombifolia L.*

Mechanical control of *S. rhombifolia* in plantation crops and pastures can be achieved by hand weeding. Hand pulling and mowing are only partially effective because the species is difficult to pull and quickly sprouts after cutting. This species is a greater problem in reduced-tillage agriculture (Holm et al. 1997). For pastures, mowing or chain slashing is recommended (FAO 2002).

Sida rhombifolia is somewhat tolerant of 2,4-D, dicamba, and triclopyr (Motooka et al. 2003). Application at the higher end of labeled rates is required. Australian sources indicate good control with wick applications of glyphosate by means of double passes at right angles. Soil-applied herbicides (pre-emergence metribuzin and imazaquin and postemergence imazethapyr) are reported to be beneficial for glyphosate-tolerant crops, reducing early-season competition from weeds, including *S. rhombifolia*, particularly those inherently more tolerant to glyphosate (Arregui et al. 2001).

Few recent attempts at biological control have been made. Because of the importance of this weed in the United States, a search has been made for insects feeding on it. Candidate control agents for all species of *Sida* include the beetles *Calligrapha feline, C. pantherina,* and *Acanthoscelides brevipes* (Coleoptera: Chrysomelidae), which are being investigated in the Northern Territory of Australia. *Calligrapha pantherina,* which feeds exclusively on three weedy species of *Sida,* has been released in New South Wales (Parsons and Cuthbertson 2001) and has provided effective biological control in Australia and Papua New Guinea (Kuniata and Rapp 2001).

3.17 *Momordica charantia L.*

Most infestations can be effectively removed by mechanical or manual means. The stale-seed-bed technique will assist in removing many just before sowing (Douglas 2002).

No chemicals are recommended or registered for control of balsam pear; more work on this subject is needed.

No biological-control agents have been identified.

3.18 *Chrysopogon aciculatus (Retz.)*

Chrysopogon aciculatus is an aggressive and noxious weed. It withstands trampling, poor soils, and mowing. Sharp seeds can penetrate flesh and work their way in, causing festering sores; it is particularly hard on dogs (Stone 1970).

No specific mechanical, chemical, or biological control methods have been developed for this weed, but methods indicated for other grass species can be applicable.

3.19 *Chamaesyce hypericifolia (L.) Millsp.*

No control methods for *Chamaesyce hypericifolia* are readily available.

3.20 *Chloris barbata (L.) Sw.*

Chloris barbata can be effectively controlled by hand weeding every 4 weeks for the first 8–12 weeks. After this period, emerging weed seedlings are not competitive against crops (Galinato et al. 1999).

The application of 1.5 kg fluchloralin/ha combined with hand weeding produced the most efficient control of weeds, including *C. barbata*, in field experiments during the summer (96.24 % control) and rainy seasons (96.73 % control) (Balasubramanian et al. 2002).

No research has been done on biological control. Cattle are said to be partial to this weed when it is young but to avoid it when the inflorescence matures. (Anonymous 1978).

Acknowledgments This project was supported by grant 08-DG-1052021-190 Forest Health Protection, Invasive Plants, from the USDA Forest Service and USDA—McIntyre-Stennis Program for forestry related research at Land Grant Universities (Project#GUA0611). We also thank David Bakke, Pesticide Specialist and Invasive Plants Program Manager, State and Private Forestry, USDA Forest Service, Vallejo, CA and Anne Marie LaRosa, Forest Health Coordinator, Institute of Pacific Islands Forestry, USDA Forest Service, Hilo, HI for the useful discussions. In accordance with federal law and USDA policy, this institution is prohibited from discrimination on the basis of race, color, national origin, sex, age, or disability. This work benefited from assistance in the field by Z. T. Cruz and R. Gumataotao.

References

Anonymous (1978) Poaceae in Flora of Taiwan. http://www.eFloras.org

Anonymous (2006) *Mimosa pudica*. http://www.tropical-biology.org/research/dip/species/Mimosa%20pudica.htm

Anonymous (2007) Guinea grass, *Panicum maximum*. Produced by Land Protection Australia (Invasive plant and animals), Queensland Government, Department of Primary Industries and Fisheries, PP82

Anonymous (2008) Weed management guide. Feathergrass and mission grass—*Pennisetum* species. CRC Weed Management, Australia

Arregui MC, Scotta R, Sánchez D (2001) Weed control in established alfalfa (Medicago sativa) with postemergence herbicides. Weed Technol 15:424–428

Balasubramanian V, Anbumani S, Nadanassababady T (2002) Studies on evaluation of weed management techniques for irrigated soybean (*Glycine max* L.). Legume Res 25:205–207

Cruz ZT, Muniappan R, Reddy GVP (2006) Establishment of *Cecidochares connexa* (Diptera: Tephritidae) in Guam and its effect on the growth of *Chromolaena odorata* (Asteraceae). Ann Entomol Soc Am 99:845–850

Daehler CC, Carino DA (2000) Predicting invasive plants: prospects for a general screen system based on current regional models. Biol Invasions 2:92–103

Daehler CC, Denslow JS, Ansari S, Kuo HC (2004) A risk-assessment system for screening out invasive plants from Hawaii and other Pacific islands. Conserv Biol 18:360–368

Douglas J (2002) Balsam pear. New Zealand Institute for Crop & Food Research Ltd. http://www.crop.cri.nz

Erasmus DJ (1988) A review of mechanical and chemical control of Chromolaena odorata in South Africa. In: Muniappan R (ed) Proceedings of the First International Workshop on Biological Control of *Chromolaena odorata*, Bangkok, Thailand, p 34

Esguerra NM (1998) The Siam weed infestation in the Federated States of Micronesia—seven years of attempting to control it. Proceedings of the Fourth International Workshop on Biological Control and Management of *Chromolaena odorata*, Bangalore, India, p 80

Esguerra NM, William WS, Smith JR (1991) Status of biological control of the Siam weed *Chromolaena odorata* (L.) R. M. King and H. Robinson on Pohnpei, Federated States of Micronesia. Ecology and management of *Chromolaena odorata*, p 49, BIOTROP special publication 44

Evans J (1992) Plantation forestry in the tropics, 2nd edn. Oxford University Press, Oxford

FAO (Food and Agriculture Organization of the United Nations) (2002) Cultural practices. http://www.fao.org/ag/AGP/AGPC/doc/Publicat/FAOGUL2/B204.htm

Galinato MI, Moody K, Piggin CM (1999) Upland rice weeds of South and Southeast Asia. International Rice Research Institute, Makati City

Groves RH, Panetta FD, Virtue JG (2001) Weed risk assessment. CSIRO, Collingwood, Australia

Hall DW, Vandiver VV, Ferrell JA (1991) Common Beggar's-tick (Hairy Beggar's-tick), *Bidens alba* (L.) DC. Florida Cooperative Extension Service, Institute of Food and Agricultural Sciences, University of Florida, Gainesville

Holm L, Doll J, Holm E, Pancho J, Herberger J (1997) World weeds. Wiley, New York

Hughes CE (1998) Leucaena, a genetic resources handbook. Oxford University Press, Oxford

Jones RJ, Brewbaker JL, Sorensson CT (1992) Leucaena leucocephala (Lamk) de Wit. In: Mannetje L, Jones RM (eds) Plant resources of South-East Asia No. 4. Forages. Pudoc Scientific Publishers, Wageningen, p 150

Kolar CS, Lodge DM (2001) Progress in invasion biology: predicting invaders. Trends Ecol Evol 16:199–204

Kuniata L, Rapp G (2001) Biocontrol of *Sida rhombifolia* in Papua New Guinea. http://www.nt.gov.au/dpif/pubcat/agnotes/542.htm

Land Protection Australia (2006a) Common sensitive plant *Mimosa pudica.* Department of Natural Resources and Water, Australia

Land Protection Australia (2006b) Snakeweed and its control. Land Protection, Queensland Government, Australia. http://www.nrm.qld.gov.au/factsheets/pdf/pest/pp52.pdf. Accessed 28 Sept 2012

Macdonald G, Ferrell J, Sellers B, Langeland K, Duperron-Bond T, Ketterer E (2008) Excerpted from invasive species management plans for Florida. University of Florida, IFAS Extension, Circular 1529

Magda D, Duru M, Huguenin J, Gleizes B (2006) Impact of shading and cutting on the demography and composition of *Mimosa pudica* L., a ligneous weed species of tropical grasslands. Grass Forage Sci 61:89–96

Mcconnell J, Muniappan R (1991) Introduced ornamental plants that have become weeds on Guam. Micronesica 3(Suppl):47–49

Motooka P, Castro L, Nelson D, Nagai G, Ching L (2003) Weeds of Hawaii's pastures and natural areas: an identification and management guide. College of Tropical Agriculture and Human Resources, University of Hawaii at Manoa

Muniappan R, Marutani M, Denton GRW (1988) Introduction and establishment of *Pareuchaetes pseudoinsulata* Rego Barros (Arctiidae) against *Chromolaena odorata* in the Western Caroline Islands. J Biol Control 2:141–142

Muniappan R, Cruz J, Bamba J (2002) Invasive plants and their control in Micronesia. Micronesica 6(Suppl):85–92

Muniappan R, Reddy GVP, Lai PY (2005) Distribution and biological control of *Chromolaena odorata*. In: Inderjit S (ed) Invasive plants: ecological and agricultural aspects. Birkhäuser Verlag, Basel, p 223

Muniappan R, Englberger K, Reddy GVP (2007) Biological control of *Chromolaena odorata* in the American Pacific Micronesian Islands. In: Lai P-Y, Reddy GVP, Muniappan R (eds) Proceedings of the Seventh International Workshop on biological control and management of

Chromolaena odorata and *Mikania micrantha.* National Pingtung University of Science and Technology, Taiwan, p 49

Muniappan R, Reddy GVP, Raman A (eds) (2009) Biological control of tropical weeds using arthropods. Cambridge University Press, Cambridge

Neal JC, Derr JF (2005) Weeds of container nurseries in the United States. http://ppwsipm.contentsrvr.net/weeds_container_nurseries.php

Nechet KLD, Barreto RW, Mizubuti ESG (2006) *Bipolaris euphorbiae* as a biological control agent for wild poinsettia (*Euphorbia heterophylla*): host-specificity and variability in pathogen and host populations. BioControl 51:259–275

Parsons WT, Cuthbertson EG (1992) Noxious weeds of Australia. Inkata Press, Melbourne

Parsons WT, Cuthbertson EG (2001) Noxious weeds of Australia, 2nd edn. Inkata Press, Melbourne

Petter FA, Procópio SO, Cargnelutti FA, Barroso ALL (2007) Herbicide management in Roundup Ready soybean crop. Planta Daninha 25:557–566

Pheloung PC, Williams PA, Halloy SR (1999) A weed risk assessment model for use as a biosecurity tool evaluating plant introductions. J Environ Manage 57:239–251

PIER (Pacific Island Ecosystems at Risk) (2003) *Pennisetum polystachion.* http://www.hear.org/Pier/species/pennisetum_polystachion.htm

PIER (Pacific Island Ecosystems at Risk) (2005) *Stachytarpheta jamaicensis.* http://www.hear.org/Pier/species/stachytarpheta_jamaicensis.htm

Prostko EP (2009) Wild Poinsettia identification and control in Peanut. University of Georgia Cooperative Extension Circular #855

Reddy GVP (2011) Survey of invasive plants on Guam and identification of the 20 most widespread. Micronesica 41:263–5274

Sankaran KV, Muraleedharan PK, Anitha V (2001) Integrated management of the alien invasive weed *Mikania micrantha* in the Western Ghats. Final report of the research project KFRI/283/97 August 1997–December 1999

Seibert TF (1989) Biological control of the weed, *Chromolaena odorata* (Asteraceae) by *Pareuchaetes pseudoinsulata* (Lepidoptera: Arctiidae) on Guam and the Northern Mariana Islands. Entomophaga 34:531–539

Sharadamma BC, Gowdam JVN, Prasad TVR (2002) Chemical weed management in crossandra (*Crossandra undulaefolia* Salisb.) cv. Soundarya. Crop Res 23:313–316

Stone BC (1970) The flora of Guam. Micronesica 6:1–659

Swarbrick JT (1997) Weeds of the Pacific Islands. South Pacific Commission, Noumea, New Caledonia

Waterhouse DF, Norris KR (1987) Biological control. Pacific prospects. Inkata Press, Melbourne

Weber E (2003) Invasive plants of the world. CABI Publishing (CAB International), Wallingford

Zachariades C, Day M, Muniappan R, Reddy GVP (2009) *Chromolaena odorata* (L.) King and Robinson (Asteraceae). In: Muniappan R, Reddy GVP, Raman A (eds) Biological control of tropical weeds using Arthropods. Cambridge University Press, Cambridge, p 130

Genetic Resources of Vegetable Crops: Indian Perspective

Anjula Pandey

1 Introduction

India is the world's largest producer of many fresh vegetables and it ranks second in production of potatoes, onions, cauliflowers, brinjal, cabbages, etc. Vegetables such as onion, okra, bitter gourd, green chillies, mushrooms and potatoes contribute largely to the vegetable export basket (FAO 2013; National Horticulture Database 2012). Concurrent developments in quality of vegetable produce and the areas of state-of the-art infrastructure have largely contributed to this (http://www.apeda.gov.in/apedawebsite/six_head_product/FFV.htm). Growing popularity of vegetables among the farming community is due to the quick plant growth to produce maximum quantity of food for the area planted and more income generation in short time as compared to other food commodities. Vegetables provide variety and flavour to the diet, making meals more appetizing. Though food value of many vegetables, especially leafy and fruit types is low because of large amount of water, low fat and calories, they are rich source of vitamins, minerals and fibre. Some vegetables such as tomato, potato, cabbage, cauliflower, onion, etc. are among the top most important and popular crops at the global level. But others of lesser importance are only regionally utilized and relished by small population or community.

Botanically diverse plant part(s) contributes to the vegetable of commerce. The edible part of a vegetable, may be a root/tuberous root (potato, radish, carrot, turnip, sweet potato, cassava, beet root), stem (knol-khol, asparagus); stem tubers (potato, taro, yams), corm (taro), bulb (onion, garlic), bud (brussels sprouts), leaf (spinach, lettuce, cabbage, parsley, chive), leaf stalk/petiole (celery, rhubarb, Swiss chard), bracts/thickened receptacle (globe artichoke), flower stalk/flower buds (broccoli, cauliflower), seeds (beans, pea), young fruits (brinjal, cucumber, sweet corn), mature fruits (tomato, chilli) and the flower petal (pumpkins, bauhinia, moringa, silk

A. Pandey (✉)
Division of Plant Exploration and Germplasm Collection, National Bureau of Plant Genetic Resources, New Delhi 110 012 , India
e-mail: anjuravinder@yahoo.com

D. Nandwani (ed.), *Sustainable Horticultural Systems,* Sustainable Development and Biodiversity 2, DOI 10.1007/978-3-319-06904-3_15

cotton). In general the term 'Vegetables' means an edible product of a herbaceous plant (a plant with a soft stem), and can be distinguished from the edible nuts and fruits produced by plants with woody stems.

Vegetable genetic resources (VGR) play a major role in Indian agriculture by providing food, nutritional and economic security. Gradual reduction in cultivable land, declining natural resources and increasing biotic and abiotic stresses is posing challenges to meet with climate change and enhance quality and productivity in vegetable crops. This chapter highlights salient points on strength of the vegetable genetic resources (VGR) in the wider perspective of India and discusses on diversity available in cultivated taxa including regionally important cultivars, less-known and potential species and wild relatives of vegetables in light of efforts made in the genetic resource management at national level. Future targets are highlighted on the basis of gap areas identified in genetic resource management.

2 Vegetable Genetic Resources

The vegetable genetic resources (VGR) with contribution of native species diversity include improved and obsolete varieties, landraces/local cultivars and breeding materials of vegetable crops developed here along with their wild and weedy relatives in the Indian region (Dhillon et al. 2001) (Fig. 1).

Terminology to define vegetable genetic resources such as major and minor vegetables is used in sense of production and consumption at commercial levels. Besides, some terms like under-utilized/underexploited, less-known, semi-domesticated, and potential vegetables are often used interchangeably for vegetables having value at regional or at local level. The terms 'cultivated', 'semi-domesticated', 'semi-wild', 'protected' and 'wild' species generally refer to their state of occurrence in an area.

2.1 VGR: To Meet with Challenge of Climate Change

Plant genetic resources are the backbone of agriculture and play a positive and unique role in development of new cultivars including the improving/restructuring of the existing ones. Genotypes which could withstand better under abiotic and biotic pressures are the keys for sustainable agriculture. Historically, plant improvement has always crossed thresholds by identifying new traits in genetic resources, and introducing them into new varieties. Gene for such traits are often available in landraces or locally adapted cultivars and wild related species. Therefore the genetic diversity in vegetables crops at their primary center of origin/domestication and secondary centers of diversity is important (Paroda et al. 1988; Arora 1991). This diversity need to be systematically collected by the explorers, characterized/evaluated for promising traits for use in crop improvement and also conserved for posterity. In recent past there has been a paradigm shift

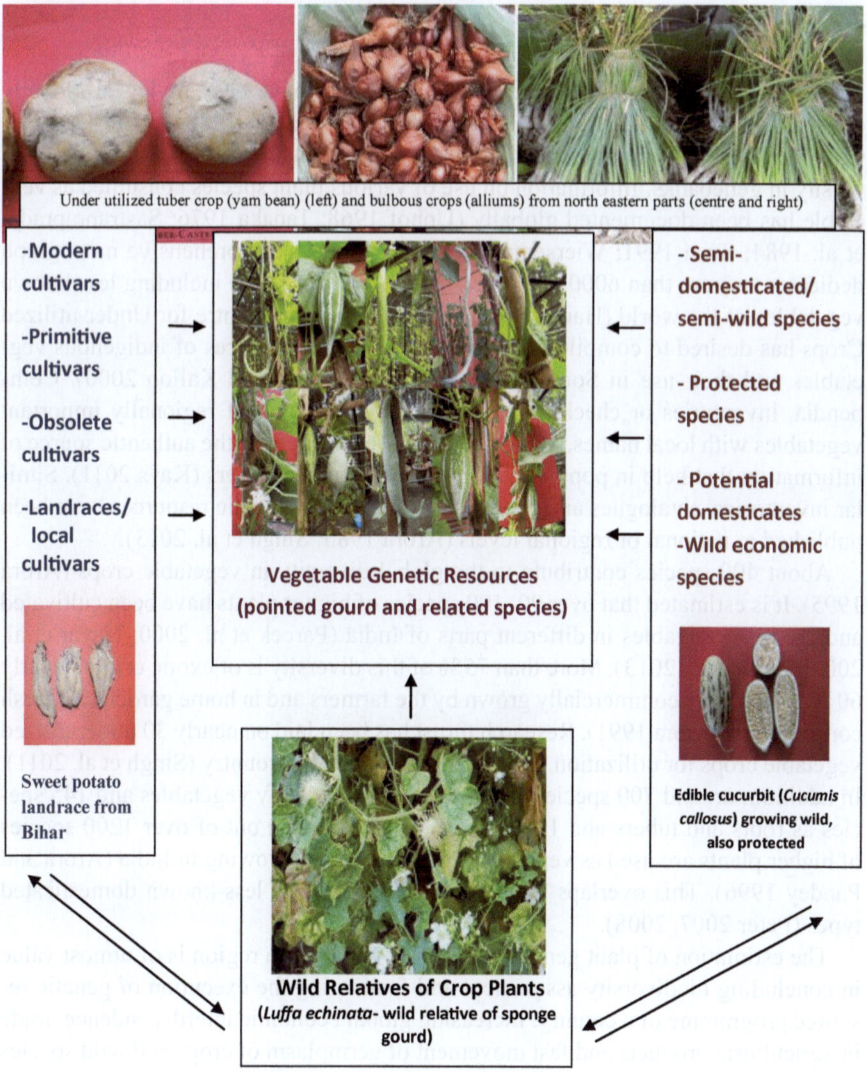

Fig. 1 Flow chart showing various components of vegetable genetic resources in India. (photos illustrate some examples)

in germplasm collection and conservation priorities from general to trait specific one (Bhandari and Pradheep 2010). Identification of the areas of concentration of diversity vis.-a-vis. availability of important traits is the most difficult but challenging task. Screening of germplasm of earlier collections from areas of availability can further be explored through fine grid surveys to find the targeted traits. Subsequent efforts towards multilocational evaluation can further help in identifying the traits under genetic resources programmes.

2.2 VGR Strength

In past several decades vegetable genetic resources have been reviewed globally (Sloten 1980; Arora 1985; Crisp and Astley 1985; Kalloo 1988; Kotlinska 1993; Cross 1998), largely emphasizing on the need to collect and conserve the genetic diversity in genebanks. Information on use of various plant species consumed as vegetable has been documented globally (Uphof 1968; Tanaka 1976; Sastraprapradja et al. 1984; Swai 1991; Wiersema and Leon 1999). A comprehensive monograph dedicates to more than 6000 cultivated species of food value including less known vegetables of the world (Hanelt 2001). The International Centre for Under-utilized Crops has desired to compile information on genetic resources of indigenous vegetables and their use in South Asian countries (Ghosh and Kalloo 2000). Compendia, inventories or checklists describing the potential of regionally important vegetables with local names, use and areas of occurrence are the authentic source of information that help in popularization of less-known cultivars (Kays 2011). Similar inventories, catalogues and checklists on vegetable genetic resources have been published at national or regional levels (Arora 1985; Singh et al. 2013).

About 400 species contribute to the global diversity in vegetable crops (Arora 1995). It is estimated that over 90–100 species of higher plants have been cultivated and used as vegetables in different parts of India (Pareek et al. 2000; Nayar et al. 2003; Singh et al. 2013). More than 75 % of this diversity is of exotic origin. Nearly 60 vegetables are commercially grown by the farmers and in home gardens for fresh consumption (Arora 1991). Research thrust has been laid on nearly 30 underutilized vegetable crops for utilization in different regions of the country (Singh et al. 2011). In addition, around 700 species (521 species used as leafy vegetables and 145 species as roots and tubers and 101 species as flower/buds) out of over 1200 species of higher plants are used as vegetable from wild plants growing in India (Arora and Pandey 1996). This overlaps with about 200 species of less-known domesticated types (Peter 2007, 2008).

The estimation of plant genetic resource diversity of a region is of utmost value in concluding biodiversity assessment and facilitating the execution of genetic resource programme of a country. Increasing global economic interdependence, trade in agricultural products and fast movement of germplasm of crops and wild species of desirable genotype regionally or globally have influenced the magnitude of this diversity to a great extent.

Various publications on vegetables genetic resource provided different updates on species richness; regionally important species are considered under the less-known species or wild edible vegetable (Arora and Pandey 1996; Peter 2007, 2008). Many other publications provide scattered account on their distribution and use (Watt 1889-1896; Ambasta et al. 1986; Rao et al. 1988; Arora and Pandey 1996; Ramachandran 1997; Roy et al. 1988; Nayar et al. 2003). In past one decade a significant increase in exotic vegetable genetic resources was recorded in the national markets thereby indicating the enhanced species number on one hand and an alarming threat to native diversity on the other hand. An update on a 'comprehensive list' of cultivated vegetable crops of India has been attempted that includes over 120 species of indigenous and exotic origin (Pandey, pers. com. 2014).

3 Centres of Diversity of Major Cultivated Vegetables

Globally the cultivated vegetables of a region reflect larger percentage of introduced diversity, that is well adapted to the cropping patterns and also been accepted by the native population (Siemonsma and Kasern 1994). Based on the richness of diversity in cultivated plants and their wild relatives the Indian region is designated to be one of the 12 megacentres of diversity in the world and a centre of origin for many domesticated vegetable species (Tab. 1; Zeven and de Wet 1982). Extent of diversity and its distribution in the Indian gene centre in different agro-climatic regions is discussed in various publications (Arora 1991; Dhillon et al. 2001; Gopalakrishnan 2007). Primitive cultivars of some native crops are still in *in situ/on farm* protection, in home gardens or grown in mix cropping.Good variability has been developed over a period of time in exotic and indigenous cultivated vegetables. Like any other crop-groups, germplasm introductions have played a major role in diversifying the vegetable crops in India. In the sixteenth century over a dozen of vegetables such as yams and taros, gourds, ivy gourd, dolichos bean, ash gourds, melons, water chestnut, etc. were commonly grown in India. Some of these are still under cultivation (Randhawa 1982). Later chilli, potato, amaranth and many others were introduced into India through different modes. Over a period of time diversity in cultivated vegetables in the Indian gene centre got further enriched to over 30 crops mainly representing the cucurbits, tuberous types, leguminous vegetables and leafy types (Zeven and de Wet 1982). Vegetables from other diversity regions of the world were well represented in this region. Major cultivated vegetables in world centres of diversity are given in Table 1, Fig. 2.

3.1 Diversity in Vegetable Crops in India

The Indian subcontinent has roughly contributed to about 20–25 species of domesticated vegetables. Over more than 20 different families of higher plants represent diversity in vegetable crops of native and exotic origin (Pandey, pers. com. 2014). Among them family Cucurbitaceae (25 crops), Brassicaceae (12 crop) and Solanaceae (6 crops) are the major contributors towards the vegetable crops (Table 2). The mustard family, Brassicaceae, includes many important vegetables. Single species *Brassica oleracea* includes the 'cole crops' such as cabbage, cauliflower, knol-khol, brussels sprouts, broccoli and kale. The family Solanaceae, in addition to common potato (*Solanum tuberosum*), provides several other vegetables like tomato, brinjal and chilli/hot pepper and sweet pepper. Among major crops the cucurbits are important for the Indian region, including 38 endemic species (Chakravarty 1982).

Table 1 World centres of diversity of major cultivated vegetables. (Modified from Zeven and de Wet 1982; Hawkes 1983.)

Gene centre	Crops in primary centers of origin/domestication; Secondary centers
Chinese-Japanese	Brinjal, ash gourd, Chinese cabbage, water spinach (kang-kong), Japanese bunching onion, leafy mustard; Water melon, amaranth
Indo-Chinese	Ash gourd, sponge gourd, ridged gourd, bottle gourd, bitter gourd, cucumber, cho-cho, sword bean, winged bean, taro, yam; Cucumber, bottle gourd, cho-cho, shallot, yam bean, yardlong bean, Chinese cabbage, amaranth, water spinach, leafy mustard
Hindustani	Brinjal, ash gourd, cucumber, ridged gourd, sponge gourd, bitter gourd, lablab bean, water spinach, okra, drumstick, rat's tail radish, cowpea, leafy mustard, taro, yams; Water melon, melon, bottle gourd, amaranths, roselle
Central Asia	Onion, garlic, mustard, spinach, carrot (Asiatic varieties), pea, radish, beet root, faba bean; Brinjal, water melon, melon, cauliflower
Near East	Onion, garlic, leek, mustard, beet, melon, faba bean, Cucumis; Okra, pea (endemic types; secondary centre)
Mediterranean	White cabbage, cauliflower, broccoli, radish, faba bean, broad bean, cabbage, lettuce, beet root; Chilli, garlic, okra
African	Brinjal, water melon, melon, bottle gourd, cowpea, okra, roselle, yams, cucumber; Onion, shallot, lima bean, mustard, amaranth
European-Siberian	Lettuce; Onion, common bean, white cabbage, cauliflower, spinach, carrot
South American	Tomato, chilli, pumpkin/squash, lima bean, French bean, cowpea, okra, roselle, sweet potato, potato, tannia, amaranth, chenopods, manioc; Common bean
Central American and Mexican	Tomato, hot pepper, pumpkin/squash, yam bean, French bean, amaranth, sweet potato; French bean, potato, chenopods
North America	Jerusalem artichoke; Tomato, brinjal, pepper, melon, water melon, pumpkin/squash, onion, lettuce, French bean, lima bean, okra

Cultivated vegetable crops belonging to Fabaceae yield minor vegetables—yam bean, yard long bean, broad bean, lima bean, etc.; other legume crops grown as pulse or multipurpose type are also consumed as for their tender fruits.

The 'Hindustani centre' or 'Indian gene centre' is one of the centres of origin and a primary centre for crops possessing rich genetic diversity in native crops like brinjal (*Solanum melongena*), cucumber (*Cucumis sativus*), ridged and sponge gourd (*Luffa* spp.) and many root crops (Zeven and de Wet 1982). This centre possesses

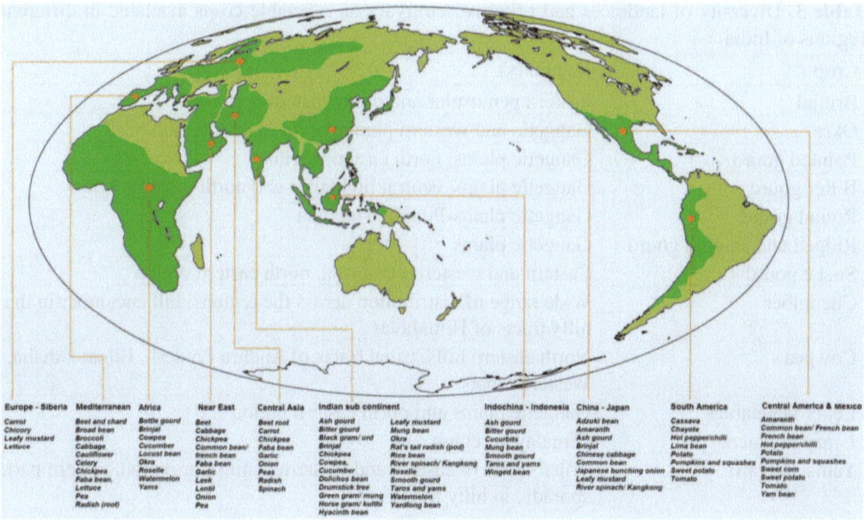

Fig. 2 World centres of diversity of major cultivated vegetables

Table 2 Diversity occurring in cultivated vegetables in India

S. no.	Diversity	Status (no. of species)
1	Botanical families	Cucurbitaceae (25), Fabaceae (13), Brassicaceae (12), Alliaceae (11), Solanaceae (6), Chenopodiaceae (5), Dioscoreaceae (5), Araceae (4), Apiaceae (3), Asteraceae (2), Convolvulaceae (2), Amaranthaceae (1), Euphorbiaceae (1), Lamiaceae (1), Liliaceae (1), Malvaceae (1), Marantaceae (1), Moringaceae (1), Poaceae (1), Basellaceae (1), Cannaceae (1)
2	Life form (annual/perennial)^	Annual (3), perennial (10), both (5)
3	Mode of propagation^	Seed (72), vegetative (28), both (14)
4	Part used^	Fruits (tender pods) (44), seeds (1), leaves/tender shoot (33), swollen stem (1), inflorescence (flower heads/buds-7), underground parts (roots-9, bulbs-9, tubers-13, rhizome-4)
5	Centre of origin^	Indian subcontinent/southeast Asia (29), Mediterranean/Europe (20), America (19), Africa (18), West Asia/central Asia (13)

^Category overlapping

rich diversity for crops like cowpea (*Vigna unguiculata*), okra (*Abelmoschus esculentus*), chilli (*Capsicum annuum*) and water melon (*Citrullus lanatus*) and the pumpkin for which it is the secondary centre.

Important indigenous crops include: brinjal, smooth gourd, ridged gourd, snake gourd, round gourd, snap melon (phut), cucumber, leafy vegetables (Malabar spinach/*Basella alba*, chenopods, leafy brassicae, amaranths, etc.), taro, yam, Elephant foot yam and ginger. Among the introduced crops, tomato, peas, French bean, scarlet runner bean, lima bean, winged bean, faba bean, cauliflower, cabbage,

Table 3 Diversity of landraces and primitive cultivars of vegetable crops available in different regions of India

Crop	Region (s)
Brinjal	Eastern peninsular and north eastern region
Okra	Gangetic and western plains; north eastern region
Pointed gourd	Gangetic plains; north eastern region
Bitter gourd	Gangetic plains, central highlands and north eastern region
Round gourd	Gangetic plains-Punjab, Haryana
Ridged and smooth gourd	Gangetic plains
Snake gourd	Eastern and western peninsula, north eastern region
Cucumber	Wide range of distribution across the regions; hill cucumber in the hilly tracts of Himalayas
Cowpea	North eastern hills, tribal tracts of Andhra Pradesh, Bihar, Odisha, Western Ghats
Leafy vegetables	Gangetic plains and north eastern region
Leafy brassicae	Himalayan region
Yams and taro	Tribal tracts of eastern and western peninsular region, eastern part; sporadic in hilly tracts

knol-kohl, turnip, radish, carrot, beta spinach, bell pepper, cho-cho, potato, meetha karela/*Cyclanthera pedata*, onion, leek, shallot, garlic/*Allium* spp., asparagus, artichoke, parsley and coriander are some major ones. Majority of vegetable crops grown in India are from the tropical American, tropical Asian and the Mediterranean regions. Table 3 lists crop diversity of some landraces and primitive cultivars of vegetable crops in different regions of India.

Vegetable crops are widely distributed across the country showing regional variation due to varied climate and physiographic condition in the Indian sub-continent. This is prevalent among different species of *Solanum* in north-eastern region, yams in Western Ghats and south-eastern states; chives, leeks and other wild *Allium* spp. in western and eastern Himalayas; cluster bean in western arid zone; lablab in Deccan plateau; cucurbits in Indo-Gangetic plains and Rajasthan, Haryana and Madhya Pradesh; leafy brassicae in western Himalaya and chilli, tuber crops (cassava, yams, colocasia/alocasia, etc.) mainly in the peninsular region including the tribal dominating areas.

North-western and eastern Himalayan region is rich in diversity of leafy types-leek, shallot, other alliums, spinach, chenopods, amaranths, bean, colocasia, parsley, cho-cho and meetha karela. North-eastern region particularly Assam is rich in underutilized solanaceous vegetables, leafy vegetables, less-known legume vegetables (winged bean, jack bean, sword bean) and cucurbits (cho-cho and meetha karela). The northern plains/Gangetic plains including tarai region is considered to exhibit variability in cucurbits and other aquatic leafy vegetables. The north western/Indus plain region is rich in diversity particularly in amaranth, chenopods, *Cucumis, Momordica* and *Citrullus.* The central region/plateau has more diversity in cucurbits like melon, bitter gourd, pointed gourd and ridged gourd. Western and eastern peninsular regions have considerable diversity of regionally important species like snake gourd, *Luffa* spp. and *Moringa.*

3.1.1 Variability in Some Vegetable Crops

Of different species of chilli, *C. annuum* var. *annuum* is the most widespread and highly variable for fruit colour, shape size and capsaicin content. Varieties with high capsaicin content are used as spice whereas large fruited forms with meagre capsaicin content are used as vegetable, salads or pickles. However, dark red types are used for vegetable colourants. Peruvian pepper (*C. frutescens* var. *baccatum*) was introduced to India from Brazil by Portuguese prior to 1585. Different types of hot peppers (derivatives of *C. annuum* var. *annuum*) distinguished mainly on the basis of fruit type are known by various common/vernacular names such as cluster pepper (vern. guchha mirch), wrinkled pepper, cherry pepper, paprika, baghi mirch, cone pepper, etc. and are grown in diverse but localized areas of India.

Tuberous crops are important in the context of Indian gene centre. In colocasia two types viz. eddoe (*C. esculenta* var. *antiquorum*—arvi for vegetable use), and dasheen (*C. esculenta* var. *esculenta*) are common (Purseglove 1975). Another the tuberous root vegetable cassava (*Manihot esculenta*) was introduced into India from Brazil by the Portuguese. Elephant foot yam (*Amorphophalus paeoniifolius*) is a very old but regionally important tuberous vegetable crop of commercial value in eastern region of India. Less-known cultivated tuberous vegetables- tannia (*Xanthosoma sagittifolium*) and yam bean (*Pachyrrhizus erosus*), both of exotic origin from south and central America respectively are cultivated on small scale in Bihar and parts of eastern region. They are now only grown as backyard cultigens in these regions. Not much variability exists in root crop- radish (*Raphanus sativus* var. *sativus*). Related minor vegetable of Indian origin *R. sativus* var. *caudatus* (rat tailed radish/mugri) which is well recognized for its tender pods used as vegetable/salad; it is under limited cultivation in north-western parts especially Punjab, Haryana, Uttar Pradesh and Rajasthan. Beet root (*Beta vulgaris*) and *B. vulgaris* var. *cicla* (beta spinach) originated from Mediterranean region was initially domesticated for root and gradually selected for edible leaves used as vegetable; it is under sporadic cultivation in different parts of the country more so in Uttar Pradesh, West Bengal, Maharashtra and Gujarat.

Cabbage and cauliflower are old introduced cole crops in India which have shown a gradual development of regional diversity suited to diverse agro-climatic conditions of the country. Knol khol (*B. oleracea* var. *gongylodes*) crop of Mediterranean origin is a minor vegetable grown in Kashmir, West Bengal, Uttar Pradesh, with limited use and preferences. Kale (*Brassica oleracea* var. *acephala*) is cultivated mainly in Kashmir, Himachal Pradesh, Uttar Pradesh, Punjab, Haryana and Nilgiris. Newer cole crops such as sprouting broccoli, asparagus broccoli, and purple cauliflower of the Mediterranean region are recent additions to the vegetable markets of India and are gaining popularity mainly in the urban areas.

Leafy types as celery (*Apium graveolens* var. *dulce*) from Mediterranean region have limited preference for use and are under restricted cultivation. *Amaranthus*, *A. hybridus*, *A. hybridus* ssp. *cruentus* and *A. lividus* are mainly used as leafy types. Leguminous vegetables exclusively grown for pods are winged bean, sword bean, mucuna bean, hyacinth bean, yard long bean, etc. These crops are regionally important. Some leguminous crops such as French bean and cow pea that are primarily cultivated for pulses/grains are also used for tender pods whereas pea has dual use (young seeds used as vegetable; mature seeds as pulse).

3.1.2 Underutilized/Underexploited and Less-known Vegetables

Apart from cultivated species, enormous diversity occurs in the form of underutilized, semi-wild and wild types (Ghosh 1984; Arora and Pandey 1996; Mal et al. 1997; Hore 2001, 2004). The genetic wealth of underutilized vegetable species is increasing as more and more native species are being domesticated in different plant diversity regions of the world (Grubben and Soetjipto 1996; Mal et al. 1997; Arora 2003; Arora et al. 2006; Chadha et al. 2007). It is estimated that out of 1600 economic species globally used, 422 species in Asia, and 107 species in South Asia are used as vegetable (Nath et al. 1987). Of a total of about 225 species used as vegetable, 100–125 are cultivated for commercial purpose and home consumption; and over 100 minor vegetables represent semi-domesticated/domesticated and plants gathered from the wild (Siemonsma and Kasern 1994). Vegetable network programmes at the global level focus on promotion, collection, conservation and use of indigenous/regional vegetables to significantly contribute to enriching this diversity involving national programmes (Mal et al. 1997; Engle and Altoveros 2000; APAARI 2006, 2007; Chadha et al. 2007; Arora et al. 2006).

Underutilized species of vegetable crops are well protected, semi-domesticated, domesticated types under cultivation in the marginal habitats, and as backyards and home gardens. Apart from their minor use as edible species, their hidden potential lies in their identification and use for specific traits in crop improvement (Arora and Pandey 1996). Diversity in this group of plant genetic resources has been discussed along with its distribution (Arora 1985; Arora and Pandey 1996). Several of them are also endemic with narrow range of distribution as cultigens, and others have relatively wider distribution and use by the native communities. In times to come these genetic resources having reservoir of genes for disease resistance, stress tolerance to extremely hot and extremely cold arid regions, water logged and tidal in coastal regions, salinity prone areas/habitats, acidic soils, etc. can meet with the climate change.

Less selection pressure has operated on underutilized cultivated vegetables due to their regional distribution and use, and thus they are likely to carry some useful traits like well adapted to adverse environmental conditions and tolerance/resistance to diseases, etc. Many of these species are hardy and can grow on soils, which are not suitable for other crops (Smartt and Haq 1997). Many underutilized vegetables are vast reservoirs of vitamins, minerals and dietary fibre. Despite being recognized for their importance, the possible reason of their poor utilization is lack

Table 4 Distribution of important wild relatives of vegetable crops in different phyto-geographical zones. (Arora and Nayar 1984)

Phyto-geographical zones: species
Western Himalaya: *Abelmoschus manihot* (tetraphyllus forms), *Cucumis hardwickii*, C. trigonus, *Luffa echinata, L. graveolens, Solanum indicum/S. lasiocarpum, Trichosanthes multiloba, T. himalensis*
Eastern Himalaya: *Abelmoschus manihot, Cucumis trigonus, Luffa graveolens, Neoluffa sikkimensis*
North-eastern region: *Abelmoschus manihot (pungens forms), Alocasia macrorhiza, Amorphophallus bulbifer, Colocasia esculenta, Cucumis hystrix, C. trigonus, Dioscorea alata, Luffa graveolens, Moghania vestita, Momordica dioica, M. cochinchinensis, M. macrophylla, M. subangulata, Solanum indicum, Trichosanthes cucumerina, T. dioica, T. khasiana, T. ovata, T. truncata*
Gangetic plains: *Abelmoschus tuberculatus, A. manihot (tetraphyllus forms), Luffa echinata, Momordica cymbalaria, M. dioica, M. cochinchinesis, Solanum incanum, S. indicum*
Indus plains: *Momordica balsamina, Citrullus colocynthis, Cucumis prophetarum*
Western peninsular region: *Abelmoschus angulosus, A. moschatus, A. manihot (pungens forms), A. ficulneus, Amorphophallus campanulatus, Cucumis setosus, C. trigonus, Luffa graveolens, Momordica cochinchinensis, M. subangulata, Solanum indicum, Trichosanthes anamalaiensis, T. bracteata, T. cuspidata, T. perottitiana, T. villosa*
Eastern peninsular region: *Amorphophallus campanulatus, Abelmoschus manihot, A. moschatus, Colocasia antiquorum, Cucumis hystrix, C. setosus, Luffa acutangula var. amara, L. graveolens, Momordica cymbalaria, M. denticulata, M. dioica, M. cochinchinesis, M. subangulata, Solanum indicum, S. melongena* (insanum types), *Trichosanthes bracteata, T. cordata, T. lepiniana, T. himalensis*

of supply and non-viable indigenous market when compared to major vegetables. The domestic potential of these species has to be identified for their selection as new crops suitable for a region (Vielmeyer 1990; Smartt and Haq 1997).

3.1.3 Wild Relatives of Vegetable Crops

Wild relatives of crop plants (WRCP) are important for their inherent resistance to one or more biotic and abiotic stresses. The wild relatives of vegetable crops belong to the categories legumes (31 spp.), vegetable (54 spp.) and spices and condiments (27 spp.) (Arora and Nayar 1984; Arora 2000). Concerted efforts have been made in collection and conservation of such genetic resources from the Indian region through mission mode approach (Pandey et al. 2005; Arora and Pandey 2008; Pandey et al. 2008). Important wild relatives and related types of vegetable crops in different phytogeographical zones of India are mainly distributed in the western and eastern Himalaya, north-eastern region, Gangetic plains, Indus plains and in the western and eastern Peninsular regions (Table 4).

Botanically the WRCP are classified into five major families (Cucurbitaceae, Malvaceae, Fabaceae/Leguminosae, Zingiberaceae and Polygonaceae) and eight minor families (Alliaceae, Apiaceae, Asteraceae, Amaranthaceae, Brassicaceae, Chenopodiaceae, Araceae and Dioscoreaceae).

4 Potential Domesticates

About 1500 wild species have been originally involved in the process of domestication of cultivated vegetable crops (Grubben 1977). Many minor vegetables are grown in tribal areas under traditional/subsistence agriculture and kitchen gardens/backyard cultigens. Some common examples are amaranths, chenopods, portulaca and Indian poke (*Phytolacca acinosa*). These vegetables occur as semi-domesticated, semi-wild/protected/gathered wild types. Tree bean (*Parkia roxburghii*) is one of the most common multipurpose tree species used for protein rich pods in Manipur and Mizoram. Velvet bean (*Mucuna pruriens*) is considered to be one of the most preferred legume vegetables used in the north eastern region (Arora and Pandey 1996; Hore 2001; Yadav et al. 2009).

Many wild species of *Momordica*, *M. subangulata* ssp. *renigera* (often confused with *M. cochinchinesis*) and *M. dioica* are semi-domesticated vegetables native of Assam-Mayanmar and eastern region. Fruit is less bitter than the bitter gourd (*M. charantia*) and leaves twigs are used as vegetable in tribal areas (Ram et al. 2002). Creeping cucumber (*Solena amplexicaulis*) is a cucurbitaceous vegetable under cultivation in the tribal pockets of Odisha otherwise found wild in disturbed habitat of tropical areas (Peter 1998, 2007, 2008). An Asian species *Colocasia gigantea* occurs wild through India is harvested for the leaves from wild plants has good domestication potential; in south east Asia it is cultivated in Hawaii for export of petioles/leaves to United States of America. It is primarily used for leaves but tubers are inedible. Many indigenous vegetable crops are predominantly available in the tribal tracts of north eastern hill region and relished by the tribal population.

Drumstick or horse radish tree (*Moringa oleifera*) locally cultivated as swanjana tree is yet another perennial vegetable tree growing wild in the lower altitude zones of Himalaya. Curry leaf tree (*Murraya koenigii*) occurs wild in tarai tracts of Uttarakhand foothills but widely cultivated all over India as backyard tree where good aroma types are especially sold in the southern region. Some less known types as basella, chekkumanis, alternanthera, etc are regionally important. Among the leafy vegetables amaranth occupies a place due to fast plant growth, tolerance to soil borne diseases, suitability to any crop rotation practice favourable response to fertilizers and low production cost, affordability by poor people (Hore 2004; Arora et al. 2006). Detailed account in this category covering wild consumable vegetable species from India along with the areas of distribution, nutritive value and potential for their domestication has been updated routinely in different publications from this region (Arora and Pandey 1986; Barrau 1989; Rao and Dora 2002; Hore 2004; Arora et al. 2006).

5 Efforts in VGR: A Glance

At the national level concerted efforts have been made in collection, characterization and conservation of vegetable genetic resources under the aegis of Indian Council of Agricultural Research (Nath et al. 1994; Rana et al. 1995; Ghosh and

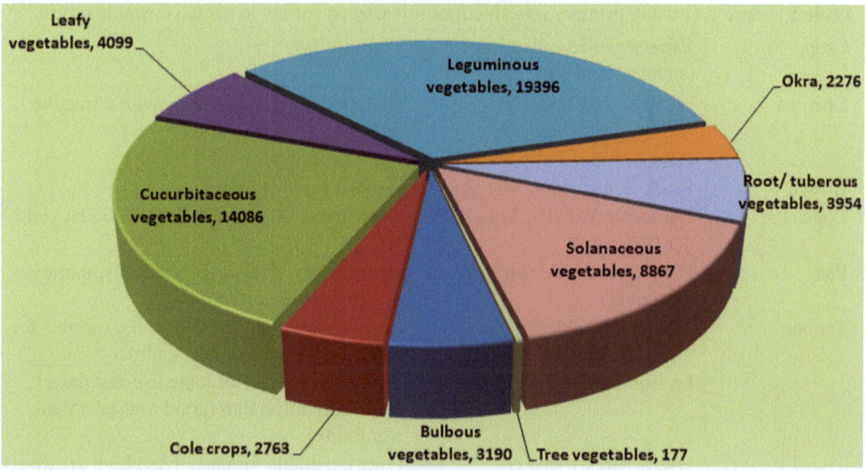

Fig. 3 Germplasm accessions collected in cultivated vegetable crop-groups

Kalloo 2000; Singh and Pandey 2004; Dhillon et al. 2005). Comprehensive augmentation of VGR has been made by National Bureau of Plant Genetic Resources (NBPGR), New Delhi through explorations in the sub-continent and sizeable number of introductions from abroad. During 1976–2013, a total of 50,863 accessions of various vegetable genetic resources were assembled from different phyto-geographical regions of the country (Fig. 2). Cucurbits, solanaceous vegetables and root/tuberous/bulbous vegetables were predominantly represented among the cultivated diversity (Fig. 3).

Introduced crop cultivars with specific traits through primary and secondary introductions have played a pivotal role in upgrading and developing indigenous diversity to a great extent (Singh et al. 2006; Table 5). During 1976–2012, introduction of these genetic resources was facilitated through global sources and with the help of several International Agricultural Research Centres (IARCs). Some of the important introductions of traits specific germplasm of vegetables crop and wild relatives were made in cowpea, tomato, onion, cabbage, vegetable brassicae, etc.

Genetic resources of vegetable crops such as solanaceous (eggplant, tomato, chilli), leguminous (cowpea, pea, beans), cucurbitaceous (pumpkin, gourds, melons, cucumber etc.), leafy (spinach, Chinese cabbage, fenugreek, etc.), bulbous crop (onion), root crops (radish, carrot, turnip) and okra are maintained as base collection/working collections in the National Gene Bank (NGB) at NBPGR, New Delhi. A total of 25,232 accessions of different vegetable crops have been kept in long-term storage (−18 °C); working collections are maintained in medium-term storage (+40 °C). Besides, over 700 accessions of vegetatively propagated allium*s* and tuberous types were conserved in-vitro through tissue culture and in field genebanks at regional stations.

Table 5 Some promising primary introductions released as variety in some vegetable crops

Crop	Variety/exotic collection (source country)	Character(s)
Cowpea	EC5000 (exRhodesia)	High pod yield, bushy type with attractive light green medium pods
	Pusa Barsati (Philippines)	Light green pods
Brinjal	Black Beauty (exUSA)	Darkest purplish fruit
Capsicum	California Wonder, Yolo Wonder (exUSA)	Deep green fruits, thick flesh, 170q/ha yield
Pea	Harbhajan (EC33866, ex Portugal)	Dwarf, early, dual purpose variety, maturing in 110 days in northern India
Tomato	Sioux (exUSA)	Early variety with smooth fruits, suitable for cultivation in winter and summer
	La Bonita (exUSA)	Dwarf, dual type (for table use and paste), fruits with thick skin (good transportation, storability)
	Dwarf Money Maker (EC108759, ex Israel)	Dwarf, suitable for paste type, high yielding, fruits deep red
	Molakai (exAustralia)	Prolific fruit bearer, good table variety, fruit large in size
	Fire Ball (exCanada)	Early maturing, suitable for high altitude areas
French bean	Contender (exUSA)	Dwarf, bushy, early type with green round pods
	Kentucky Wonder (exUSA)	Viny habit, round podded, stringless, seeds light brown
	EC108101 (exUSA)	Bushy, Vegetable type, round pods
	Giant Stringless (exUSA)	Stringless, resistant to mosaic and powdery mildew
Meetha Karela	EC36910 (exGermany)	High yielding, fruit green, almost smooth, narrow basal part
Cauliflower	Early Snow Ball	Early variety, curd white
	Improved Japanese (exIsrael)	Curd white, compact, maturity late-November—mid December
	Snow Ball (EC12013, exHolland)	Medium duration variety
Cabbage	Golden Acre (exDenmark)	Early variety, with compact round white head
	Drumhead (exJapan)	Late variety, with flat compact head
	Express (exDenmark)	Medium type variety, very popular in Himachal Pradesh
	September (exGermany)	Compact head, better shelf life
Carrot	Nantes (ex France)	Temperate type, cylindrical roots, sweet
	Zeno (ex Germany)	Suited to cultivation in Nilgiris
Water melon	Asahi Yamato (exJapan)	Fruit with deep pink flesh, mid-season type
	Sugar Baby (ex USA)	Fruits round, fine textured, attractive dull-green skin; flesh uniform deep red, very sweet, 10–12 % TSS
	Improved Shipper (exUSA)	High TSS
Cucumber	Poinsette (exUSA)	Resistance to anthracnose, mosaic
Onion	Pusa Ratnar, Early Grano (exUSA)	Large deep red bulbs; large yellow coloured bulbs with mild pungency
Radish	China Red (ex China)	Root red, early type (40–45 days)

Systematic work on evaluation of germplasm and improvement of vegetables in a networking mode through All India Coordinated Vegetable Improvement Project, Indian Agricultural Research Institute, Project Directorate, Vegetable Research, Varanasi, Indian Institute of Vegetable Research, National Bureau of Plant Genetic Resources, Indian Institute of Horticultural Research, and various State Agricultural Universities dealing with the VGR improvement programmes has helped in identifying and utilization of potential germplasm with promising traits (Mishra et al. 2006). Characterization and preliminary evaluation was carried out mainly in New Delhi and the regional stations of NBPGR. Large number of accessions were evaluated for yield, fruit setting, quality and other traits and promising donors were identified in vegetable crops. Concerted efforts on crop improvement in solanaceous crops have resulted in release of varieties and hybrids in tomato, brinjal and chilli (Kalloo et al. 2005). In India, tomato is one of the best examples of utilization of wild species in crop improvement. Tomato variety H24 tolerant to tobacco leaf curl virus was developed using *L. hirsutum* f. *glaberatum* as a donor. Characterization and evaluation of germplasm resulted in identification of donors and genetic stocks have been registered at NBPGR.

Direct selection of germplasm, or recombination and through mutation breeding resulted in release of varieties of cucurbits. In okra promising germplasm accessions for high number of fruits, branches with attractive smooth green fruits, resistance to biotic and abiotic stresses were identified and utilization of these donors has resulted in release of large number of varieties (Sirohi et al. 2005; Dhankhar et al. 2005). In water melon, musk melon, cucumber, pumpkin, bottle gourd, sponge and ridged gourd, bitter gourd, pointed gourd, snake gourd, ash gourd selections from the local germplasm were used to develop improved cultivars with promising traits (Sirohi et al. 2005). Multi-location evaluation has been done in some selected crops; okra and brinjal were evaluated for important agronomic traits, biotic and abiotic stresses and quality parameters in collaboration with Project Directorates/Project Coordinators and state agricultural universities and promising accessions have been identified and documented. In preliminary evaluation under field conditions a total of 2376 accessions of various crops were screened namely okra (192 accessions) against YVMV, tomato (582 accessions) and brinjal (200 accessions) against fruit borers. Some of the underutilized vegetable crops such as kankoda (*Momordica dioica*), winged bean (*Psophocarpus tetragonolobus*), and faba bean (*Vicia faba*) were evaluated for promising traits under the All India Coordinated Research Network on Underutilized Crops (AICRP on UC), New Delhi.

Based on characterization and preliminary evaluation data, germplasm accessions were catalogued to facilitate breeders in selecting the material of their interest. More than two dozen of crop catalogues have been published on various vegetable crops like tomato, okra, chilli, winged bean, cowpea, etc. Bureau maintains strong linkages with several international and national institutes/state agricultural universities in respect to various programmes for vegetable crops. To meet with the need of researchers and other users seeds of promising lines were supplied on request basis.

6 Factors Responsible for Richness of Diversity

Intensity of cultivation of different vegetable crops in an area is influenced by consumers preference (taste, habit), climate suitable for cultivation, physiological adaptation, marketability and the socio-economic factors. Leafy vegetables, spinach/*Spinacea oleracea*, *Chenopodium album* and others are more popular in northern plains, whereas *Brassica oleracea* var. *rugosa* (elephant ear brassica, hill brassica), purselane/*Portulaca oleracea* are favourite among diets of the local communities in Himalayan region. In southern and coastal areas there are large number of other leafy types- *Amaranthus*, Malabar spinach/*Basella rubra*, *Ipomoea aquatica*, in the western plains of Bihar, Odisha and Bengal. Wider adaptability in many of these crops has extended their distribution to diverse habitats. All this has influenced the pattern of distribution of a species.

Total vegetable crop diversity in India represents a major share of the introduced crops. The introduced crops have developed a lot of variability in traits of use in the region of introduction mainly based on adaptability and climatic factors. Species native to other regions of the world are now widely cultivated in India and have great commercial importance in the Indian region. The cauliflower, introduced by European colonists from temperate regions of Western Europe, has been acclimatized to warm and humid conditions of north India, such that it has been transformed according to tropical and sub-tropical conditions. The temperate long-day onion with perennial habit has been adapted to as annual sub-tropical and near tropical short-day conditions of Maharashtra and Karnataka respectively. Likewise temperate garden pea is cultivated in the warmer and humid conditions of India after diversification. In chilli, tomato and potato wide range of diverse forms have been developed as compared to the types that were introduced in historical times.

Direct or indirect plant introductions were made in the past for food, commercial or research purposes. The many crops originated in Africa had reached India around 3000 years ago (Purseglove 1968). Crop introduction expanded the range cultivation of vegetable crops such as bottle gourd (*Lagenaria siceraria*) and sweet potato (*Ipomoea batatas*) to the Old as well as the New Worlds before 1492. After the discovery of the New World by Columbus, there was a rapid exchange of crops between the New and Old Worlds.

7 Genetic Erosion in VGR

The diversity under cultivation is also considered as an important genetic resource itself, and thus need to maintained, enhanced and conserved for posterity (Crisp and Astley 1985). The spread of high yielding varieties leading to cultivar replacement is reported to be one of the main causes of genetic erosion. The risk of genetic erosion due to introduction of newer crops to a region is difficult to predict. Data on genetic erosion cannot be measured quantitatively; they can only be visualized through a trend of variety turnover and displacement data in the recent past (Maggioni 2004). Large-scale cultivation of improved cultivars, particularly in widely grown crops, has resulted in genetic vulnerability to landraces rich in traits relating to biotic or abiotic stress. In okra, cultivar Pusa Sawani released in 1959–1960, swept off the native varieties/local landraces out of cultivation in India (Seshadri 1987). Fast genetic erosion was reported of valuable variability in heat tolerant cauliflower that was selected/developed in India after introduction of the temperate types. This may happen in other vegetable crops like garden pea, water melon, etc.

Leading causes of genetic erosion in vegetable genetic resources are due to the following:

- *Loss of genetic diversity*: due to replacement of varieties landraces/primitive cultivars with high yielding varieties, and introduction of newer crops; over-exploitation of species; pressures on farmers to take up monoculture of an improved cultivar.
- *Loss of habitat*: loss of vegetable growing areas due to natural calamities (cyclones, flood, and drought) or man made changes (land clearing, constructions of buildings, dams, overgrazing and changing agricultural systems/land degradation, etc.); shrinking areas under vegetable cultivation.
- *Economic causes*: promotion of crops with high commerce value (potato, soybean, new cole crops—lettuce, broccoli) thus decreasing areas under vegetable cultivation under traditional/native vegetables.
- *Low input conditions*: due to poor rainfall/irrigation facility.
- *Cost of cultivation*: high cost of sustenance and increased cost of production.

8 Prioritization of VGR

On a global level, efforts of the International Board for Plant Genetic Resources (also IPGRI, now Bioversity International) to conserve vegetable crop germplasm began in 1980, with prioritization of crops for conservation on the basis of their importance for rural development and their economic value for farmers in the tropics (Sloten 1980). The list included okra and related species, onion and related species, *Amaranthus* spp., *Brassica* spp., *Capsicum* spp., *Cucurbita* spp., tomato, bitter gourd and related species and brinjal. Thrust on identification of prioritized genetic diversity in vegetables and its collection led to major enhancement in conservation activity (Plucknett 1983).

Landraces are more sensitive and prone to genetic erosion than those of the wild relatives of vegetables and are likely to be replaced faster. Native crops like brinjal, pointed gourd, Indian round gourd, bitter gourd, ridged gourd and nor ridge and smooth gourd, snake gourd, cucumbers and leafy amaranths and leafy brassicae have long history of cultivation in India and are at the risk of erosion of the primitive types/landraces. Areas of cultivation of these VGR need to be identified for collection of these germplasm with useful traits.

The national priority for collection and conservation of targeted vegetable crops/crop-group(s) are set as per guidelines of the executing institute or VGR management programme. This is primarily based on assessment of: (i) loss of genetic diversity, (ii) economic importance, (iii) nutritional value, and (iv) level of research. In priority list *Allium* spp. can be rated highest for general economic importance, level of research and level of genetic erosion, despite having low nutritional value. Similarly solanaceous vegetables—tomato, chilli and brinjal and their related species rank moderate for collection as these species were among the top priority for collection in the past three decades. Based on research thrust and level of genetic erosion, only trait specific germplasm and wild relatives need to be given more thrust. Vanishing landraces or primitive types in some of the native crops—brinjal, pointed gourd and others such as okra, cauliflower, etc. need to be collected. However, need based modifications in the priority list may be done if desired so.

In general native diversity in vegetables such as cucumber and other *Cucumis* species, pointed gourd, bitter gourd and their related species, yam, taro, and others like chilli, musk melon, lablab bean are to be considered for collection and conservation as compared to less popular or regionally important vegetable crops like drumstick, velvet bean, winged bean, cho-cho, tree tomato and newly introduced cole crops. Vegetable crops like *Amaranth*, leafy brassicae, although higher in nutritional value and economic importance but have been designated at lower national priority vegetables at the level of genetic erosion and research. In general wild relatives fall under the top national priority due to meagre representation in the genebank holdings, high research priority as well as threat of genetic erosion due to habitat loss.

Areas need to be identified for native species under cultivation along with their wild relatives growing in vicinity; the germplasm developed through introgression of this may have intermediate forms (easy to utilize) resulting from natural hybridization. If desired special missions need to be executed to explore, collect and conserve the germplasm having desirable traits. Difficult/inaccessible areas, tribal dominating tracts particularly in eastern and north-eastern region are rich in landraces of vegetable crops and thus figure among the top target regions for collection of unique or rare types.

9 Future Thrust

Indian gene center exhibits tremendous diversity in the vegetable crops, landraces, primitive cultivars and their wild relatives. There are outstanding examples of direct use of introduced germplasm like tomato and chilli, particularly sweet pepper (Singh et al. 2006). Whereas others have been utilized in crop improvement and used from cultivated and wild native diversity. Keeping in view the above mentioned points on vegetable genetic resource perspective for the Indian region, following thrust areas have been identified:

- Working out gap areas for germplasm collection and conservation of vegetable and wild relatives: difficult terrains, under-explored/inaccessible/tribal diversity rich areas (ethnically and culturally)/diversity-rich pockets; disturbed habitats/ insurgency areas through special missions.
- Collecting and conservation of wild relatives, endangered, endemic and potential species including less-known vegetables of nutraceutical importance in collaboration with concerned crop-based institutes
- Collecting and conservation of trait-specific germplasm mainly for biotic and quality parameters; working out characterization and evaluation and bio-risk management to identify these traits in germplasm of vegetables collected earlier;
- Strengthening linkages through network approach for management of vegetable genetic resources

During germplasm collection, the traits preferred in the foreign markets should be given due attention. Tremendous genetic variability present in vegetable crops needs to be utilized with greater vigour for the development of native cultivars. Future prospects for collection of germplasm are through explorations (a) within the country (for indigenous germplasm), their documentation, evaluation, conservation and use, and (b) to other countries (especially to Africa, Latin America, and South East Asia) designated primary centres of origin and diversity of many vegetable crops that are presently grown in India.

Conclusions

There has been national and international concern for systematic collection, conservation, characterization and documentation of the potential wealth of a region on priority basis. During CBD regimes when introduction of the trait specific germplasm has become difficult, there is an urgent need to identify indigenous vegetable genetic resources with desirable traits using native germplasm/diversity. Existing genetic diversity in VGR at their primary center of origin/domestication and secondary centres of diversity are important and therefore, need to be collected, conserved, characterized/evaluated and documented. Inventorization of new plant species identified from tribal areas, need special attention especially for their cultivation/popularization. Landrace diversity, local types, underutilized vegetables, less-known types, wild relatives and wild economic species about which there is scanty information need to be documented at top most priority. This is achievable by the dedicated efforts of the team working in VGR programme with support of the stakeholders.

Acknowledgements The author expresses sincere thanks to the Director, NBPGR for valuable guidance and support for undertaking this work. Acknowledgements are also due to Dr DC Bhandari, former Head, Plant Exploration and Germplasm Collection, NBPGR and my colleagues Dr ER Nayar, Dr. K Pradheep Ms Rita Gupta, Sh OP Dhariwal and Sh Shasikant Sharma for providing help in all respect.

References

Ambasta SP, Ramachandran K, Kashyap K, Chand R (1986) The useful plants of India. Publication and Information Directorate, Council for Scientific and Industrial Research, New Delhi

APAARI (2006) Fifteen years of APAARI: a retrospective. Asia-Pacific Association of Agricultural Research Institutions, Bangkok

APAARI (2007) Expert consultation to review progress of agricultural research networks and consortia in Asia-Pacific, Hyderabad, India. In: Proceedings Asia-Pacific Association of Agricultural Research Institutions, Bangkok, Thailand, 8–9 Oct

Arora RK (1985) Genetic resources of less known cultivated food plants. Science monograph no 9. National Bureau of Plant Genetic Resources, New Delhi

Arora RK (1991) Plant diversity in the Indian Gene Centre. In: Paroda RS, Arora RK (eds) Plant genetic resources: conservation and management. International Board for Plant Genetic Resources, Regional Office, New Delhi, pp 25–54

Arora RK (1995) Ethnobotanical studies in plant genetic resources—national efforts and concern. Ethnobotany 7:125–136

Arora RK (2000) Wild relatives of cultivated plants. In: Flora of India. Introductory volume, part II. Botanical Survey of India, Calcutta, pp 218–234

Arora RK (2003) Genetic resources of underutilized vegetables and their regions of diversity. Veg Sci 30(1):1–9

Arora RK, Nayar ER (1984) Wild relatives of crop plants in India. Science monograph no 7. National Bureau of Plant Genetic Resources, New Delhi

Arora RK, Pandey A (1996) Wild edible plants of India. National Bureau of Plant Genetic Resources, New Delhi

Arora RK, Pandey A (2008) Conservation of crop wild relatives in India: country reports. The Newsletter of the Crop Wild Relative Specialist Group, Bioversity International Sub-regional in Crop Wild Relative, Issue 6, Jan 2008. http://www.cwrsg.org/publications/newsletters/crop%20wild%20relative%20Issue%206.pdf

Arora RK, Sharma GD, Joshi V, Phogat BS, Bhatt KC, Rana JC (2006). Underutilized crops. In: Dhillon BS, Saxena S, Agrawal A, Tyagi RS (eds) Plant genetic resources: food grain crops. Narosa Publishing House, New Delhi, pp 320–342

Barrau JF (1989) The possible contribution of ethnobotany to the development of new crops for food and industry. In: Wickens GE, Haq N, Das P (eds) New crops for food and industry. Chapman and Hall, London, pp 402–410

Bhandari DC, Pradheep K (2010, January 5–16) Trait-specific germplasm collection: what ecology can offer? In: Training manual on molecular markers for efficient management and enhancing utilization of plant genetic resources. National Bureau of Plant Genetic Resources, New Delhi, pp 35–41

Chadha ML, Kuo G, Gowda CLL (eds) (2007) Proceedings of 1st international conference on indigenous vegetables and legumes: prospects for fighting poverty, hunger and malnutrition, Hyderabad, 12–15 Dec 2006. Acta Horticult 752:622

Chakravarty HL (1982) Cucurbitaceae. Fascicle 11. Botanical Survey of India, Calcutta

Crisp P, Astley D (1985) Genetic resources in vegetables. In: Russel GE (ed) Progress in plant breeding-1. Butterworths, London, pp 281–310

Cross RJ (1998) Review paper: global genetic resources of vegetables. Plant Var Seeds 11:39–60

Dhankhar BS, Mishra JP, Bisht IS (2005) Okra. In: Dhillon BS, Tyagi RK, Saxena S, Randhawa GJ (eds) Plant genetic resources: horticultural crops. Narosa Publishing House, New Delhi, pp 59–74

Dhillon BS, Varaprasad KS, Srinivasan Kalyani, Singh Mahendra, Archak S, Srivastava U, Sharma GS (eds) (2001) National Bureau of Plant Genetic Resources: a compendium of achievements. National Bureau of Plant Genetic Resources, New Delhi

Dhillon BS, Tyagi RK, Saxena S, Randhawa GJ (eds) (2005) Plant genetic resources: horticultural crops. Narosa Publishing House, New Delhi

Engle LM, Altoveros NC (eds) (2000) Collection, conservation and utilization of indigenous vegetables. Proceedings of a workshop. Asian Vegetable Research and Development Centre, Shanhua, Tainan, Taiwan, 16–18 Aug 1999

FAO (2013) Food and Agriculture Organisation of the United Nations. http://faostat.fao.org/site/339/default.aspx. Accessed Aug 2014

Ghosh SP, Kalloo G (2000) Genetic resources of indigenous vegetables and their uses in south Asia. Technical Bulletin 4, Indian Institute of Vegetable Research, Varanasi, Uttar Pradesh

Gopalakrishnan TR (2007) Vegetable crops. Horticulture science series, vol 4. New India Publishing Agency, New Delhi

Gosh SP (1984) Horticulture in north east region. Associated Publishing Company, New Delhi

Grubben GJH (1977) Tropical vegetables and their resources. International Board on Plant Genetic Resources, Rome

Grubben GIH, Soetjipto P (eds) (1996) Plant resources of south east Asia, no 10, cereals. Backbuys Publications, Leiden

Hanelt P (ed) (2001) Mansfeld's encyclopedia of agricultural and horticultural crops. Institute of Plant Genetics and Crop Plant Research, Springer, Berlin

Hawkes JG (1983) The diversity of crop plants. Harvard University Press, Cambridge.

Hore DK (2001) North east India—a hot-spot for agrodiversity. Summer school on agriculture for hills and mountain ecosystem, pp 361–362

Hore DK (2004) Leafy vegetables of north east India. Indian J Plant Genet Resour 17(1):10–16

Kalloo G (1988) Vegetable breeding, vol III. CRC, Boca Raton

Kalloo G, Srivastava U, Major S, Sanjeet K (2005) Solanaceous vegetables. In: Dhillon BS, Tyagi RK, Saxena S, Randhawa GJ (eds) Plant genetic resources: horticultural crops. Narosa Publishing House, New Delhi, pp 19–33

Kays SJ (2011) Cultivated vegetables of the world: a multilingual onomasticon. Wageningen Academic, Chicago

Kotlinska T (1993) Vegetable crops genetic resources conservation and utilization integration of conservation strategies of plant genetic resources in Europe. In: Begemann F, Hammer L (eds) Proceedings of an international symposium on plant genetic resources in Europe, 6–8 Dec 1993, Gatersleben, Germany

Mal B, Paroda RS, Kochhar S (1997) Underutilized crops and their implications in farming systems in India. In: Smartt J, Haq N (eds) Domestication, production and utilization of new crops. International Centre for Underutilised Crops, Southampton, pp 30–45

Maggioni L (2004) Conservation and use of vegetable genetic resources: a European perspective. In: McCreight JD, Ryder EJ (eds) Proc. XXVI IHC-advances in vegetable breeding. Acta Horticult 637, ISHS 2004

Mishra SK, Singh M, Kumar A, Rana JC (2006) Germplasm characterization, evaluation and utilization. In: Singh AK, Srinivasan K, Saxena S, Dhillon BS (eds) Hundred years of plant genetic resources in plant genetic resource management in India. National Bureau of Plant Genetic Resources, New Delhi, pp 133–70

Nath, Sudari V, Singh DP (1994) Vegetables for tropical region. Indian Council of Agricultural Research, New Delhi

Nath, Badra T, Danton I (1987) Germplasm resources and genetic improvement of vegetable crops in the tropics. National Horticultural Research Institute, Abaidan

National Horticulture Database (2012) National Horticulture Board, Ministry of Agriculture, Govt. of India. http://iasri.res.in/agridata/13data/chapter4/db2013tb4_7.pdf

Nayar ER, Pandey A, Venkateswaran K, Gupta R, Dhillon BS (2003) Checklist of crop plants of India. National Bureau of Plant Genetic Resources, New Delhi

Pandey A, Bhandari DC, Bhatt KC, Pareek SK, Tomer AK, Dhillon BS (2005) Wild relatives of crop plants in India: collection and conservation. National Bureau of Plant Genetic Resources, New Delhi

Pandey A, Tomer AK, Bhandari DC, Pareek SK (2008) Towards collection of wild relatives of crop plants in India. Genet Resour Crop Evol 55(2):187–202. doi:10.1007/s10722–007-9227-4

Pareek SK, Bisht IS, Bhatt KC, Kumar A, Koppar MN, Gupta PN, Mittal SK, Singh R (2000) Manual on exploration and collection of plant genetic resources and related indigenous knowledge. Jai Vigyan National Science and Technology Mission on Conservation of Agro-biodiversity (Plant Genetic Resources). Agro-biodiversity (PGR) 5, National Bureau of Plant Genetic Resources, New Delhi

Paroda RS, Arora RK, Chandel KPS (eds) (1988) Plant genetic resources: Indian perspective. Proceedings, National Symposium on Plant Genetic resource, National Bureau of Plant Genetic Resources, New Delhi

Peter KV (1998) Genetics and breeding of vegetable crops. Indian Council of Agricultural Research, New Delhi

Peter KV (ed) (2007) Underutilized and under exploited horticultural crops, vol 1(2) New India Publishing Agency, New Delhi

Peter KV (ed) (2008) Underutilized and under exploited horticultural crops, vol 3(4) New India Publishing Agency, New Delhi

Plucknett DL (1983) Taxonomy of the genus *Colocasia*. In: Wang JK (ed) Taro—a review of *Colocasia esculenta* and its potentials. University of Hawaii, Honolulu, pp 14–14

Purseglove JW (1968) Tropical crops—dicotyledons, vol 1(2) Longmans Green, London

Purseglove JW (1975). Araceae. Tropical crops: monocotyledons, vol 1(2) Longmans Green, London, pp 58–66

Ram D, Kalloo G, Banerjee MK (2002) Popularizing kakrol and kartoli: the indigenous nutritious vegetables. Indian Horticult (Oct–Dec):6–9

Ramachandran K (1997) Compilation of wealth of India. Proc Indian Natl Sci Acad B63 (3):139–150

Rana RS, Gupta PN, Mathura Rai, Kochhar S (1995) Genetic resources of vegetable crops: management conservation and utilization. National Bureau of Plant Genetic Resources, New Delhi

Randhawa MS (ed) (1982) A history of agriculture in India. Eighth to eighteenth century, vol II. National Printing Works, New Delhi

Rao Srinivasa M, Dora DK (2002) Lesser known vegetables for nutritious, health and economic security—Indian context. International Conference on Vegetables, Bangalore, pp 610–17

Roy B, Hardev RC, Pal DC (1998) Plants for human consumption in India. Botanical Survey of India, Calcutta

Sastraprapradja S, Hambali GG, Prana TK (1984) Edible *Amorphophallus* and its related species. In: Chandra R (ed) Edible aroids. Clarendon Press, Oxford

Swai REA (1991) Genetic resources of horticultural crops in Tanzania. In: Shao FM, Magingo FSS, Minjas AN, Bitanyi H, Mahunnah R (eds) Proceedings of the first national workshop on plant genetic resources and biotechnology. Arusha, Tanzania, 16–20 January 1990, Benedictine Publications, Ndanda, pp 251–266

Seshadri VS (1987) Genetic resources and their utilization in vegetable crops. In: Paroda RS, Arora RK, Chandel KPS (eds) Genetic resources: Indian perspective. National Bureau of Plant Genetic Resources, New Delhi, pp 335–343

Siemonsma JS, Kasern P (eds) (1994) Plant resources of south east Asia, no 8. Vegetables. PROSEA, Bogor

Singh AK, Pandey C (2004) Vegetable germplasm diversity in Asia and its use in India. In: Proceedings of the first Indian horticulture congress, New Delhi, pp 62–81

Singh RV, Chand D, Tyagi V, Gupta S (2006) Exchange of crop germplasm: achievements and challenges. In: Singh AK, Srinivasan K, Saxena S, Dhillon BS (eds) Hundred years of plant genetic resources in plant genetic resource management in India. National Bureau of Plant Genetic Resources, New Delhi, pp 63–82

Singh B, Singh M, Rai AB (2011) Vision 2030. The Indian Institute of Vegetable Research (IIVR), Varanasi

Singh AK, Rana RS, Bhag Mal, Singh B, Agrawal RC (2013) Cultivated plants and their wild relatives in India—an inventory. Protection of Plant Variety and Farmers Right, New Delhi

Sirohi PS, Gunjeet Kumar, Munshi AD, Behera TK (2005) Cucurbits. In: Dhillon BS, Tyagi RK, Saxena S, Randhawa GJ (eds) Plant genetic resources: horticultural crops. Narosa Publishing House, New Delhi, pp 34–58

Sloten DH van (1980) Vegetable genetic resources. Plant Genet Res Newslet 44:20–25
Smartt J, Haq N (eds) (1997) Domestication, production and utilization of new crops. International Centre for Under-utilized Crops, Southampton, Colorine Printers, Dhaka
Tanaka Y (1976) Cyclopedia of edible plants of the world. Keigata, Tokyo
Uphof JCTh (1968) Dictionary of economic plants, 2nd Edn Cramer, Lehre
Vielmeyer ND (1990) The new crops era. In: Advances in new crops. In: Janick J, Simon J (eds) Proceeding of the first national symposium on new crops—research, development, economic. Tirober Press, Indianapolis, 22 Oct 1988
Watt G (1889–1896) A dictionary of economic products of India, vol 7, Suppl 2. Publications and Information Directorate, Council of Scientific and Information Directorate, New Delhi
Wiersema JH, Leon B (1999) World economic plants: a standard reference. CRC, New York
Yadav RK, Deka BC, Sanwal SK (2009) Genetic resources of vegetable crops of north eastern Himalayan region. ENVIS Bulletin: Himalayan Ecology, Envis centre, GBPIHED, Uttarakhand
Zeven AC, de Wet JMJ (1982) Dictionary of cultivated plants and their regions of diversity. Centre of Agricultural Publishing and Documentation, Wageningen

Plant Disease Prevention and Management in Sustainable Agricultural Systems

Koon-Hui Wang and Janice Uchida

1 Introduction

Plant disease management strategies have been dominated by the use of "silver bullet" products to control plant disease outbreaks. Unfortunately, most of these therapeutic strategies are effective only for a short time. Long-term plant disease preventative strategies that are based on the ecological understanding of pathogens have been explored and studied extensively, but its implementation is restricted by economic consideration for large scale farming. Crop losses due to arthropods, diseases, and weeds have increased on a world basis from 34.9% in 1965 to 42.1% in 1990 (Oerke et al. 1994; Lewis et al. 1997). Despite the intensification of developing modern technologies for pest control, today's crop loss caused by arthropods, diseases, and weeds remains at 20–40% (Oerke 2006). Worldwide, herbicide use ranked highest among all the pesticide usage in the agriculture sector, followed by insecticide/miticide and fungicide use (Grube et al. 2011). Based on recent pesticide usage statistics conducted by the United States Environmental Protection Agency, nematicide or fumigant usage is actually second to herbicide use (200 mil kg), with 49 mil kg being used in agriculture sectors in the U.S. (Grube et al. 2011), higher than insecticide (29.5 mil kg) and fungicide (20 mil kg) use. One positive sign of this statistic is that the amount of organophosphate insecticides used in the U.S. has declined more than 60% since 1990, from an estimated 38.6 million kg in 1990 to 15 million kg in 2007 (Grube et al. 2011). This reduction is somewhat encouraging as organophosphates are among the most acutely toxic pesticides still used.

K.-H. Wang (✉)
Department of Plant and Environmental Protection Sciences, College of Tropical Agriculture and Human Resources, University of Hawaii at Manoa, 3050 Maile Way, Gilmore 310, Honolulu, HI 96822, USA
e-mail: koonhui@hawaii.edu

J. Uchida
Department of Plant and Environmental Protection Sciences, College of Tropical Agriculture and Human Resources, University of Hawaii at Manoa, 3190 Maile Way, St. John 304, Honolulu, HI 96822, USA

D. Nandwani (ed.), *Sustainable Horticultural Systems*, Sustainable Development and Biodiversity 2, DOI 10.1007/978-3-319-06904-3_16

Much of these acute toxic pesticides are now slowly replaced by biologically based pesticides. At the end of 2001, there were approximately 195 biopesticide active ingredients and 780 products registered in the U.S. (i.e. 25 % are biopesticides). Today, there are 330 registered biopesticide active ingredients out of 1100 products registered in the U.S. (i.e. 30 % are biopesticides) (Meister and Sine 2012). However, this shift in pesticide use still did not avoid a common paradigm of those "therapeutic pesticides" used. Two of the main concerns of therapeutic pesticides are the buildup of pesticide resistant pest populations that might lead to the pesticide treadmill, and broad-spectrum effects of pesticides that will harm non-target beneficial organisms and humans.

A total systems approach for developing sustainable pest management strategies for plant disease management was thoroughly reviewed more than a decade ago (Cook et al. 1995; Cook and Baker 1983; Hornby 1990). Most of these literature reviews focused on managing bacteria, fungi and nematodes. To achieve sustainable pest management, we must go beyond replacing toxic chemicals with sophisticated, biologically based agents. This chapter focuses on sustainable pest management approaches for tropical plant diseases through environmentally friendly and renewable strategies. Particularly, we are focusing on managing nematodes, fungi, bacteria and insect vectored plant viruses. Based on ecological knowledge about targeted pests/pathogens, their associated natural enemies, and their interactions with hosts, we recommend integrating five sustainable approaches for plant disease management. These approaches are: (1) enhancing high biological diversity through polyculture instead of the conventional preference of monoculture cropping systems; (2) increasing ecosystem community stability by promoting natural enemies of multiple pests and pathogens; (3) stimulating inherent plant defenses; (4) improving plant health by maintaining nutrient cycling and energy flow; and (5) targeting vulnerable stages of pests or pathogens through the understanding of their ecology. The use of pesticides or even biocontrol agents as 'therapeutic' approaches are unsustainable and should be the last resort in developing a sustainable plant disease management program. While managing multiple pathogens in an agroecosystem involves multi-disciplinary studies and would be very comprehensive, we chose to start by reviewing sustainable pest management approaches recommended for plant-parasitic nematodes, followed by how these approaches can be extended to or complement plant fungal and viral disease management.

2 Sustainable Pest Management for Plant-Parasitic Nematodes

The life cycle of most plant-parasitic nematodes is rather typical, consisting of the egg, four juvenile stages, and the adults, usually parasitizing plant roots. One might suggest that root infecting plant-parasitic nematodes can be managed by the same approach across genera. In reality, management strategies for plant-parasitic nematodes could differ significantly based on the amount of time they spend inside the plant roots vs outside of roots (i.e. endoparasitic vs ectoparasitic), whether they are

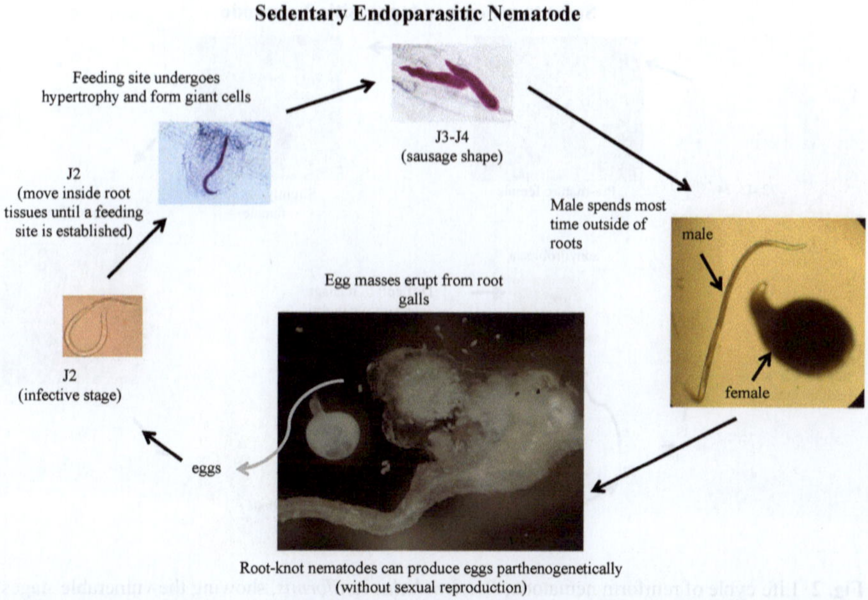

Fig. 1 Life cycle of root-knot nematode, *Meloidogyne javanica*, showing the vulnerable stage of the nematode being the vermiform second stage juvenile (J2). Other stages of the nematode are buried inside the root system, and are thus less prone to chemical exposure, parasitism or predation by its natural enemies. Eggs of root-knot nematodes though exposed outside of the root gall are protected by a gelatinous matrix, making them less prone to predation but can often be parasitized by nematode egg parasites. (Pictures by M. Davilla and K.-H. Wang)

sedentary or migratory after infecting the roots, if they produce eggs protected by egg masses or not, and if they have the capability to transform into an anhydrobiotic stage. Examples of some of the most economically damaging and commonly found plant-parasitic nematodes with distinct life cycle patterns are root-knot nematode (*Meloidogyne spp.*) (Fig. 1), reniform nematode (*Rotylenchulus reniformis*) (Fig. 2), lesion (*Pratylenchus* spp.), and burrowing nematode (*Radopholus similis*) (Fig. 3). Knowing the nematode distribution in the plants and their survival strategies help to determine the best management approaches.

A thorough review on managing plant-parasitic nematodes in sustainable and subsistence agriculture in the tropics and subtropics has been published by Bridge (1996). In that review, Bridge described how to prevent the introduction and spread of nematodes by the use of nematode-free planting materials through heat treatment, production of seedlings in nematode-free seedbeds, surface soil burning of plant debris, soil solarization, adjust planting time, flooding, postharvest removal of infected crop residues, and soil cultivation between crops. Bridge (1996) also provided some insight on encouraging naturally occurring nematode antagonistic organisms through the use of soil amendments and designing multiple cropping and multiple cultivars in tropical and subtropical climates to increase plant resistance or

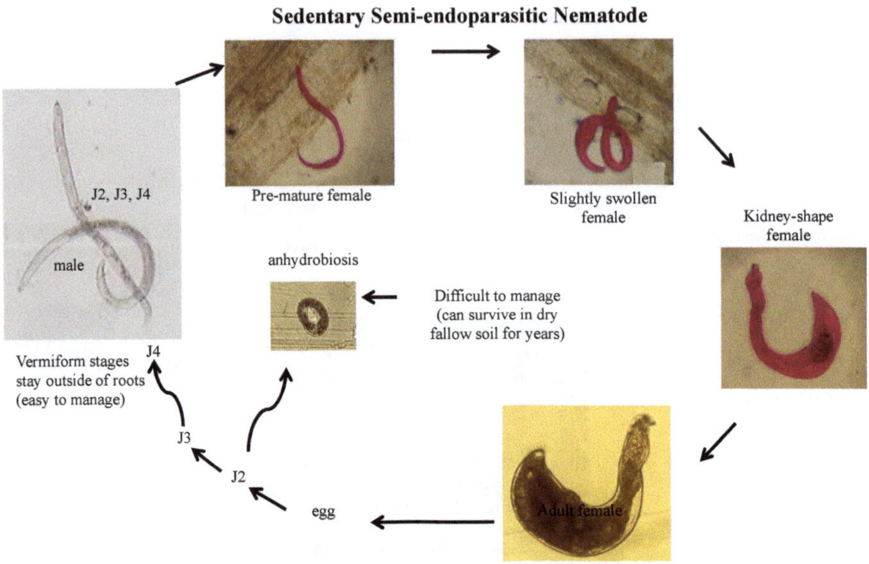

Fig. 2 Life cycle of reniform nematode, *Rotylenchulus reniformis*, showing the vulnerable stages of the nematode: vermiform juvenile stages (J2, J3, and J4) as well as the male. Females of the nematode partially penetrate the roots. Although most life stages of reniform nematodes are prone to chemical exposure or parasitism and predation by its natural enemies in the soil, this nematode can survive in harsh environments without the presence of a host plant in a form of anhydrobiosis. During anhydrobiosis, the nematodes curl up their bodies and remain dormant. (Pictures by K.-H. Wang)

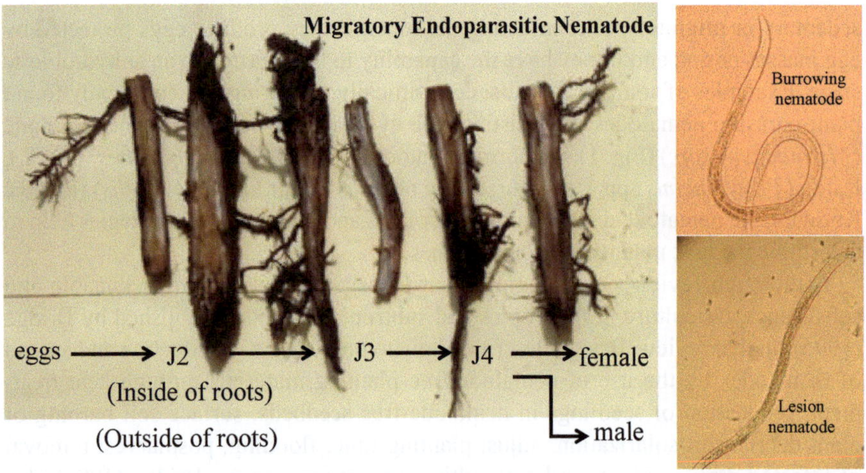

Fig. 3 Life stages of burrowing (*Radopholus similis*) and lesion nematodes (*Pratylenchus* spp.). These migratory endoparasitic nematodes burrow into root tissues soon after hatching, spend most of their life cycle inside the roots of the host plant. Only male nematodes migrate out of the roots when mature. Females lay eggs inside of roots. (Pictures by K.-H. Wang)

tolerance to nematodes. The current chapter will focus on four sustainable farming approaches for nematode management.

2.1 Increasing Host Plant Diversity in a Cropping System

Many plant-parasitic nematodes commonly found in tropical climates tend to have a wide host range, making crop rotation practices rather challenging to design. In addition, host plant resistance against nematode pests for minor crop production in the tropics is also rare. Thus, planting cover crops with allelopathic compounds against nematode pests offer a good alternative strategy. Integrating cover crops into conventional cropping systems for managing plant-parasitic nematodes has been extensively researched and has shown great potential for crops commonly grown. For example, rye (*Secale cereal*), wheat, sorghum (*Sorghum bicolor*), hairy vetch (*Vicia villosa*), sunn hemp (*Crotalaria juncea*), cowpea (*Vigna unguiculata*), and marigold (*Tagetes* spp.) effectively suppressed root-knot nematodes (*Meloidogyne* spp.), one of the most important plant-parasitic nematodes of vegetable and field crops in tropical and subtropical climates (McSorley et al. 1994; Ploeg and Maris 1999). In general there are four possible mechanisms that may play a role in nematode suppression by cover crops: (1) non or poor host effect; (2) nematicidal or nematostatic effect in which the cover crop produces volatile and nonvolatile toxic compounds; (3) enhancement of nematode antagonists (e.g., parasites, predators); and (4) "dead end" trap crop effect, in which cover crop roots are penetrated by nematodes but the nematodes are not capable of reproducing (Gardner and Caswell-Chen 1994; Ploeg 2000; Ploeg and Maris 1999; Wang et al. 2001).

A list of the most well-known cover crops with allelopathic properties against specific plant-parasitic nematodes is summarized in Table 1. Some of these cover crops suppressed multiple nematode pests. For example, sunn hemp was found to significantly reduce root-knot (*Meloidogyne* spp.) (McSorley et al. 1994), reniform (*Rotylenchulus reniformis*) (Charchar and Huang 1981; Wang et al. 2001), lance (*Hoplolaimus* spp.) (Charles 1995), and burrowing (*Radopholus similis*) (Birchfield and Bristline 1956) nematodes. Sunn hemp was found to suppress plant-parasitic nematodes when incorporated into soil. Similarly, marigold suppresses a wide range of plant-parasitic nematodes (up to 14 genera) (Suatmadji 1969). However, the nematicidal effect of marigold varies according to marigold and nematode species, cultivar, and soil temperature (Ploeg and Maris 1999). *Tagetes patula* 'Single Gold' consistently suppressed a diverse range of plant-parasitic nematodes. Only living marigold root systems exhibit nematicidal properties. Incorporation of 'Single Gold' residues into the soil would not suppress root-knot nematode (Ploeg 2000; Suatmadji 1969; Jagdale et al. 1999). The critical stage for the marigold suppressive effect to occur is during its growth. Thus, to maximize its nematicidal effect, marigolds should not be terminated until late after establishment. Sipes and Arakaki (1997) found that *T. patula* was the most effective cover crop for improving

Table 1 Cover crops with allelopathic compounds against plant-parasitic nematodes in tropical and subtropical regions

Cover crop	Common name	Effect on nematode pests
Brassicacea	Rapeseed Mustard Oil radish	When incorporated into soil, their residues contain glucosinolates that will break down into isothiocyanates and nitriles that suppress nematodes. These cover crops were known to suppress stubby root (*Paratrichodorus* sp.), lesion (*Pratylenchus* sp.), and root-knot (*M. incognita*) nematodes (Brown and Morra 1997)
Sorghum bicolor × *Sorghum arundinaceum* var. *sudanense*	Sorghum × Sudan grass	When incorporated into soil, releases dhurrin that degrades into hydrogen cyanide, which is nematicidal (Widmer and Abawi 2000)
Sesamum indicum	Sesame seeds	As a rotation crop with cotton, peanut, and soybean, it suppressed peanut root-knot (*M. arenaria*) and southern root-knot (*M. incognita*) nematodes but not Javanese root-knot (*M. javanica*) (Starr and Black 1995). It is made into commercial products Dragonfire™ (oil), Ontrol™ (seed meal) (Poulenger, USA), and Nemastop™ (ground up sesame plant) (Natural Organic Products)[a]
Crotalaria juncea	Sunn hemp	Act as a poor host of root-knot, reniform, soybean cyst (*Heterodera glycine*) nematodes, etc. When incorporated into soil, it releases monocrotaline that is nematostatic (immobilizes the movement of nematodes) (Wang et al. 2001; Warnke et al. 2008)
Tagetes spp.	Marigold	There are 14 genera of plant-parasitic nematodes suppressed by marigold, among which root-knot and lesion nematodes are most consistently suppressed (Hooks et al. 2010). Gommers and Bakker (1988) suggested that marigold as a standing cover crop (not after incorporation), releases α-terthienyl (nematocidal compound) when the roots are penetrated by nematodes. Marigold roots also enhance activity of endophytic bacteria that might be responsible for nematode suppression (Sturz and Kimpinski 2004). Unfortunately, marigold is good host to many ectoparasitic nematodes include sting (*Belonolaimus*), stubby root (*Paratrichodorus*), and lance (*Hoplolaimus*) nematodes
T. erecta	African marigold ('Cracker Jack')	Suppressed lesion nematodes when in rotation with potato (Ball-Coelho et al. 2003). However, it is a good host to reniform nematodes (Wang et al. 2003)
T. patula	French marigold ('Single Gold')	Suppressed many root-knot nematode species (Ploeg 2002; Ploeg and Maris 1999)
T. minuta		More tolerant of warm summer temperatures in Florida than the more commonly used marigold species

Table 1 (continued)

Cover crop	Common name	Effect on nematode pests
Tagetes hybrid 'Polynema'		Suppressed many root-knot nematode species if soil temperature is below 30 °C (Ploeg and Maris 1999)
Mucuna pruriens var. *utilis*	Velvetbean	Velvetbean has been shown to suppress some weed species in tropical production systems. In addition, it also releases nematicidal compounds against many plant-parasitic nematodes including root-knot nematodes (Zasada et al. 2006)
Canavalia ensiformis	Horsebean	It contains lectins, such as concanavalin A, which may disrupt nematode behaviors in host-finding (Marban-Mendoza et al. 1989). Co-cultivation of *C. ensiformis* with tomato reduced root galling caused by *M. incognita* and *Nacobbus aberrans* (Marban-Mendoza et al. 1989)

[a] Mention of a trade product does not imply a recommendation by the University of Hawaii

taro (*Colocastia esculenta* L.) yields among the 22 cover crops tested in *M. javanica* infested fields in Hawaii.

On the other hand, brassica crops such as rapeseed (*Brassica napus*), oilseed and mustard are used as green manure to suppress root-knot and lesion nematodes. When green manure is incorporated into soil, glucosinolates in tissues of brassica break down into isothiocyanates and nitriles that suppress nematodes. Thus, rapeseed should be incorporated prior to cash crop planting to produce allelopathic compounds against plant-parasitic nematodes (Cardwell and Ingham 1996). In Bangladesh, populations of the rice root-knot nematode (*Meloidogyne graminicola*) can be reduced significantly on deep water rice by growing rice after oilseed crops, such as mustard and sesame (Rahman 1990).

Velvetbean (*Mucuna pruriens*) is another leguminous cover crop suitable for use as a green manure to reduce some important nematode species such as *Meloidogyne* spp. and *Heterodera glycines* (Weaver et al. 1998; Ritzinger and McSorley 1998). Two alcohols, which inhibited *M. incognita* hatching, were isolated from the velvetbean (Nogueira et al. 1996). Another leguminous plant known to be suppressive to plant-parasitic nematodes in the tropics is horsebean (*Canavalia ensiformis*). It contains lectins, such as concanavalin A, which may disrupt the capability of nematodes to locate a host (Marban-Mendoza et al. 1989). Co-cultivation of *C. ensiformis* with tomatoes reduced root galling caused by *M. incognita* and *Nacobbus aberrans* (Marban-Mendoza et al. 1989). For more information about other plant extracts capable of suppressing plant-parasitic nematodes, please refer to Oka (2010). Most recently, a new biopesticide registered by Monterey AgResources as a nematicide was extracted from the soap bark tree, *Quillaja saponaria*, and has been used for controlling plant-parasitic nematodes in vineyards, orchards, field crops, turf and ornamentals (Meister and Sine 2012).

Beside cover crops, another organism known to produce allelopathic compounds suppressive to plant-parasitic nematodes is the oyster mushroom (*Pleurotus ostreatus*). The mushroom exudes a toxin from the fungal hyphae, known as trans-2-decenedioic acid (Kwok et al. 1992). This toxin paralyzes the nematode on contact, which allows the hyphae to move into the nematode to colonize and digest the nematode. Studies on the effects of nematodes have been predominantly in vitro. Thorn and Barron (1984) screened 27 species of mushrooms for toxicity against nematodes in vitro and found that 5 species of *Pleurotus*, 5 species of *Hohenbuehelia*, and one species of *Resupinatus*, all of which were classified in the family of Pleurotaceae, were capable of destroying nematodes. Complete control of *M. incognita* by *P. osteatus* compost amended at 0.5 % in soybean pots has been reported in Nigeria (Okorie et al. 2011). More research is needed to investigate the potential of using mushroom compost waste for managing plant-parasitic nematodes in agriculture production systems.

However, performance of cover crops or mushroom compost to suppress plant-parasitic nematodes could vary based on cover crop or mushroom compost quality and quantity, cultural practices, history of the crop site, and time of planting. Several approaches can be used to integrate cover crops or mushroom compost into a cropping system: (1) incorporating mushroom compost or cover crops as a soil amendment or green manure prior to cash crop planting; (2) strip-till cover cropping system, where only the rows for planting the cash crop are tilled under and the remaining cover crop remains on the soil surface as a living or hay mulch; (3) no-till system where the cover crop is destroyed by chemical or physical means (e.g. flail mower, cover crop roller), prior to planting the cash crop with a no-till planter; or (4) as a dying mulch in which the cash crop is planted into a senescing cover crop. Understanding the mechanism of action on how a particular cover crop suppresses plant-parasitic nematodes is critical in deciding how the cover crop or compost should be integrated into a cropping system.

Based on a review of research using rotation crops and cover crops for root-knot nematode management in the Southern United States, McSorley (2011) concluded that the performance of rotation crops or cover crops was similar to clean fallow in most studies. This review suggested that rehabilitation of heavy nematode infested sites is difficult, and could require several years of crop rotation to achieve economic benefits. These results are discouraging for sustainable farming systems that rely heavily on crop rotation and cover cropping. Marahatta et al. (2012a) provided evidence that the effects of the allelopathic cover crops suppressing plant-parasitic nematodes are more efficient if targeting the vulnerable stage of the nematodes. For example, marigold suppressed *Meloidogyne* spp. if planted when these nematodes are in their active stage. This means that marigold would suppress *Meloidogyne* more effectively if planted right after a susceptible crop rather than planted in fields that had been dry fallow for some time. Similarly, Marahatta et al. (2012b) also documented that soil incorporation of *C. juncea* suppressed *R. reniformis* more efficiently in soil previously planted with a susceptible host (e.g. cowpea, *Vigna unguiculata*) than in soil being dry fallowed for 3 months. This is most likely due to the fact that dry fallow can trigger *R. reniformis* to transform into their survival or anhydrobiotic stage, and caused the eggs of *Meloidogyne* to remain unhatched. Future work should examine how to manipulate cropping systems to allow effective use of nematode antagonistic cover crops so as to target the vulnerable stage of the nematodes.

2.2 *Increasing Ecosystem Community Stability by Promoting Antagonists of Nematodes*

Nematode antagonists encompass diverse organisms that include natural enemies such as parasites and predators, but also organisms that produce antibiotics, extracellular enzymes, or induce host plant systemic resistances (Stirling 2011). Numerous organisms that are capable of reducing populations of plant-parasitic nematodes include egg parasitic fungi, endoparasitic fungi, fungal endophytes, parasitic bacteria, predatory nematodes, actinomycetes, and plant growth promoting rhizobacteria. A thorough review for each of these groups of nematode antagonists has been published in the book "Biological control of plant-parasitic nematodes: progress, problems and prospects" (Stirling 1991). Although nematode suppressive soils do occur naturally (Westphal and Becker 1999), it often takes time to build up without initial economic yield loss.

Stirling (2011) reminds us that "nematode biological control is a normal part of a properly functioning soil ecosystem, with plant-parasitic nematodes only becoming pests when they are no longer constrained by the biological buffering mechanisms that normally keep them in check". Common agricultural practices that rely on soil tillage, application of pesticides for various pest control, and synthetic fertilizer are generally unfavorable for the establishment of nematode suppressive soil. Conventional farming practices that attempt to introduce nematode antagonists of plant-parasitic nematodes back into human intervened intensive farming systems tend to conduct augmentative biological control where biocontrol agents are isolated, mass produced and introduced to a targeted area. Despite the high numbers of nematode antagonists of plant-parasitic nematodes that have been commonly isolated and identified (Stirling 2011), they are only four bacteria or fungi that were registered as biopesticides for nematode control by the United States Environmental Protection Agency as listed on the Crop Protection Handbook of 2012 (Meister and Sine 2012). Among which, *Myrothecium verrucaria* is prepared as fermented materials with killed mycelium, whereas *Paecilomyces lilacinus*, *Bacillus firmus,* and *Streptomyces saraciticus* are prepared as live biological control agents. *Pasteuria* is another nematode biological control agent that is commercialized, but with specificity against target species. Worldwide, there are more commercialized nematode biological control agents, but they are all challenged by the same dilemma, i.e. their inconsistent performance in the field.

Therefore, recently there has been more work focused on approaches that can be used to restore, maintain or enhance the natural nematode suppressive mechanisms that should operate in all agricultural soils (Timper 2014). These types of practices that enhance naturally occurring biocontrol agents through cultural practices are known as conservation biological control. A thorough review on the use of the conservation biological control approach for nematode management is recently published by Timper (2014). In summary, these approaches include: (1) providing a supplemental food source for nematode antagonists, (2) identifying the host plants that are compatible with the colonization or establishment of nematode antagonists, (3) integrating the host plants that are favorable for the expression of antibiotics from nematode antagonistic bacteria, and (4) reducing cultural practices that could disturb the establishment of nematode antagonists.

Stirling (2011) categorized the supplemental food source for nematode antagonists into two distinct groups: carbon (C) or nitrogen (N) dominated organic matter. Organic matter that are N dominated such as animal manures, oil-cakes, and residues from leguminous crops (Muller and Gooch 1982; Rodriguez-Kabana 1986; Stirling 1991) are thought to release nematicidal levels of ammonia during the decomposition process, killing nematodes in a short period of time. However, the effect is usually short lived, and requires frequent application if the nematode problem persists. Accumulation of nitrite and nitrate would be a great concern when relying on N dominated organic matter. Oka and Pivonia (2003) found that adding chitin or cottonseed amendments to N dominated soil organic matter could serve as a nitrification inhibitor by slowing the oxidation of ammonia to nitrite and nitrate and allowing ammonia concentrations to build up for an extended period, and thus extending the nematode suppressive period.

The second group of organic matter is dominated by C content. In contrary to N dominated compounds, nematode suppressive effects provided by wood chips, yard waste, or grassy crop residues (such as sugarcane residues) tend to build up slowly (McSorley and Gallaher 1996; Stirling et al. 1995). Often taking years, the mechanisms involved are usually biological based, and the effects often last much longer than the N dominated soil amendments (Stirling et al. 2005). The nematode-trapping fungi (NTF) and several genera of wood-decaying basidiomycetes are commonly found in habitats rich in cellulose and lignin and are thought to have evolved the capacity to scavenge for additional N in low N environments by preying on nematodes (Barron 1992).

Exceptional responses of nematode suppressiveness by organic matter with different C:N ratios do occur. For example, Wang et al. (2002) demonstrated that when the leguminous cover crop sunn hemp (*Crotalaria juncea*) was grown prior to a pineapple crop, population densities of nematode-trapping fungi increased significantly compared to two non-leguminous cover crops, marigold (*Tagetes erecta*) and rapeseed (*Brassica napus*), and most significantly to a weed fallow treatment. Later, Ching and Wang (2012) compared sunn hemp (SH) to oats (O) (*Avena sativa*), and a mix of sunn hemp and oats (SH+O) as preplant cover crops prior to a no-till planting of eggplant. They monitored NTF over a 10 month period after eggplant was planted, and found that SH (whether SH alone or SH+O) had higher numbers of NTF propagules per gram of root than treatments without SH (O alone and bareground). This is probably due to the fact that SH alone contains tissues that are made of different C:N ratios, ranging from C:N of 11 in the leaves to 48 in stem tissues (Marshall 2002). However, although SH could enhance NTF significantly in this system, abundance of NTF was only positively correlated ($r=0.26$, $P=0.05$) with total abundance of plant-parasitic nematodes, indicating a density dependent relationship between nematode prey and NTF, and not necessary resulted in suppression of the nematode pests.

Clearly, integrating augmentative and conservation biological control strategies together might offer a more versatile approach to promote nematode-antagonistic microorganisms in a sustainable agriculture system. For example, Stirling (2008) integrated crop rotations, reduced tillage, residue retention, more frequent cover cropping, and regular inputs of animal manures and organic wastes into a sugarcane farming system and found that damage caused by *M. javanica* and *Pratylenchus zeae* was reduced due to enhancement of natural biological control mechanisms against these pests (Stirling 2008). More work will be needed to develop integrated approaches for different cropping systems specific for different regions. Thus, the era of the "silver bullet" approach needs to be adjusted to allow for more efficient and sustainable use of natural enemies.

2.3 Stimulating Inherent Plant Defenses Against Nematode Pests

The availability of natural host plant resistance against nematode pests relies on traditional plant breeding or through genetic recombination, both of which are time consuming. These technologies are mostly available for highly profitable or intensive farming crops. Most of the minor crops grown in the tropics might not be economically significant. One alternative is to induce resistance in the host plant. Systemic induced resistance is an enhanced defensive capacity throughout the plant that is triggered by a specific stimulus such as a chemical inducer, a pathogen or insect, or a non-pathogenic microorganism (van Loon et al. 1998). There are two recognized types of induced resistance, systemic acquired resistance (SAR) and induced systemic resistance (ISR) that are differentiated by their signal transduction pathways (van Loon et al. 1998). These induced resistances often lead to the systemic expression of a broad spectrum and long lasting disease resistance that is efficient against bacteria, fungi, Oomycetes, nematodes, viruses (Durrant and Dong 2004; Zinovieva et al. 2013) as well as insect transmitted viruses (Zehnder et al. 2000). ISR occurs not only in place of an elicitor exposition or a penetration of the pathogen (local resistance), but also in remote areas of the plant induced systemic resistance. ISR in plants activate the same defense mechanisms that operate in genetically determined resistance, but unlike this degree of protection, usually, it does not exceed 30% of the regular host plant resistance (Zinovieva et al. 2013). None-the-less, it still provides a form of plant disease management that would reduce risk to the environment and be more compatible with sustainable pest management.

Although ISR is dependent on jasmonic acid and ethylene signaling in the plant, ISR differs from SAR in that resistance can be induced by the rhizobacterium, particularly *Pseudomonas* and *Bacillus* (Kloepper et al. 2004; Van Loon 2007). This type of rhizobacteria is referred to as plant growth-promoting rhizobacteria (PGPR). PGPR are naturally occurring soil bacteria that colonize plant roots and benefit plants by promoting plant growth and/or reducing disease or insect damage. Lugtenberg and Kamilova (2009) recently reviewed works on PGPR and listed all the mechanisms responsible for PGPR to promote plant growth and protect plants from pest damage. In recent years, increasing numbers of PGPR species have been identified. *Azotobacter, Pseudomonas, Azospirillum, Klebsiella, Enterobacter, Alcaligenes, Arthrobacter, Burkholderia, Bacillus*, and *Serratia* are among the most commonly found (Pathma and Sakthivel 2013). These bacteria either induced ISR, produce siderophore bacteria or antibiotics, improve nutrient acquisition of plants, or produce phytohormones.

One sustainable agriculture approach that can enhance PGPR and subsequently lead to ISR is through the use of vermicompost tea extract. Vermicompost is the product of accelerated bio-degradation of organic matter by earthworms through mesophilic decomposition. It generally has higher concentrations of plant available nutrients (NO_3^-, exchangeable Ca, P and soluble K) and significantly larger and more diverse microbial populations than thermophilic compost (Tognetti et al. 2005). Vermicompost tea (VCT) is a water-based extract of vermicomposts through aerated and non-aerated procedures. Aerated VCT is usually prepared by adding vermicompost into a porous container, then suspending it in a water containing vessel, typically 1 part compost to 10–50 parts water. Constant mechanical energy input is used to provide aeration either by air injection directly into the water or by re-circulation of the water, typically for 12–24 h. Non-aerated compost tea is prepared by mixing 1 part compost with 3–10 parts water in an open container, where it remains with or without daily stirring, for at least several days (often for 1–3 weeks) (NOSB 2004). VCT may supply microbial biomass, fine particulate organic matter (POM), organic acids, plant growth regulator like substances and soluble mineral nutrients to plant surfaces (Edwards et al. 2006; Scheuerell and Mahaffee 2002). These properties of vermicompost tea might serve as a means to enhance the soil food web structure that eventually leads to a plant tolerance to stress (Arancon et al. 2004). VCT increased seedling vigor and plant growth, as well as N content, total carotenoids, and total glucosinolates in plant tissue as compared to no VCT treatment (Pant et al. 2011). In addition, they also measured dehydrogenase activity and soil respiration and found that increased VCT concentration increased total soil microbial activities.

Vermicompost tea extract has been shown to suppress plant pathogens including nematodes (Edwards et al. 2007). Wang and Radovich (2012) found that weekly application of chicken manure based VCT (10%) suppressed root-knot nematode population densities initially and improved crop yields of squash. Besides direct suppression of plant-parasitic nematodes, Wang and Radovich (2012) also reported that drenching of VCT resulted in a significant increase of predatory nematode numbers within one cropping system of zucchini (<30 days after transplanting), thus showing an improvement in soil food web structure in a short period of time. Many have documented that the microbial population found in compost teas or extracts is responsible for suppressing soil-borne pathogens (El-Masry et al. 2002; Scheuerell and Mahaffee 2002; Zmora-Nahum et al. 2008). Future work needs to examine the performance of different formulations of VCT for targeted nematode pests.

2.4 *Improving Plant Tolerance to Plant-Parasitic Nematodes by Maintaining Soil Nutrient Cycling and Energy Flow*

Plant-parasitic nematodes cause damage to plant roots, resulting in root systems which are less able to take up nutrients and water. Enhancing soil nutrient availability not only supplies nutrients for plant uptake, but also provides plants with materials needed to grow functional roots, thus increasing the plant's tolerance to nematode damage. On the other hand, a great resource in most soil ecosystems for suppressing plant-parasitic nematodes is the pool of natural enemies of nematodes in the soil. Thus, maintaining a complex soil food web would increase plant tolerance to nematode pests.

When organic matter is first added into the soil, it is in a form that is unavailable for plant uptake until it is decomposed by bacteria or fungi. After initial decomposition, some organic matter will be converted into an inorganic form that plants can utilize. However, these same bacteria or fungi can immobilize nutrients in the soil until they are grazed by bacterivorous and fungivorous nematodes. However, overgrazing by these nematode groups can reduce the overall activity of decomposers. Fortunately, in the hierarchy of the soil food web, predators such as predatory nematodes and mites, feed on these bacterivorous and fungivorous nematodes, thus allowing more nutrients to be released into an inorganic form for plant uptake. Thus, an increase in predatory nematodes may contribute to increased nutrient mineralization. Ferris et al. (2012) recently reviewed the studies of nematode ecology since the beginning of 1980s, and revealed the importance of free-living nematodes in the soil food web (Fig. 4).

Availability of nutrients from soil organic matter to plants relies on the mineralization (release) of nutrients from the organic matter (immobilized forms). Nematodes have higher respiration rates, lower N needs, and lower growth efficiency than bacteria. Therefore they can release a majority of both the C and N that they consume from bacteria into nutrients available for plants. Grazing of nematodes may contribute up to 19 % of soluble N in soil (Neher 2001), much higher than that contributed by bacteria in soil ecosystems. In general, increases in bacterial and fungal feeding nematodes are associated with higher N availability to plants.

In addition to N mineralization, populations of free-living nematodes, especially bacterivorous, omnivorous, and predatory nematodes, have also been found to correlate with concentrations of most of the other soil nutrient elements including P, K, Na, Ca, Mg, Fe, Cu, Mn, Zn, and cation exchange capacity (Wang et al. 2003). This suggests that nematodes are also responsible for the mineralization of other soil nutrients. Therefore, nematodes play important roles in soil nutrient cycling.

Unfortunately, nematode communities in an agroecosystem, especially those in the higher hierarchy of the soil food web, are often too disturbed by human intervention, such as frequent tillage, pesticide applications, etc. (McSorley et al. 2007). Several attempts have been made to use cover crops in combination with conservation tillage practices to reduce soil disturbances and increase the abundance of soil organisms higher in the soil food web hierarchy (DuPont et al. 2009;

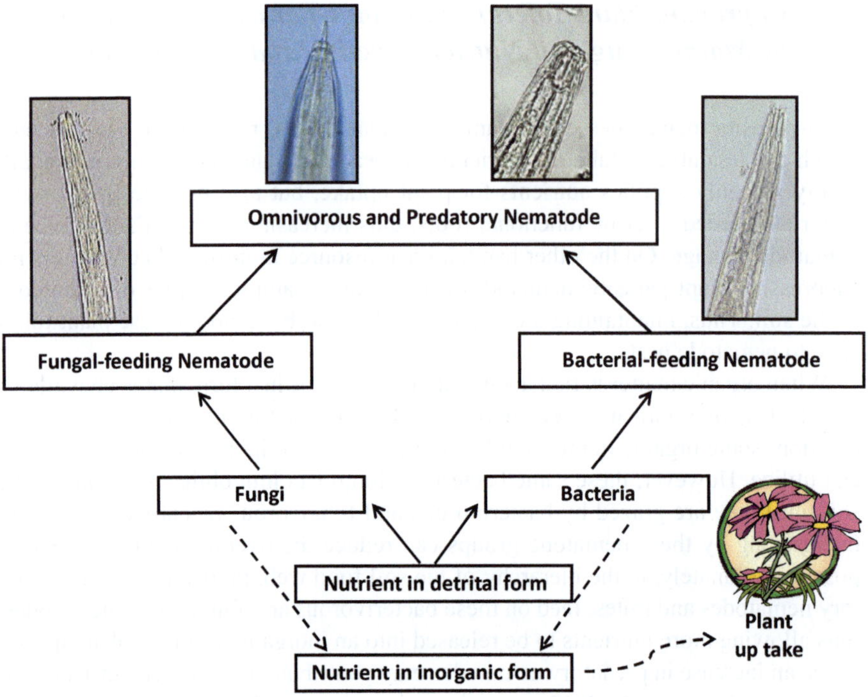

Fig. 4 Functional groups of free-living nematodes in a soil food web in relation to soil nutrient cycling. (Pictures by K.-H. Wang)

Sánchez-Moreno and Ferris 2007). However, these studies generally found that long term conservation tillage (more than 2 years) is required before enhancement of omnivorous or predatory nematodes can be observed. Alternatively, adding resources such as organic matter into the soil fuels the food source for free-living nematodes, and thus, can also enhance nematodes that are of higher hierarchy in the soil food web. Using these concepts, growing sunn hemp (*Crotalaria juncea*) as a leguminous cover crop in a strip-till cover cropping system (STCC) followed by periodic clipping of the living mulch intercropping between vegetable cropping rows and added as organic surface mulch (SM) provide an approach to reduce soil disturbance, while fueling organic materials to the soil food web over a longer period of time than a conventional cover cropping system (Wang et al. 2011).

Another approach to improve the soil food web structure in short-term agroecosystems is to enrich the soil with vermicompost tea extract, as mentioned in Sect. 2.3. VCT enhanced microbial biomass and improved soil structure (Edwards et al. 2006; Scheuerell and Mahaffee 2002). These properties of vermicompost tea might be responsible for the significant increase of abundance in predatory nematodes within one cropping cycle of zucchini, drenched weekly with chicken manure based VCT (Wang and Radovich 2012). Other approaches to enrich crop rhizosphere with microbial biomass and improve soil structure might obtain similar results and should be integrated into sustainable pest management planning.

3 Sustainable Pest Management for Plant Fungal and Oomycete Pathogens

Similar to management of other groups of plant pathogens in a sustainable agriculture system, sanitation of planting materials or growing media should be the first line of a preventative measure against fungal or Oomycete caused diseases. This chapter will mainly focus on biological and cultural practices that are compatible with the five approaches listed in the introduction section. Natural enemies of fungal or Oomycete pathogens are mostly fungi or bacteria. Some of the most commonly found fungal and Oomycete biocontrol agents are *Trichoderma, Gliocladium, Acremonium, Geotrichopsis, Pythium, Verticillium, Coniothyrium, Piptocephalis, Kuzuhaea, Melanospora*, and many others. However, most of these are not commercially available. In general the pathogenic relationships that have evolved between the biocontrol antagonist and the fungal pathogen are often highly specific. Through the understanding of the mechanisms of how these potential biocontrol agents suppress fungi or Oomycetes, perhaps better ecologically based management strategies could be developed. In general, five mechanisms could be involved between the fungal biological control antagonist and the plant pathogen as reviewed by Jeffries and Young (1994).

Toxin producers: Production of extracellular metabolites (*e.g.* toxins) that diffuse from the antagonist that harm the plant pathogen. For example, *Stereum* species that attack plums are inhibited by the antagonist, *Hypomyces aurantius.* When the *Hypomyces* is grown near the *Stereum*, the *Hypomyces* releases substances that cause lipid (component of fat, waxes, etc) bodies to accumulate, or that lead to destruction of the membranes in the energy producing mitochondria, or that cause membrane disruption of the endoplasmic reticulum, needed for internal transport, and this causes the cell membrane to withdraw from the cell wall (Kellock and Dix 1984). The cell membrane is crucial to the survival of all cells and regulates what enters and leaves the cell. Thus disruption of these systems in the *Stereum* results in cell death of the pathogen. In another example, fungal pathogen destruction was observed when grown with *Trichoderma* species that are common biocontrol agents. Thus between the use of *Hypomyces* and *Trichoderma* biocontrol species, plum diseases caused by *Stereum* can be controlled.

Hyphal cell interference: Ikediugwu and Webster (1970) described hyphal interference as any process that occurs when two different fungi are grown close together and the plant pathogen is harmed by the antagonist, when substances diffuse from the antagonist to the pathogen and reduce its growth rate and cause cytoplasm disruption. When studied using an electron microscope, a marked change in the permeability of the cell membrane of the pathogen can be observed. There is loss of turgor pressure with granulation and vacuole formation in the pathogen cytoplasm. These cells then die. The substances are able to diffuse across a cellophane membrane and this implies that the substance is a small molecule and not a large enzyme molecule (Ikediugwu and Webster 1970).

This process has been documented for the ability of *Phanerochaete gigantea* to control infection of the pine tree by *Heterobasidion annosum*. When pine trees are harvested, the stumps are rapidly infected with *Heterobasidion*, the bracket fungus. This highly destructive fungus can grow from one tree to another using the connected root systems, infecting hundreds of trees. If the *Phanerochaeta* is sprayed on the freshly cut stumps, infection by *Heterobasidion* is prevented. Research has shown that the *Phanerochaeta* occupies the sites needed by the *Heterobasidion* for infection and that the interference reaction prevents respiration (energy metabolism) of the *Heterobasidion* (Ikediugwu 1976). This has been a highly effective biocontrol for the pine forest industry.

Rhizoctonia solani is a very common worldwide pathogen of roots and some foliar disease (thread blights). It can be controlled by the biological control agent *Arthrobotrys oligospora*, another fungus common in soil (Persson and Baath 1992). When the *Arthrobotrys* hyphae grows adjacent to *Rhizoctonia* hyphae, the *Arthrobotrys* forms numerous membranous structures indicating that following contact, serious metabolites are formed by the *Arthrobotrys* that trigger the lyses and cell death of *Rhizoctonia*.

Haustoria formers: In these cases the mycelium of the antagonist remains on the outside of the pathogen but grows closely with it. Appressoria, or cushion shaped hyphal tips are formed and firmly anchored to the plant pathogen. An infection peg is produced from the appressoria and it enters the pathogen hyphae. Inside the hyphae, it inflates to form a balloon like structure, the haustoria. This haustoria releases enzymes that disrupts the cells contents and cause them to degrade and die. Nutrients are released and the haustorium absorbs these nutrients and transports the nutrients back to the antagonist mycelium for further growth. Haustoria formers are common in biological control of fungi. Commonly, several methods of attack occur. For example, pathogen control of onion diseases uses interference principles, and then enzymes are formed for wall degradation and growth of the antagonist in the mycelium of the *Botrytis*, that destroys the pathogen. The cells of the plant pathogen *Botrytis alli*, which attacks onion, are killed when the biocontrol *Gliocladium roseum* are grown with it. The series of events that occur are: (1) the close growth of the *Gliocladium* and *Botrytis* hyphae or threads, (2) the *Gliocladium* produces toxins that diffuse into the *Botrytis*, (3) the cells of the *Botrytis* are killed, (4) cell wall degrading enzymes are then produced by the *Gliocladium* which enters the *Botrytis* and feeds on it (Pachenari and Dix 1980).

Invasive necrotrophs: These biological control agents penetrate the hyphae of the pathogens they invade. Following penetration, there is a huge amount of lyses of the cell walls and destruction of the pathogen cytoplasm and finally death of the pathogen. These necrotrophs can invade the entire pathogen structure but are more commonly associated with the hyphae, spores, or surviving sclerotia. A well-known necrotroph, *Schizophyllum commune*, lives on stumps of felled trees. It attacks the nematode destroying fungus, *Arthrobotrys oligospora*, parasitizing *Rhizoctonia solani* and species of *Cunninghamella, Rhizopus,* and *Zygorhynchus*. *Schizophyllum* coils around the pathogen and other fungi it infects, develops appressoria and pegs, and penetrates the host. The parasitized hypha collapses and is destroyed (Tzean and Estey 1978).

Pythium species can be plant pathogens but some are also mycoparasites, which have been well studied. Mycoparasitic species include *P. acanthicum, P. mycoparasiticum, P. nunn, P. oligandrum* and *P. periplocum* (Deacon et al. 1991). *Pythium oligandrum* has been frequently isolated from agricultural soils and is known to be an aggressive pathogen of significant plant pathogens (Deacon 1976; Deacon and Henry 1978). The fungal pathogens differ in susceptibility with *Trichoderma aureoviride* and *Fusarium oxysporum* as highly susceptible, *Botrytis cinerea, Fusarium culmorum, Rhizoctonia solani* and *Botryotrichum piluliferum* as moderately susceptible and *Pythium graminicola* and *P. vexans* as highly resistant (Laing and Deacon 1990). Susceptible pathogens are rapidly lyses with total destruction of the pathogen while those that are more resistant degraded at slower rates and lyses are not consistent.

Intracellular Biotrophs: Although many types of relationships exist, one that is unique is the total movement of the biological control agent into the host. This occurs as the zoospore of the biological control agent becomes an endobiotic body within the pathogens cytoplasm. Several species of the *Oomycota* and *Chytridiomycota* are involved. The invading biotroph does not seem to cause damage to the pathogen initially, and there appears to be a close relationship of the parasite wall and the endoplasmic reticulum of the host (Powell 1982).

However a few successes in using fungal biocontrol agents against fungal or Oomycete pathogens have been developed and more are in the process of being improved. For detailed examples, please visit Jeffries and Young (1994). In general, fungal biocontrol agents that are resistant or compatible with fungicide application are recognized as effective biocontrol agents. For example, *Pythium oligandrum* is being considered for commercial production now that isolates resistant to Benlate were found (Lewis et al. 1989). Effective control of damping-off was attained with the use of *P. oligandrum* and lower rates of the fungicides. Another example is *Trichoderma. Trichoderma* was predicted to be a good source of biocontrol because it grows rapidly in soil, is able to use low levels of nutrients, and is able to grow in acidic and alkaline soils (Jeffries and Young 1994). It was also found that some strains were resistant to fungicides used to control pathogens (Papavizas 1985). Today there are nine species/strains of *Trichoderma* listed as biocontrol agents, and these include *Trichoderma harzianum* (no specific strain; some mixed with *T. viridie*, available as Ecosom, Trisan, others), *T. harzianum* (strain T-22, Rootshield), *T. harzianum* (strainT-39, Trichodex) *T. virens* (SoilGard), *T. viride* (Bioderma, Ecfosom, Bio-Cure, and Tricho-shield), *T. harzianum/polysporum* (BINAB T) (Meister and Sine 2012). These types of biocontrol agents could be used in Integrated Pest Management Programs to control plant diseases.

Bacteria as biocontrol agents: In addition to the use of fungi to control fungal diseases, bacterial species have also been employed to control fungal rots. Three species of *Bacillus* are used to control fungal diseases. For example *B. subtilis* is used to control fungal pathogens on seeds of barley, peanuts, wheat, cotton, soybean and other leguminous crops, while *B. pumilus* controls mildew and rust on cereals, roses, strawberries, and vegetables. *Bacillus pumilus* is sold in the USA as Ballad or Yield Shield, and available in Brazil as Sonata. *Bacillus subtilis* is widely available from many companies and is sold as Serenade, Kodiak, Companion and others. *Streptomyces candidus* is used to control *Phytophthora* and *Pythium* on fruits and vegetables. It is sold as BioAId and Sun Mycan. Other *Streptomyces* species are used to control fungal diseases such as damping-off, root rots and wilt of herbs, vegetables, ornamentals, and landscape plants. For *S. lydicus*, it is marketed as Actinovate AG, Actinovate Sp, or others. *Pseudomonas fluorescens* can be used to control root rots, banana wilt, diseases of chickpeas, soybean, tomatoes, and sheath bight on rice, many other diseases of cereals, cotton fruits and vegetables. It is available as Bio-cure-B and Biomonas. *Pseudomonas syringae* is employed to control post-harvest diseases of citrus, cherries and pomes fruits. It can be purchased as Bio-Save 10LP and should be kept refrigerated (Meister and Sine 2012).

None-the-less, introducing biocontrol agents is still considered as a therapeutic approach for pest management. Integrating multiple species of biological control agents to complement each other would be an effort to increase biodiversity to keep pathogenic fungi or Oomycete in check.

3.1 Increasing Ecosystem Community Stability by Promoting Antagonists of Fungal and Oomycete Pathogens

In reality, root systems of agricultural crops are often exposed to multiple soil-borne or foliar fungal pathogens. Introducing biocontrol agents that specifically target a single pathogen will be too time consuming and costly. Amending soil with organic matter could enhance multiple beneficial indigenous soil organisms that might lead to natural suppressive soils. For example, broad-spectrum control of *Pythium*, *Phytophthora* and *Rhizoctonia* was reported in peat and compost-based soilless container media (Hoitink et al. 2001); *Pythium* was suppressed in Mexican fields following the application of large quantities of organic matter over many years (Lumsden et al. 1987); *Phytophthora* root rot of avocado in Australia was suppressed by the use of cover crops, organic amendments and mulches (You and Sivasithamparan 1994, 1995); similarly, suppression of the same disease with eucalyptus mulch was reported in California, USA (Downer et al. 2001).

One drawback of using organic amendments is that when pathogens are good saprophytes but poor competitors (e.g. *Pythium* and *Fusarium*), they may multiply on organic amendments before being suppressed. Similarly, use of organic amendments is also problematic in *Rhizoctonia* infested soil. This is because *Rhizoctonia* has a high capacity to degrade cellulose as well as simple sugars, making it a good saprophyte that can proliferate in soil with high organic matter. Thus, organic-matter mediated general suppression might not be sufficient to achieve control and specific antagonists may also be required (Stone et al. 2004).

However, van Bruggen and Termorshuizen (2003) compared disease severity in organic and conventional farming systems from various studies conducted in the U.S. and Europe, root diseases are generally less severe in organically than conventionally managed soils. Although specific mechanisms of this phenomenon are not totally understood, sometimes positive correlation between soil N availability and disease severity was found (Tamis and van den Bink 1998; Daamen et al. 1989; El Titi and Richter 1987). Another possible reason for lower disease severity in organic farming systems compared to conventional farming systems is the higher diversity and abundance of non-pathogenic fungi with antagonisitic properties. For example, the density of non-pathogenic *Fusarium* species with antagonistic properties towards *F. culmorum* was significantly higher in an established organic farm than in a neighbouring conventional farm (Knudsen et al. 1999). Bulluck and Ristaino (2002) documented that southern blight (caused by *Sclerotium rolfsii*) was suppressed in plots receiving organic amendments rather than synthetic fertilizers, and they attribute this to higher microbial biomass and activity in organic plots. In addition, van Bruggen and Termorshuizen (2003) also concluded that crop protection in organic farming is generally not directed at controlling particular pathogens or pests but at management of the environment so that plants are able to withstand potential attacks.

Similar to the sustainable approaches for nematode management, sustainable farming practices to promote soil suppressiveness against fungal and Oomycete pathogens are long, balanced crop rotations, organic amendments and reduced tillage, all geared towards maintenance of the soil organic matter content and fertility. Unfortunately, conversion from a disturbed, conventional farming system to disease suppressive organic farming system often takes times. Roget (1995) demonstrated that after conversion from regularly tilled to no-till wheat production, *Rhizoctonia* root rot increased initially, but this increase was followed by a decline in this disease after about 5 years of no-till.

3.2 Stimulating Inherent Plant Defenses against Fungal or Oomycete Pathogens

Examples of ISR to suppress fungal or oomycete pathogens is also present. For example, *Pseudomonas* sp. strain WCS417r was found to induce ISR against Fusarium wilt of carnation (Wei et al. 1991). In addition, induction of systemic resistance of cucumber to *Colletotrichum orbiculare* by selected strains of plant growth-promoting rhizobacteria (PGPR) have also been reported (Wei et al. 1991). Commercially available chemicals are also available to induce ISR against this group of pathogens. For example, Acibenzolar-S-methyl (ASM), which is commercially available as Actigard, contains a Harpin protein, has been used successfully in some ornamental crops such as Fuchsia (Titus 2012) and is sold as Messenger or Employ. It was effective in boosting the health of plants, which were more vigorous following treatment, and grew at a faster rate. Harpin protein is obtained from the bacterium *Erwinia amylovora* which activates plant physiology. Better growth and vigor is the major effect but minor, weak pathogens may be inhibited as well (see bacterial biocontrol with Actigard).

3.3 Plant Extracts to Control Fungal Diseases

A few compounds have been extracted from plants which demonstrate disease control characteristics. An extract from *Reynoutria sachalinensis*, sold as Regelia in the U.S. A., has been commonly cited as a control for powdery mildew and leaf spots on greenhouse grown ornamentals, strawberries and vegetables. Much research is still needed for specific crops, as well as, the optimal methods of application and the timing of these extracts to be applied. Clove oil is reported to reduce silver scurf and sprouting of potatoes. It is sold as Matratec, Matran and Biox. Other oil based compounds used are anise, canola, jojoba, rosemary, wintergreen, lemongrass, mint, orange, rapeseed, soybean, thyme and others. These are sold as Armorex, Deter, Ecotec, EcoTrol, Organocide, and Pest Out (Meister and Sine 2012). Many compounds have some effect but do not completely prevent or "cure" diseases. Still they offer growers alternatives to harsh chemical controls.

4 Sustainable Pest Management for Insect-Vectored Plant Viruses

In terms of plant viruses, breeding for virus resistant hybrids is available on several crops, but these approaches are time consuming. This is especially challenging when dealing with crops that are susceptible to multiple plant viruses, as well as, other pathogens. Managing insect vectors such as aphids, thrips, and whiteflies that transmit plant viruses serve as a preventative measure for this type of virus disease. Rear and release of biological control agents against virus insect vectors has become of great interest to farmers that are considering the sustainable pest management approach. A wide range of therapeutic bioinsecticides are available (Meister and Sine 2012), and more are being developed with modern technology. Therapeutic approaches play a valuable role in ecologically based pest management strategies, but they could potentially be disruptive to natural enemies. For example, although spinosad is considered a reduced-risk insecticide, it is harmful to *Trichogramma* wasp (*T. exiguum*), and slightly harmful to a predatory mite, *Iphiseius degenerans* (Charles et al. 2000; Van Driesche et al. 2006) and bumble bees (Johnson 2004; Morandin et al. 2005). Azadirachtin (neem) reduced life table parameters of *T. exiguum* (Saber et al. 2004), and harmed a soil predatory mite, *Hypoaspis aculeifer*, when applied to soil (Thoeming and Poehling 2006). Upon contact, neem seed extract reduced the population densities of parasitoids, *Encarsia* sp. and *Aleurodiphilus* sp. (Price and Schuster 1991). Both natural and synthetic pyrethrins are harmful to green lacewing (*Chrysoperla carnea*) and multicolored Asian lady beetle (*Harmonia axyridis*) (Huerta et al. 2003; Kraiss and Cullen 2008). Insecticidal soap is moderately lethal to first and third instars of *H. axyrids* (Kraiss and Cullen 2008). Another U.S. National Organic Program (NOP) compliant pesticide, sulfur, is moderately harmful to the predatory mite *Phytoseiid* sp. (James 2005) and harmful to *Trichogramma cacoeciae* (Grutzmacher et al. 2004). Frequent use of

Bacillus thuringiensis (Bt) on diamondback moth (*Plutella xylostella*) has resulted in resistance developing in that population (Liu et al. 1996). *Beauvaria,* a biocontrol agent with minimal non-target impact, is unfortunately very costly and requires frequent application.

A thorough review on developing sustainable pest management strategies against insect pests can be found in an article by Lewis et al. (1997). This article will again focus on some of the key sustainable pest management approaches.

4.1 Enhancing High Biological Diversity and Promoting Natural Enemies of Insect Pests

Intercropping cash crops with trap crops or insectary plants is compatible with the approaches on enhancing high biological diversity and promoting natural enemies of multiple pests. Insectary plants are plants that attract beneficial insects by providing nectar and pollen sources for many predatory insects (Cowgill et al. 1993; Lavandero et al. 2005; Hogg et al. 2011); and supplemental food for spiders (Taylor and Pfannenstiel 2008). Conventional farming practices that involve intensive tillage, monoculture, and extensive weed control, or even some innovative alternative farming practices such as high tunnel shade house crop production, hydroponic and aquaponic practices often lack the insectary plant resources beneficial to agroecosystems. Introducing insectary plants into these systems could be one approach to attract beneficial insect allies back into our farming systems.

Natural enemies of insect pests that are of interest to attract into agroecosystems to protect plants from insect transmitted viruses include hoverflies, various insect parasites or parasitoids (e.g. Tachinid fly, Trichogramma, Braconid wasp, etc.), lady beetles, lacewings, spiders, assassin bugs, minute pirate bugs, and ground beetles. Careful timing of the planting of insectary plants can help enhance population densities of predatory insects or insect parasitoids that could result in a reduction of insect transmitted viral diseases (Mandanhar et al. 2009).

Some criteria on selecting insectary plants include: (1) attractiveness to beneficial insects; (2) an early and long blooming period; (3) low potential to host plant viruses; (4) ability to out-compete weeds; (5) low potential to become a weed; (6) low attractiveness to pest species; and (7) low cost of seed and establishment (Hogg et al. 2011).

Families of plants commonly found to be attractive to various beneficial insects mentioned above include, but are not limited to, Apiacea (Umbelliferae), Asteracea (Compositae), Lamiaceae and Fabaceae. Apiacea includes fennel (*Foeniculum vulgare*), dill (*Anethum graveolens*), cilantro or coriander (*Coriandrum sativum*), carrot (*Daucus carota* subsp. sativus), wild carrot or Queen Anne's-lace (*Daucus carota* subsp. Carota). These Apiacea are excellent insectary plants as they provide great numbers of tiny flowers required by parasitic wasps.

Asteracea known to be attractive to parasitoids include zinnia (*Zinnia peruviana*), creeping zinnia (*Sanvitalia speciosa*), marigold (*Tagetes* spp.), sunflower (*Helianthus annuus*), etc. These plants produce showy composite flowers that are

Table 2 Natural enemies of insect pests or beneficial arthropods and plants that attract them (insectary plants). (Wang 2012)

Beneficial arthropods	Insectary plants
Various parasitoids and predators	Fennel, dill, coriander (cilantro), parsley, carrot, wild carrot (Queen Anne's-lace), angelica, yarrow (milfoil), sow thistle, dandelion, zinnia, tansy, marigold, sunflowers
Predatory wasps, hoverflies	Sweet alyssum, buckwheat, mustard, Cuban oregano, sage, salvia, lavender, oregano, thyme, marjoram, perilla
Lady beetles	Dill, marigold, Mexican tea, morning glory, oleander, yarrow
Lacewing	Carrot, oleander, red cosmos, wild lettuce, tansy
Minute pirate bug	Carrot, Mexican tea, oleander, sunn hemp, cowpea
Ground beetles	Low-growing plants: thyme, rosemary, mint or mulches
Spider	Marigold, yellow-sweet clover, white clover

favorable for many parasitoids as well as predatory insects. Lamiaceae (mint family) includes many herbs such as basil (*Ocimum basilicum*), mint (*Mentha* spp.), rosemary (*Rosmarinus officinalis*), sage (*Salvia officinalis*), marjoram (*Origanum marjorana*), oregano (*Origanum vulgare*), thyme (*Thymus* spp.), lavender (*Lavandulla spica*), and perilla (*Perilla frutescens*). These plants attract wasps, hoverflies and other beneficials. Planting these insectary plants may provide additional economic incentive for farmers to set aside land for harvestable herbs. In addition, low-growing Lamiaceae such as thyme, rosemary, or mint also provides shelter for ground beetles that are predators of insect pests or weed seeds. Among Fabaceae, sunn hemp (*Crotalaria juncea*), cowpea (*Vigna unguiculata*), white clover (*Trifolium repens*), and yellow sweet clover (*Melilotus officinalis*) are most commonly used as cover crops in Hawaii. White clover and yellow sweet clover has been documented to increase the abundance of spiders when planted as living mulch, intercropping with broccoli (Hooks and Johnson 2004). At the flowering stage, sunn hemp and cowpea increased numbers of beneficial insects such as the Trichogramma wasp and the minute pirate bug when planted as living mulch in corn fields (Manandhar, personal communication). In addition to serving as insectary plants, this plant family is the most favorable cover crop group as it fixes nitrogen, reducing the need for additional nitrogen inputs into the agroecosystem.

One factor to take into consideration when selecting for insectary plants, is the length of time needed for plants to reach blooming. Buckwheat (*Fagopayum esculentum*) and mustard are early blooming plants, but their flowering period is short. On the other hand, wild arugula and tansy phacelia might take a longer time to bloom, but their flowering period lasts for a longer time (Hogg et al. 2011). Table 2 summarizes natural enemies of insect pests that are attracted to but not limited to these insectary plants.

Understanding the benefits of enhancing various beneficial organisms can provide incentive for farmers to incorporate cover cropping into their farmscapes. Many approaches to integrate cover cropping with cash crop production have already been practiced. These include planting cover crops as a (1) border crop or barrier crop, (2) living mulch intercropping with cash crop, (3) undersown ground cover in an orchard system, (4) pre-plant rotation crop followed by conservation tillage, i.e.

serve as surface mulch or organic mulch after mowing, crimping or natural die back of cover crop, (5) preplant cover crop followed by strip tilling, and clipping for surface mulch, and (6) trap crop. Farmers can select cover cropping practices based on target insect vectors, occurrence of beneficial organisms, and compatibility of the cover crop to their cash crop production practices.

Various research projects conducted in the tropics on the use of cover crops to manage insect transmitted viruses have shown promising results. Hooks et al. (1998) has demonstrated that intercropping buckwheat (*Fagopayum esculentum*) with zucchini reduced population densities of whiteflies and aphids, thus reducing silver leaf symptoms and aphid transmitted viruses such as papaya ring spot virus on zucchini. They supported that this was partly due to the "virus sink theory" (Mandanhar and Hooks 2011) where insect vectors that transmitted non-persistent, non-propagative viruses lost the virus after visiting a border crop. In addition, cover crops could also provide niches for generalist-type of insect predators. For example, when sunn hemp or marigold (*Tagetes patula*) were intercropped with cucumber in a strip-till system as a living mulch followed by clipping to provide organic mulch (Wang et al. 2011), abundance of spiders in the agroecosystem were higher in these cover crop plots than cucumber planted in the bare ground plots.

4.2 *Stimulating Inherent Plant Defenses Against Insect Pests*

Plants grown with high levels of N supplements resulted in larger infestations of insect pests (Dixon 1969; House 1965). Conversely, inadequate N availability increased consumption rates of plant tissues by insects (Hamilton and Moran 1980). Radovich and Arancon (2011) suggested that vermicompost provides plants with balanced nutrients through gradual decomposition of organic matter and slower mineralization rates of nutrients, and could suppresses various insect pests and diseases in greenhouse and field conditions. Vermicomposts are produced by mesophilic decomposition and stabilization of organic matter by certain earthworms and microorganisms. Solid vermicompost produced significant suppression of mealy bug infestations (*Pseudococcus* sp.) on cucumbers and tomatoes, two-spotted spider mite attacks (*Tetranychus urticae*) on bush beans and eggplants, and aphids (*Myzus persicae*) on cabbages at low application rates (Arancon and Edwards 2004; Arancon et al. 2005a, b). Yardim et al. (2006) reported the suppression of tomato hornworms and cucumber beetles by solid vermicomposts.

When Radovich and Arancon (2011) applied food-waste based vermicompost water extract at 10 and 20% every week, they found that aphid populations on tomatoes and cucumbers decreased with time as compared to water control. Several mechanisms have been proposed to explain the pest resistance induced by vermicompost. Organic and inorganic nutrition affects plant growth such as the onset of senescence, lignification of the epidermal cells, increased sugar concentrations in the apoplasts, amino-N in phloem sap, and levels of secondary plant compounds (Patriquin et al. 1995). Water-soluble phenols in VCT extracted during the brewing have been proposed to be the most likely ingredients in VCT to contribute to insect pest suppression (Arancon et al. 2004; Arancon and Edwards 2004). Several soil microorganisms associated with vermicompost including *Trichoderma hamatum, Pseudomonas fluorecens, P. putida, Xanthomonas maltophilia, Bacillus subtilis* etc. (Hoitink and Fahy 1986; Dowling et al. 1996; O'Sullivan and O'Gara 1992) have also been shown to induce ISR by secretion of hydrolytic enzymes (chitinase, protease, and β-1,3 glucanase).

5 Sustainable Pest Management Approaches Against Bacterial Pathogens

Given a highly susceptible host, a bacterial disease is nearly impossible to control with conventional methods. The task of using sustainable methods becomes even more challenging. The best method is to prevent introduction of the disease. This requires knowledge of the severe bacterial diseases of the crop being grown. Great efforts need to be made to purchase seeds and plants that are free of bacterial disease. For many horticultural crops, stocks that are free of bacterial (and other diseases) are available. Investment in these clean crops is worth the cost. Once the crop is contaminated, it is nearly impossible to significantly reduce the level of disease, especially in a tropical environment. In Hawaii, after the bacterial blight on anthurium entered the State, growers refused to discard diseased anthurium, and over 75 % of the industry was lost. Thus, prevention is the best approach. For a few crops that are grown even with the bacterial pathogen present, options such as the following are available: growing the crop under solid cover while using minimal irrigation, use of drip systems and avoiding splash irrigation that wet the foliage. Applying copper sulfate (Phyton-27), copper hydroxide (Kocide, Champ), and other compounds that reduce bacterial reproduction and spread. In Florida, Actigard with mode of action to induce SAR provide leaf spot control on tomato but the crop yield was not increased. Sprays needed to be made weekly and begun early in the crop cycle (Vallad 2013; Vallad and Goodman 2004). Cool environments are not conducive to bacterial diseases. Any plant with resistance is also beneficial. Antibiotics can be applied but overall they are expensive. For bacterial diseases, a common sense approach using prevention, sanitation, dry environments, and clean stock are the best approaches.

6 Future Prospects of Sustainable Pest Management for Plant Diseases

The sustainable pest management approaches emphasized in this chapter focus on ecosystem management, induced host plant resistance, while finding alternatives to replace conventional therapeutic plant disease management approaches. The use of therapeutic approaches, whether biological, chemical, or physical, is not compatible with the five fundamental approaches for sustainable pest management. More efforts should be devoted to developing farming practices that can: (1) enhance high biological diversity through polyculture instead of the conventional preference of the monoculture cropping system; (2) increase ecosystem community stability by promoting natural enemies of multiple pests and pathogens; (3) stimulate inherent plant defenses; (4) improve plant health by maintaining nutrient cycling and energy flow; and (5) target the vulnerable stage of pests or pathogens through the understanding of their ecology. Successful examples have been observed for each of the key plant pathogens that met these approaches. While there are more sustainable pest management approaches, this chapter only listed some that are known to be successful. An article on "Research on Plant Disease and Pest Management is Essential to Sustainable Agriculture" published by Cook et al. (1995) about two decades ago mentioned that while new technologies are needed, traditional methods for plant disease management need to be strengthened. Decades later, this statement still holds true. We now have more documentation on the importance of maintaining ecosystem biodiversity, sustaining the function and integrity of the soil food web, and ensuring the continuation of natural suppressiveness in our agroecosystems. Yet, more work remains to be done to reveal the nature of the myriad of interactions between pathogens and their antagonists, before biological control can be successful in managing targeted pathogens. The study of biological control of insect pests has provided much of the theory and principles available to understanding and predicting the predatory-prey relationships in the environment. To encourage biological control agents to be more widely used for plant pathogen management, sustainable approaches reviewed by Lewis et al. (1997) should be referenced. In addition, the study of plant disease management for sustainable agricultural systems should integrate with other disciplines. For example, information from the study of plant disease epidemics has provided much theory for predicting genotype changes of microorganisms. This could address the issues of constant resistance buildup of plant pathogens against pesticides or resistant cultivars. Research on a single disease in a monoculture system, or under artificial laboratory conditions is not applicable for sustainable crop production. Thus, future work should look into developing cultural practices that can be suppressive to multiple pathogens concurrently in an agroecosystem while protecting our environmental health, economic profitability, and social and economic equity.

Disclaimer Neither, the University of Hawaii, the College of Tropical Agriculture and Human Resources, the State Department of Agriculture, or the United States Department of Agriculture, or the authors, shall be liable for any damages resulting from the use of the information contained in this article or information omitted in this article. Reference to any company or recommendation of any product is for examples only and do not imply approval beyond those that are not mentioned. Recommendation of any product is not an exclusive endorsement of those products. For growers using pesticides or any product mentioned, each must be responsible for following the direction on the label. All label directions must be followed.

Acknowledgement This research was supported, in part, by USDA NIFA WIPM Programs (2011–00623, 2013-04430), and in part by University of Hawaii College of Tropical Agriculture and Human Resources (CTAHR) HATCH project (HAW09022-H).

References

Arancon NQ, Edwards CA (2004) Vermicomposts can suppress plant pest and disease attacks. BioCycle 45:51–53

Arancon NQ, Edwards CA, Atiyeh R, Metzger JD (2004) Effects of vermicomposts produced from food waste on the growth and yields of greenhouse peppers. Bioresour Technol 93:139–144

Arancon NQ, Edwards CA, Bierman P, Metzger JD, Lucht C (2005a) Effects of vermicomposts produced from cattle manure, food waste and paper waste on the growth and yields of peppers in the field. Pedobiologia 49:297–306

Arancon NQ, Galvis PA, Edwards CA (2005b) Suppression of insect pest populations and damage to plants by vermicomposts. Bioresour Technol 96:1137–1142

Ball-Coelho B, Bruin AJ, Roy RC, Riga E (2003) Forage pearl millet and marigold as rotation crops for biological control of root-lesion nematodes in potato. Agron J 95:282–292

Barron GL (1992) Lignolytic and cellulolytic fungi as predators and parasites. In: Carroll GC, Wicklow DT (eds) The fungal community, its organization and role in the ecosystem. Marcel-Decker, New York

Birchfield W, Bristline F (1956) Cover crop in relation to the burrowing nematode, *Radopholus similis*. Plant Dis Report 40:398–399

Bridge J (1996) Nematode management in sustainable and subsistence agriculture. Ann Rev Phytopathol 34:201–225

Brown PD, Morra MJ (1997) Control of soil-borne plant pests using glucosinolate-containing plants. Adv Agron 61:167–231

Bulluck LR, Ristaino JB (2002) Effect of synthetic and organic soil fertility amendments on southern blight, soil microbial communities, and yield of processing tomatoes. Phytopathology 92:181–189

Cardwell D, Ingham R (1996) Management practices to suppress Columbia Root-Knot nematode. Pac Northwest Sustain Agr 6

Charchar JM, Huang CS (1981) Host range of *Pratylencus brachyurus*: 3 different species. Fitopathol Bras 6:469–474

Charles JSK (1995) Effect of intercropping antagonistic crops against nematodes in banana. Ann Plant Prot Sci 3:185–187

Charles P, Suh C, Orr DB, Van Duyn JW (2000) Effects of insecticides on *Trichogramma exiguum* (Trichogrammatidae: Hymenoptera) preimaginal development and adult survival. J Econ Entomol 93:577–583

Ching S, Wang K-H (2012) Developing bioassay method for detecting nematode-trapping fungi in field soils with different cultural practices. CTAHR Student Research Symposium, Honolulu, Hawaii (Abstract #60)

Cook RJ, Baker KF (1983) The nature and practice of biological control of plant pathogens. Am Phytopathol Soc, St. Paul, MN. p 539

Cook RJ, Gabriel CJ, Kelman A, Tolin S, Vidaver AK (1995) Research on plant disease and pest management is essential to sustainable agriculture. Bioscience 45:354–357

Cowgill SE, Wratten SD, Sotherton NW (1993) The selective use of floral resources by the hoverfly *Episyrphus balteatus* (Diptera: Syrphidae) on farmland. Ann Appl Biol 122:223–231

Daamen RA, Wijnands FG, van der Vliet G (1989) Epidemics of diseases and pests of winter wheat at different levels of agrochemical input. A study on the possibilities of designing an integrated cropping system. J Phytopathol 125:305–319

Deacon JW (1976) Studies on *Pythium oligandrum*, an aggressive parasite of other fungi. Trans Br Myco Soc 66:383–391

Deacon JW, Henry CM (1978) Mycoparasitism by *Pythium oligandrum* and *Pythium acanthicum*. Soil Biol Biochem 10:409–415

Deacon JW, Laing SAK, Berry LA (1991) Pythium mycoparasiticum sp. nov., an aggressive mycoparasite from British soils. Mycotaxon 42:1–8

Dixon AFG (1969) Quality and availability of food for a sycamore aphid population. In: Watson TF (ed) Animal populations in relation to the food resources. Blackwell, Oxford

Dowling DN, Sexton R, Fenton A, Delany I, Fedi S, McHugh B, Callanan M, Moënne-Loccoz Y, Gara F (1996) Iron regulation in plant-associated pseudomonas flourescens M114: implications for biological control. In: Nakazawa K, Furukawa K, Haas D, Silver S (eds) Molecular biology of pseudomonads. ASM, Washington, DC

Downer AJ, Menge JA, Pond E (2001) Association of cellulytic enzyme activities in eucalyptus mulches with biological control of phytophthora cinnamomi. Phytopathology 91:847–855

DuPont SP, Ferris H, Van Horn M (2009) Effects of cover crop quality and quantity on nematode-based soil food webs and nutrient cycling. Appl Soil Ecol 41:157–167

Durrant WE, Dong X (2004) Systemic acquired resistance. Ann Rev Phytopathol 42:185–209

Edwards C, Arancon NQ, Greytak S (2006) Effects of vermicompost teas on plant growth and disease. BioCycle 47:28–31

Edwards C, Arancon NQ, Emerson E, Pulliman R (2007) Suppression of plant-parasitic nematodes and arthropod pests with vermicompost tea. Biocycle 48:38–39

El-Masry MH, Khalil AI, Hassouna MS, Ibrahim HAH (2002) In situ and in vitro suppressive effect of agricultural composts and their water extracts on some phytopathogenic fungi. World J Microbiol Biotechnol 18:551–558

El Titi A, Richter J (1987) Integrierter Pflanzenschutz im Ackerbau: Das Lautenbach Projekt. III. Schadlinge und Krankheiten 1979–1983. Z. Pflkrankh Pflschutz 94:1–13

Ferris H, Griffiths BS, Porazinska DL, Powers TO, Wang KH, Tenuta M (2012) Reflections on plant and soil nematode ecology: past, present and future. J Nematol 44:115–126

Gardner J, Caswell-Chen EP (1994) Raphanus sativus, Sinapis alba, Fagopyrum esculentum as hosts to Meloidogyne incognita, Meloidogyne javanica, and Plasmodiophora brassicae. Suppl J Nematol 26:756–760

Gommers FJ, Bakker J (1988) Physiological diseases induced by plant responses or products. In: Poinar GO Jr, Jansson H-B (eds) Diseases of nematodes, vol 1. CRC, Boca Raton, pp 3–22

Grube A, Donaldson D, Kiely T, Wu L (2011) Pesticides industry sales and usage: 2006 and 2007 market estimates. U.S. Environmental Protection Agency, Washington, DC

Grutzmacher AD, Zimmermann O, Yousef A, Hassan SA (2004) The side-effects of pesticides used in integrated production of peaches in Brazil on the egg parasitoid *Trichogramma caeco-ecia* Marchal (Hym. Trichogrammatidae). J Appl Entomol 128:377–383

Hamilton WD, Moran N (1980) Low nutritive quality as a defese against herbivores. MicrobiolJ Theor Biol 86:247–254

Hogg BN, Bugg RL, Daane KM (2011) Attractiveness of common insectary and harvestable floral resources to beneficial insects. Biol Control 56:76–84

Hoitink HA, Fahy P (1986) Basis for the control of soilborne plant pathogens with composts. Ann Rev Phytopathol 24:93–114

Hoitink HAJ, Krause MS, Han DY (2001) Spectrum and mechanisms of plant disease control with composts. In: Stoffella PJ, Kahn BA (eds) Compost utilization in horticultural cropping systems. Lewis Publishers, Boca Raton, FL. pp. 263

Hooks CRR, Johnson M (2001) Broccoli growth parameters and level of head infestations in simple and mixed plantings: Impact of increased flora diversification. Assoc Appl Biol 138:269–280

Hooks CRR, Johnson M (2004) Using undersown clovers as living mulches: effects on yields, lepidopterous pest infestations, and spider densities in a Hawai'ian broccoli agrosystem. Int J Pest Manage 50:115–120

Hooks CRR, Valenzuela H, Defrank J (1998) Incidence of pests and arthropod natural enemies in zucchini grown with living mulches. Agric Ecosyst Environ 69:217–231

Hooks CRR, Wang K-H, Ploeg A, McSorley R (2010) Using marigold (*Tagetes* spp.) as a cover crop to protect crops from plant-parasitic nematodes. Appl Soil Ecol 46:307–320

Hornby D (1990) Biological control of soil-borne plant pathogens. CAB International, Wallingford

House HI (1965) Insect nutrition. In: Rockstein ME (ed) Physiology of the insecta, vol 2. Academic Press, London, pp 769–858

Huerta A, Medina P, Smagghe G, Castanera P, Vinuela E (2003) Tropical toxicity of two acetonic fractions of *Trichilia havanensis* Jacq. and four insecticides to larvae and adults of *Chhrysoperla carnea* (Stephens) (Neuroptera: Chrysopidae). Commun Agric Appl Biol Sci 68:277–286

Ikediugwu FEO (1976) The interface in hypal interference by peniophora against *Heterobasidion annosum*. Trans Br Myco Soc 66:291–296

Ikediugwu FEO, Webster J (1970) Antagonism between *Coprinus heptemerus* and other coprophilous fungi. Trans Br Myco Soc 54:181–204

Jagdale GP, Reynolds B, Ball-Coelho B, Potter J (1999) Nematicidal activity of marigold plant parts against root-lesion nematodes (*Pratylenchus penetrans*). J Nematol 31:546–547

James DG (2005) Further evaluation of synthetic herbivore-induced plant volatiles as attractants for beneficial insects. J Chem Ecol 31:493–507

Jeffries P, Young TWK (1994) Interfungal parasitic relationships. CAB International, Cambridge

Johnson M (2004) The impact of organic insecticides on beneficials. California conference features pest control and organicproduction. http://www.rodaleinstitute.org/200483/lotter

Kellock LM, Dix NJ (1984) Antogonism by hypomyces aurantius. II. Ultrastructural studies of hyphal disruption. Trans Br Myco Soc 82:332–338

Kloepper JW, Ryu C-M, Zhang S (2004) Induced systemic resistance and promotion of plant growth by Bacillus spp. Phytopathology 94:1259–1266

Knudsen IMB, Hockenhull DK, Jensen DF, Elmholt S (1999) Suppressiveness of organically and conventionally managed soils towards brown foot rot of barley. Appl Soil Ecol 12:61–72

Kraiss H, Cullen EM (2008) Efficacy and non-target effects of reduced-risk insecticides on *Aphis glycines* (Hemiptera: Aphididae) and its biological control agent *Harmonia axyridis* (Coleoptera: Coccinellidae). J Econ Entomol 101:391–398

Kwok OCH, Plattner R, Weisleder D, Wicklow DT (1992) A nematicidal toxin from *Pleurotus ostreatus* NRRL 3526. J Chem Ecol 18:127–136

Laing SAK, Deacon JW (1990) Aggressiveness and fungal host ranges of mycoparasitic *Pythium* species. Soil Biol Biochem 22:905–911

Lavandero B, Wratten S, Shishehbor P, Womer S (2005) Enhancing the effectiveness of the parasitoid *Diadegma semiclausum* (Helen) movement after use of nectar in the field. Biol Control 34:152–158

Lewis DH, Whipps JM, Cooke RC (1989) Mechanisms of biological disease control with special reference to the case study of *Pythium oligandrum* as an antagonist. In: Whipps JM, Lumsden RD (eds) Biotechnology of fungi for improving plant growth. Cambridge University, Cambridge, pp 191–217

Lewis WJ, van Lenteren JC, Phatak SC, Tumlinson JH III (1997) A total system approach to sustainable pest management. Proc Natl Acad Sci U S A 94:12243–12248

Liu YB, Tabashnik BE, Pusztai-Carey M (1996) Field-evolved resistance to *Bacillus thuringiensis* toxin CryIC in diamondback moth (Lepidoptera: Plutellidae). J Econ Entomol 89:798–804

Lugtenberg B, Kamilova F (2009) Plant-growth-promoting rhizobacteria. Annu Rev Microbiol 63:541–556

Lumsden RD, García-E R, Lewis JA, Frías-T GA (1987) Suppression of damping-off caused by *Pythium* spp in soil from the indigenous Mexican Chinampa agricultural system. Soil Biol Biochem 19:501–508

Mandanhar R, Hooks CRR (2011) Using protector plants to reduce the incidence of papaya ring spot virus-watermelon strain in zucchini. Environ Entomol 40:391–398

Mandanhar R, Hooks CRR, Wright MG (2009) Influence of cover crop and intercrop systems on *Bemisia argentifolli* (Hemiptera: Aleyrodidae) infestation and associated squash silverleaf disorder in zucchini. Environ Entomol 38:442–449

Marahatta SP, Wang K-H, Sipes BS, Hooks CRR (2012a) Effects of Tagetes patula on active and inactive stages of Root-Knot nematodes. J Nematol 44:26–30

Marahatta SP, Wang K-H, Sipes BS, Hooks CRR (2012b) Effects of *Crotalaria juncea* on anhydrobiotic state of *Rotylenchulus reniformis*. Nematropica 42:34–40

Marban-Mendoza N, Dicklow MB, Zuckerman BM (1989) Evaluation of control of *Meloidogyne incognita* and *Nacobbus aberrans* on tomato by two leguminous plants. Rev Nematol 12:409–412

Marshall AJ (2002) Sunn hemp (*Crotalaria juncea*) as an organic amendment in crop production. MS Thesis, University of Florida, Gainesville, FL

McSorley R (2011) Overview of organic amendments for management of plant-parasitic nematodes, with case studies from Florida. J. Nematol. 43:69–81

McSorley R, Gallaher RN (1996) Effect of yard waste compost on nematode densities and maize yield. J Nematol 28:655–660

McSorley R, Dickson DW, de Brito JA (1994) Host status of selected tropical rotation crops to four populations of root-knot nematodes. Nematropica 24:45–53

McSorley R, Wang K-H, Church G (2007) Suppression of root-knot nematodes in natural and agricultural soils. Appl Soil Ecol 39:291–298

Meister RT, Sine C (2012) Crop protection handbook, vol 98. Meister Pro, Willoughby, OH. p 768

Morandin LA, Winston ML, Franklin MT, Abbott VA (2005) Lethal and sub-lethal effects of spinosad on bumble bees (*Bombus impatiens* Cresson). Pest Manage Sci 61:619–626

Muller R, Gooch PS (1982) Organic amendments in nematode control. An examination of the literature. Nematropica 12:319–326

Neher D. (2001) Role of nematodes in soil health and their use as indicators. J. Nematol. 33:161–168

Nogueira MA, de Oliveira JS, Ferraz S, Dos Santos MA (1996) Nematicidal constituents in *Mucuna aterrima* and its activity on *Meloidogyne incognita* race 3. Nematol Mediterr 24:249–252

NOSB (2004) National organic standards board: compost tea task force final report. http://www.ams.usda.gov/nosb/meetings/CompostTeaTaskForce-FinalReport.pdf. Accessed 22 Dec 2013

Oerke EC (2006) Crop losses to pests. J. Agricul. Sci. 144:31–43

Oerke EC, Dehne HW, Schonbeck F, Weber A (1994) Crop production and crop protection: Estimated losses in major food and cash crops. Elsevier, Amsterdam, p 808

Oka Y (2010) Mechanisms of nematode suppression by organic soil amendments-a review. Appl Soil Ecol 44:101–115

Oka Y, Pivonia S (2003) Effect of a nitrification inhibitor on nematicidal activity of organic and inorganic ammonia-releasing compounds against the root-knot nematode *Meloidogyne javanica*. Nematology 5:505–513

Okorie CC, Ononuju CC, Okwujiako IA (2011) Management of *Meloidogyne incognita* with *Pleurotus ostreatus* and *P. tuberregium* in soybean. Int J Agric Biol 13:401–405

O'Sullivan DJ, O'Gara F (1992) Traits of fluorescents *Pseudomonas* spp. involved in suppression of plant root pathogens. Microbial Rev 56:662–676

Pachenari A, Dix NJ (1980) Production of toxins and wall degrading enzymes by *Gliocladium roseum*. Trans Br Myco Soc 74:561–566

Pant A, Radovich TJK, Hue NV, Arancon NQ (2011) Effects of vermicompost tea (aqueous extract) on pak-choi yield, quality, and on soil biological properties. Compost Sci Util 19:279–292

Papavizas GC (1985) *Trichoderma* and *Gliocladium*: biology, ecology, and potential for biocontrol. Ann Rev Phytopathol 23:23–54

Pathma J, Sakthivel N (2013) Molecular and functional characterization of bacteria isolated from straw and goat manure based vermicompost. Appl Soil Ecol 70:33–47

Patriquin DG, Baines D, Abbound A (1995) Diseases, pests, and soil fertility. In: Cook HF, Lee HC (eds) Soil management in sustainable agriculture. Wye College Press, London, pp 161–174

Persson Y, Baath E (1992) Quantification of mycoparasitism by the nematode trapping fungus *Arthobotrys oligospora* on *Rhizoctonia solani* and the influence of nutrient levels. FEMS Microbiol Ecol 101:11–16

Ploeg AT (2000) Effects of amending soil with *Tagetes patula* cv. Single gold on *Meloidogyne incognita* infestation of tomato. Nematology 2:489–493

Ploeg AT (2002) Effect of selected marigold varieties on root-knot nematodes and tomato and melon yields. Plant Dis 86:505–508

Ploeg AT, Maris PC (1999) Effect of temperature on suppression of *Meloidogyne incognita* by *Tagetes* cultivars. J Nematol 31:709–714

Price JF, Schuster DJ (1991) Effects of natural and synthetic insecticides on sweetpotato whitefly *Bemisia tabaci* (Homoptera: Aleyrodidae) and its hymenopterous parasitoids. Flo Entomol 74:60–68

Powell MJ (1982) Ultrastructure of the host-parasite interface between *Allomyces javanicus* and its endoparasite *Catenaria allomyces*. Bot Gazelle 143:176–187

Radovich T, Arancon N (2011) Tea Time in the tropics: a handbook for the compost tea production and use. CTAHR, University of Hawaii, Honolulu

Rahman ML (1990) Effect of different cropping sequences on Root-Knot nematode, *Meloidogyne graminicola*, and yield of deep water rice. Nematol Mediterr 18:213–217

Ritzinger CHSP, McSorley R (1998) Effect of castor and velvetbean organic amendments on *Meloidogyne arenaria* in greenhouse experiments. J Nematol 30:624–631

Rodriguez-Kabana R (1986) Organic and inorganic nitrogen amendments to soil as nematode suppressants. J Nematol 18:129–135

Roget DK (1995) Decline in root rot (*Rhizoctonia solani* AG-8) in wheat in a tillage and rotation experiment at Avon, South Australia. Aus J Exp Agric 35:1009–1013

Saber M, Hejazi M, Hassan SA (2004) Effects of azadirachtin/neemazal on different stages and adult life table parameters of *Trichogramma cacoeciae* (Hymenoptera: Trichogrammatidae). J Econ Entomol 97:905–910

Sánchez-Moreno S, Ferris H (2007) Suppressive service of the soil food web: effects of environmental management. Agri. Ecosyst. Environ. 119:75–87

Scheuerell S, Mahaffee W (2002) Compost tea: Principles and prospects for plant disease control. Compost Sci Util 10:313–338

Sipes BS, Arakaki AS (1997) Root-knot nematode management in dryland taro with tropical crops. Suppl J Nematol 29:721–724

Starr JL, Black MC (1995) Reproduction of *Meloidogyne arenaria, M. incognita*, and *M. javanica* on sesame. Suppl J Nematol 27:624–627

Stirling GR (1991) Biological control of plant parasitic nematodes: progress, problems and prospects. CAB International, Wallingford

Stirling GR (2008) The impact of farming systems on soil biology and soilborne diseases: examples from the Australian sugar and vegetable industries-the case for better integration of sugarcane and vegetable production and implications for future research. Australas Plant Pathol 37:1–18

Stirling GR (2011) Biological control of plant-parasitic nematodes: an ecological perspective, a review of progress and opportunities for further research. In: Davies K, Spiegel Y (eds) Biological control of plant-parasitic nematodes: building coherence between microbial ecology and molecular mechanisms. Progress in Biological Control, vol 11. Springer, London

Stirling GR, Dullahide SR, Nikulin A (1995) Management of lesion nematode (*Pratylenchus jordanensis*) on replanted apple trees. Aus J Exp Agric 35:247–258

Stirling GR, Wilson EJ, Stirling AM et al (2005) Amendments of sugarcane trash induced suppressiveness to plant-parasitic nematodes in sugarcane soil. Australas Plant Pathol 34:203–211

Stone AG, Scheuerell SJ, Darby HM (2004) Suppression of soilborne diseases in field agricultural systems: organic matter management, cover cropping, and other cultural practices. In: Magdoff F, Weil RR (eds) Soil organic matter in sustainable agriculture. CRC, Boca Raton, pp 131–177

Sturz AV, Kimpinski J (2004) Endoroot bacteria derived from marigold (*Tagetes* spp.) can decrease soil population densities of root-lesion nematodes in the potato root zone. Plant Soil 262:241–249

Suatmadji RW (1969) Studies on the effect of *Tagetes* species on plant parasitic nematodes. H. Veenman en Zonen, Wageningen, p 132

Tamis WLM, van den Brink WJ (1998) Inventarisatie van ziekten en plagen in wintertarwe in gangbare, geintegreerde en ecologische teeltsystemen in Nederland in de periode 1993-1997 IPO-DLO Rapport nr. 98-01. Wageningen

Taylor RM, Pfannenstiel RS (2008) Nectar feeding by wandering spiders on cotton plants. Environ Entomol 37:996–1002

Thoeming G, Poehling HM (2006) Integrating soil-applied azadirachtin with *Amblyseius cucumeris* (Acari: Phytoseiidae) and *Hypoaspis aculeifer* (Acari: Laelapidae) for the management of *Frankliniella occidentalis* (Thysanoptera: Thripidae). Environ Entomol 35:746–756

Thorn BG, Barron GL (1984) Carnivorous mushrooms. Science 224:76–78

Timper P (2014) Conserving and enhancing biological control of nematodes. J Nematol (in press)

Titus S (2012) Inducing systemic acquired resistance: helping plants help themselve. Greenhouse Grower, 5 Sept 2012. http://www.greenhousegrower.com/crop-inputs/inducing-Systemic-acquired-resistance-helping-plants-help-themselves/

Tognetti C, Laos F, Mazzarino MJ, Hernández MT (2005) Composting vs. vermicomposting: a comparison of end product quality. Compost Sci Util 13:6–13

Tzean SS, Estey RH (1978) *Schizophyllum commune* Fr. as a destructive mycoparasite. Can J Microbiol 24:780–784

Vallad G (2013) Systemic acquired resistance: from basic concepts to applied reality. Seminar presentation for growers and other researchers. gcrec.ifas.ufl.edu

Vallad G, Goodman RM (2004) Review and Interpretation: systemic acquired resistance and induced systemic resistance in conventional agriculture. Crop Sci 44:1920–1934

van Bruggen AH, Termorshuizen AJ (2003) Integrated approaches to root disease management in organic farming systems. Australas Plant Pathol 32:141–156

Van Driesche RG, Lyon S, Nunn C (2006) Compatibility of spinosad with predaceious mites (Acari: Phyoseiidae) used to control weatern flower thrips (Thysanoptera: Thripidae) in greenhouse crops. Flo Entomol 89:396–401

Van Loon LC (2007) Plant responses to plant growth-promoting bacteria. Eur J Plant Pathol 119:243–254

Van Loon LC, Bakker PAHM, Pieterse CMJ (1998) Systemic resistance induced by rhizosphere bacteria. Ann Rev Phytopathol 36:453–483

Wang K-H (2012) Cover crops as insectary plants to enhance above and below ground beneficial organisms. Hānai'Ai Newsletter March-April-May 2012. http://www.ctahr.hawaii.edu/sustainag/news/articles/V11-Wang-insectary-covercrops.pdf

Wang K-H, Radovich T (2012) Cover cropping system and compost tea treatment for management of nematode, whiteflies, and pollinators. J Nematol 44:496 (abstract)

Wang K-H, Sipes BS, Schmitt DP (2001) Suppression of *Rotylenchulus reniformis* by *Crotalaria juncea*, *Brassica napus*, and *Target erecta*. Nematropica 31:237–251

Wang K-H, Sipes BS, Schmitt DP (2002) Management of *Rotylenchulus reniformis* in pineapple, *Ananas comosus*, by intercycle cover crops. J Nematol 34:106–114

Wang K-H, McSorley R, Gallaher RN (2003) Effect of *Crotalaria juncea* amendment on nematode communities in soil with different agricultural histories. J Nematol 35:294–301

Wang K-H, Hooks CRR, Marahatta SP (2011) Can using a strip-tilled living mulch system enhance organisms higher up in the soil food web hierarchy? Appl Soil Ecol 49:107–117

Warnke SA, Chen SY, Wyse DL, Johnson GA, Porter PM (2008) Effect of rotation crops on hatch, viability, and development of *Heterodera glycines*. Nematology 10:869–882

Weaver DB, Rodrı'guez-Ka'bana R, Carden EL (1998) Velvetbean and bahiagrass as rotation crops for management of *Meloidogyne* spp. and *Heterodera glycines* in soybean. J Nematol 30:563–568

Wei G, Kloepper JW, Tuzun S (1991) Induction of systemic resistance of cucumber to *Colletotrichum orbiculare* by select strains of plant growth-promoting rhizobacteria. Phytopathology 81:1508–1512

Westphal A, Becker JO (1999) Biological suppression and natural population decline of *Heterodera schachtii* in a California field. Phytopathology 89:434–440

Widmer TL, Abawi GS (2000) Mechanism of suppression of *Meloidogyne hapla* and its damage by a green manure of sudangrass. Plant Dis 84:562–568

Yardim EN, Arancon NQ, Edwards CA, Oliver TJ, Byrne RJ (2006) Suppression by vermicomposts of tomato hornworm (*Manduca quinquemaculata*) and cucumber beetle (*Acalymma vittatum*) populations and damage. Pedobiologia 50:23–29
You MP, Sivasithamparan K (1994) Hydrolysis of fluorescein diacetate in an avocado plantation mulch suppressive to Phytophthora cinnamomi and its relationship with certain biotic and abiotic factors. Soil Biol Biochem 26:1355–1361
You MP, Sivasithamparan K (1995) Changes in microbial populations of an avocado plantation mulch suppressive to Phytophthora cinnamomi. Appl Soil Ecol 2:33–43
Zasada IA, Klassen W, Meyer SLF, Codallo M, Abdul-Baki AA (2006) Velvetbean (*Mucuna pruriens*) extracts: impact on *Meloidogyne incognita*survival and on *Lycopersicon esculentum* and *Lactuca sativa* germination and growth. Pest Manage Sci 62:1122–1127
Zehnder GW, Yao C, Murphy JF, Sikora EJ, Kloepper JW (2000) Induction of resistance in tomato against cucumber mosaic virus by plant growth-promoting rhizobacteria. Biol Control 45:127–137
Zinovieva SV, Vasyukova NI, Udalova ZV, Gerasimova NG. (2013) The participation of salicylic and jasmonic acids in genetic and induced resistance of tomato to *Meloidogyne incognita* (Kofoid and White, 1919). Biol Bull 40:297–303
Zmora-Nahum S, Danon M, Hadar Y, Chen Y (2008) Chemical properties of compost extracts inhibitory to germination of *Sclerotium rolfsii*. Soil Biol Biochem 40:2523–2529

Index

D. Nandwani (ed.), *Sustainable Horticultural Systems*, Sustainable Development and Biodiversity 2, DOI 10.1007/978-3-319-06904-3

C

MIX
Papier aus verantwortungsvollen Quellen
Paper from responsible sources
FSC® C105338

If you have any concerns about our products,
you can contact us on
ProductSafety@springernature.com

In case Publisher is established outside the EU,
the EU authorized representative is:
Springer Nature Customer Service Center GmbH
Europaplatz 3, 69115 Heidelberg, Germany

Printed by Libri Plureos GmbH
in Hamburg, Germany